Laboratory Manual

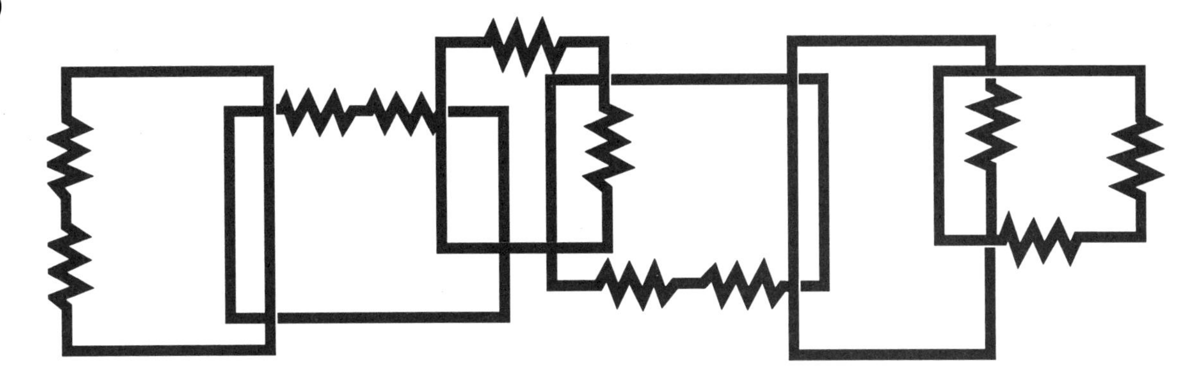

to Accompany

INTRODUCTORY DC/AC ELECTRONICS

FIFTH EDITION

and

INTRODUCTORY DC/AC CIRCUITS

FIFTH EDITION

BY NIGEL P. COOK

Gary A. Lancaster
Nigel P. Cook

Upper Saddle River, New Jersey   Columbus, Ohio

Vice President and Editor in Chief: Stephen Helba
Acquisitions Editor: Scott J. Sambucci
Assistant Editor: Kate Linsner
Production Editor: Rex Davidson
Design Coordinator: Karrie Converse-Jones
Cover Designer: Rod Harris
Cover Photo: Copyright PhotoDisc, Inc.
Production Manager: Pat Tonneman

The book was set in Times Roman by The Special Projects Group and was printed and bound by Banta Book Group. The cover was printed by Phoenix Color Corp.

---

10 9 8 7 6 5 4 3 2 1

ISBN 0-13-034031-6

Preface

This lab manual contains laboratory experiments designed specifically to translate the theory of Nigel Cook's *Introductory DC/AC Electronics, Fifth Edition*, and *Introductory DC/AC Circuits, Fifth Edition*, into practical action. Each experiment is preceded by a set of objectives, an overview, a list of materials required, an introduction, and a list of safety precautions to be aware of while conducting each particular experiment. A step-by-step procedure walks you through each experiment, and interspersed throughout the procedure are troubleshooting examples describing problem symptoms and solutions. Discussions are also found at key points throughout the experiment to summarize what has just been achieved.

Acknowledgments

This laboratory manual is a composite of the efforts of many individuals: Hugh Scriven, who set the pattern and created the manual laboratory exercises; Dail Cooper, who drafted all of the illustrations for the manual laboratory exercises; Nigel Cook, whose text this manual supports; and the creative and friendly editorial staff at Prentice Hall. I extend my thanks to each and every one for their contribution to the creation and success of this manual.

I also wish to acknowledge the valuable assistance of the following reviewers: Joe Gryniuk, Lake Washington Technical College; Gerald Schickman, Miami Dade Community College; George Sweiss, ITT Technical Institute; and Bradley Thompson, Alfred State College School of Technology.

Gary Lancaster

Michael Faraday was the electrical wizard of the nineteenth century. As an aspiring lecturer and great experimenter he was always very popular with his students, who found his presentations dynamic, enlightening, and entertaining. Commenting on experimentation, his advice was to "Let your imagination go, guiding it by judgment and principle, but holding it in and directing it by experiment. Nature is your best friend and critic in experimental science if you only allow her intimations to fall unbiased on your mind. Nothing is so good as an experiment which, while it sets an error right, gives you as a reward for your humility an absolute advance in knowledge."

This laboratory manual is dedicated to you, the aspiring Electrical and Electronics Technician or Engineer.

The opportunities for exciting careers in the Electricity and Electronics field today are more varied and more abundant than ever before. Use this manual well and you will be prepared for great success.

Contents

Introduction xi
Parts List xi
Equipment List xiii
Tools xiii

MANUAL EXERCISES

Experiment Number		
1	***The Prototyping Board—Buses*** An introduction to the bus structure of the protoboard.	1
2	***The Prototyping Board—Connections*** An introduction to the connection point aspects of the board.	8
3	***Circuit Wiring—Continuity*** An experience in creating a simple electrical circuit.	13
4	***Circuit Wiring—Measurements*** Basic series circuit voltage and current measurements.	17
5	***Using Ohm's Law*** Obtaining a predetermined current by varying resistance.	25
6	***The Series Circuit—Measurements*** More complex circuit resistance, voltage and current measurements.	29
7	***Series Circuit Application—Illumination*** Using series resistance to vary lamp intensity.	34
8	***Resistors—Values and Tolerance*** Measuring actual resistor values to establish tolerance.	39
9	***The Smoke Theory*** An exercise in the destructive effects of excess current flow.	43

Experiment
Number

10 ***The Electric Battery*** 47
Making a simple dc cell and measuring its characteristics.

11 ***Fuses for Protection*** 54
An application where a fuse is used to protect a lamp.

12 ***Troubleshooting Series Circuits*** 59
Introduction to techniques of analyzing circuit problems.

13 ***Series Circuit Voltage Division*** 65
An in-depth study of multi-component series circuits.

14 ***Series Circuits—Power Distribution*** 69
Analyzing power distribution characteristics of the series circuit.

15 ***Troubleshooting Series Circuits*** 73
In-depth analysis of problem recognition and correction.

16 ***Parallel Circuits—Measurements*** 81
Making resistance, voltage and current measurements.

17 ***Parallel Circuit Applications*** 87
Further study of parallel resistor combinations.

18 ***Troubleshooting Parallel Circuits*** 91
Using analytical techniques to recognize and correct problems.

19 ***Series-Parallel Circuits*** 98
Measuring resistance, voltage and current characteristics.

20 ***Loaded Voltage Dividers*** 104
Analyzing uses for the resistive voltage divider.

21 ***Bridge Circuits—Applications*** 108
Analyzing the bridge connection.

22 ***Measurement Limitations—The Multimeter*** 114
A study of measurement error and compensation.

23 ***AC Series Circuit Measurements*** 119
Using the multimeter to measure ac voltage and current.

24 ***Waveforms—Using the Oscilloscope*** 124
Viewing various waveforms and distinguishing ac from dc.

25 ***Capacitor Charge and Discharge*** 130
Examining dc charge and discharge characteristics.

Experiment
Number

26	***Series Capacitive Circuits—Voltage Distribution*** Analyzing both dc and ac voltage distribution.	**135**
27	***Parallel Capacitors*** A demonstration of the additive capacity of such circuits.	**139**
28	***Integrators and Differentiators*** A study of the characteristics and uses of these circuits.	**144**
29	***The Electromagnet*** Hand winding an electromagnet and testing its characteristics.	**149**
30	***The Electric Generator*** An experience which demonstrates how generators work.	**154**
31	***The Electric Relay*** Making a simple relay and testing its operation.	**159**
32	***Application of an Electromagnet*** Modifying the relay to make it into a buzzer.	**163**
33	***Listening to Alternating Current*** Making a simple loudspeaker to hear ac tones of different waves.	**167**
34	***The Electromagnet Revisited*** Using ac to power the electromagnet and noting characteristics.	**171**
35	***Inductive Kick*** A demonstration of inductive kick.	**176**
36	***RL Circuit Measurements*** Voltage and current measurements and impedance verification.	**180**
37	***Transformers—Stepping Up*** Using the transformer to step up a low voltage to a much higher voltage.	**185**
38	***Impedance Matching*** Attempting maximum power transfer.	**191**
39	***The RC Circuit—AC Characteristics*** Measurement of voltage and current characteristics.	**195**
40	***Impedance Measurement*** Verifying by measurement the impedance of RC circuits.	**201**
41	***Phase Angle*** Measurement of phase shift with the oscilloscope.	**205**
42	***The RLC Circuit*** Characteristics of a non-resonant RLC circuit.	**212**

Experiment Number

43	***Resonance—Characteristics*** Examining the characteristics of both series and parallel resonant circuits.	**217**
44	***Quality*** Evaluating the quality of a resonant circuit.	**221**
45	***Semiconductor Diodes*** Comparing the properties of different semiconductor diodes.	**225**
46	***Semiconductor Stress*** The harmful effects of stress on semiconductor diodes.	**230**
47	***Rectification*** Constructing rectifier circuits to investigate the characteristics of the resultant dc.	**236**
48	***The Reference Diode*** Electrical characteristics of the zener diode.	**242**
49	***The Filtered Power Supply*** Examining ac to dc power conversion by constructing a simple dc power supply.	**247**
50	***Full Wave, Dual Polarity, and Doubling Supplies*** Evaluating two dc power supply circuits which utilize capacitor filters.	**253**
51	***Transistor Operation—Cutoff and Saturation*** Examining the extreme conditions of transistor operation, cutoff, and saturation.	**259**
52	***Thyristors—AC and DC Power Control*** Introduction to the Silicon Controlled Rectifier and the TRIAC.	**266**
53	***The Transistor as an Amplifier*** Analyzing the amplification characteristics of the transistor and transistorized amplifier circuits.	**277**
54	***Amplifier Applications*** Constructing two electronic circuits which rely upon amplifiers to make them work.	**287**
55	***Operational Amplifiers*** Constructing a two stage circuit to demonstrate the versatility of the operational amplifier.	**297**
56	***Op Amps as Comparators*** Op Amp comparator circuits used to compare voltages.	**303**
57	***Series Circuit Measurements*** (Computer Exercise) Resistance, voltage and current measurements in a series circuit.	**309**

COMPUTER SIMULATION EXERCISES

Experiment
Number

58 ***Troubleshooting Series Circuits 1*** (Computer Exercise) **313**

59 ***Troubleshooting Series Circuits 2*** (Computer Exercise) **317**

60 ***Parallel Circuit Measurements 1*** (Computer Exercise) **321**
Resistance, voltage and current measurements in a parallel circuit.

61 ***Troubleshooting Parallel Circuits 2*** (Computer Exercise) **325**

62 ***Troubleshooting Parallel Circuits 3*** (Computer Exercise) **329**

63 ***Series-Parallel Circuit Measurements 1*** (Computer Exercise) **333**
Resistance, voltage and current measurements in a series-parallel circuit.

64 ***Troubleshooting Series-Parallel Circuits 2*** (Computer Exercise) **337**

65 ***Troubleshooting Series-Parallel Circuits 3*** (Computer Exercise) **341**

66 ***Loaded Voltage-Divider Circuit Measurements 1*** (Computer Exercise) **345**
Resistance, voltage and current measurements in a loaded voltage-divider.

67 ***Troubleshooting Loaded Voltage-Divider Circuits 1*** (Computer Exercise) **349**

68 ***Troubleshooting Loaded Voltage-Divider Circuits 2*** (Computer Exercise) **353**

69 ***RC Time Constants*** (Computer Exercise) **357**
Calculating and measuring RC time constants.

70 ***RL Circuits*** (Computer Exercise) **361**
Calculating and measuring RL circuits.

71 ***RLC Circuits*** (Computer Exercise) **365**
Impedance, voltage and current measurements in RLC circuits.

72 ***Troubleshooting RLC Circuits 1*** (Computer Exercise) **372**

73 ***Troubleshooting RLC Circuits 2*** (Computer Exercise) **376**

74 ***The Bridge Rectifier*** (Computer Exercise) **380**

75 ***Troubleshooting Power Supplies 1*** (Computer Exercise) **384**

76 ***Troubleshooting Power Supplies 2*** (Computer Exercise) **388**

77 ***Troubleshooting Power Supplies 3*** (Computer Exercise) **392**

78 ***Troubleshooting a Common-Emitter Amplifier*** (Computer Exercise) **396**

79 ***Troubleshooting a Common-Collector Amplifier*** (Computer Exercise) **400**

80 ***Troubleshooting a Common-Base Amplifier*** (Computer Exercise) **404**

VOCABULARY EXERCISES

81	*Vocabulary Exercise–Chapter 1*	408
82	*Vocabulary Exercise–Chapter 2*	411
83	*Vocabulary Exercise–Chapters 3 and 4*	414
84	*Vocabulary Exercise–Chapters 5 through 10*	417
85	*Vocabulary Exercise–Chapters 11 and 12*	420
86	*Vocabulary Exercise–Chapters 13 and 14*	423
87	*Vocabulary Exercise–Chapters 16 and 17*	426
88	*Vocabulary Exercise–Chapters 19 through 22*	429
89	*Vocabulary Exercise–Chapter 23*	432

Introduction

Parts List
The following table lists all of the components used in this lab manual.

Component Type		Quantity	Value	Description
RESISTORS		1	10Ω ±5%	1 watt
	Fixed	2	15Ω, ±5%	½ watt
		2	18Ω, ±5%	1 watt
		2	22Ω, ±5%	1 watt
		2	33Ω, ±5%	1 watt
		2	47Ω, ±5%	1 watt
		4	120Ω, ±5%	½ watt
		1	150Ω, ±5%	½ watt
		2	220Ω, ±5%	1 watt
		3	330Ω, ±5%	½ watt
		2	470Ω, ±5%	½ watt
		2	560Ω, ±5%	½ watt
		2	820Ω, ±5%	½ watt
		3	1kΩ, ±5%	½ watt
		1	1.5kΩ, ±5%	½ watt
		1	2.2kΩ, ±5%	½ watt
		2	4.7kΩ, ±5%	½ watt
		2	5.6kΩ, ±5%	½ watt
		3	10kΩ, ±5%	½ watt
		1	47kΩ, ±5%	½ watt
		1	470kΩ, ±5%	½ watt
		3	10MΩ, ±5%	½ watt
	Variable	2	1kΩ/single turn	Potentiometer, 5 watt
		2	5kΩ/single turn	Potentiometer, 5 watt
		1	1MΩ	Potentiometer

Component Type	Quantity	Value	Description
SEMICONDUCTOR	1	IN34	Germanium diode
(These or equivalent)	1	IN4002	Silicon diode
	1	IN4733A	Zener diode, 5.1V, 1 watt
	1	IN5819	Schottky diode
	1	Any visible color	Light-emitting diode
	4	IN4001	Rectifier diodes
	1	IN914	Silicon diode
	1	IN4739A	Zener Diode
	2	2N3904	*NPN* transistor
	1	S401E	SCR, 400V, 1A
	1	Q401E3	Triac, 400V, 1A
	1	MPF102	N-Channel JFET
	2	LM741CN	Op-amp chips
	1	Red	LED
	1	Green	LED
	1	Yellow	LED
CAPACITORS	3	0.01μF	50V, ceramic disk
	1	0.5μF	50V, ceramic disk
	1	1.0μF	50V, paper or mylar
	3	0.01μF	50V, ceramic
	2	20μF	25V, Electrolytic
	1	1000μF	25V, Electrolytic
	1	100μF	25V, Electrolytic
	4	10μF	25V, Electrolytic
MAGNETIC	1	120V–12.6V (300mA) min.	Center-tapped transformer
	1	1 ½ inch	Donut shaped permanent transformer
	100 feet	#30	Magnet wire
	1	120V–12.6VAC, 510mA min.	Step-down transformer
TRANSDUCERS	1	#7373 or equivalent	Light bulb (14V/100mA)
	4	#7382 or equivalent	Light bulbs (14V/80mA)
	2	#7328 or equivalent	Light bulbs (6V/200mA)
	3	Red, green, or yellow	Light emitting diode (LED)
	1	NE-2 or NE2H	Neon lamp
	1	8Ω to 40Ω (100mW)	2 ¼-inch loudspeaker
	1	10kΩ at 25°C	Thermistor
	1	60W-120V lamp	With test leads attached
	1	10kΩ dark resistance	Photoresistor
MISC.	10		Alligator clip test leads
	1		Prototyping board
	4 feet	#24 insulated	Hookup wire
	1	⅛ amp	Fuse
	1	DC permanent magnet	Miniature electric motor
	1	Momentary contact	Pushbutton switch
	2	1-inch by 3-inch strips	Aluminum foil
	1	Plastic	Margarine dish and top
	2	Plastic or glass	Containers
	1	½-full bottle	Household chlorine bleach
	2	Sewing machine bobbins	Steel
	1	4" by 6"	Plastic tray
	1	Tap water	Small bottle
	1	Roll	Electrical tape
	Several	½ and 1-inch	Roundhead/Flathead screws
	Several	½" to ¾" as necessary	Nails
	2 strips	Iron 1" × 4" and ¾" × 4"	Cut from a tin can
	Some	Tissue paper	

Equipment List

In order to realize the full learning potential of the following experiments, you should have access to the following equipment.

Equipment	Quantity	Specifications
DC Power Supply	1	Variable, dual output 0–25V (approx.) 0–500mA, current-limiting.
Digital multimeter	1	Standard
Function generator	1	Sine, square, triangular waveforms with dc offset
Dual-trace oscilloscope	1	10Mhz or greater
Oscilloscope probes	2	1× or 1×, 10× switchable
Test leads	10	6" to 8" length with booted alligator clips.
Test leads for power supply, DMM, and function generator	4	Standard
Optional: Analog VOM	1	Min 20,000 Ω/V

Computers and Software
486–100 minimum, 8 M RAM, 12 M of available space on the hard drive, CD-ROM.
Windows 95/98 or higher
MultiSim circuit simulation software or EWB software
Student disk to accompany Laboratory Manual.

Tools

All of the tools illustrated below will be needed at various times as you perform the experiments.

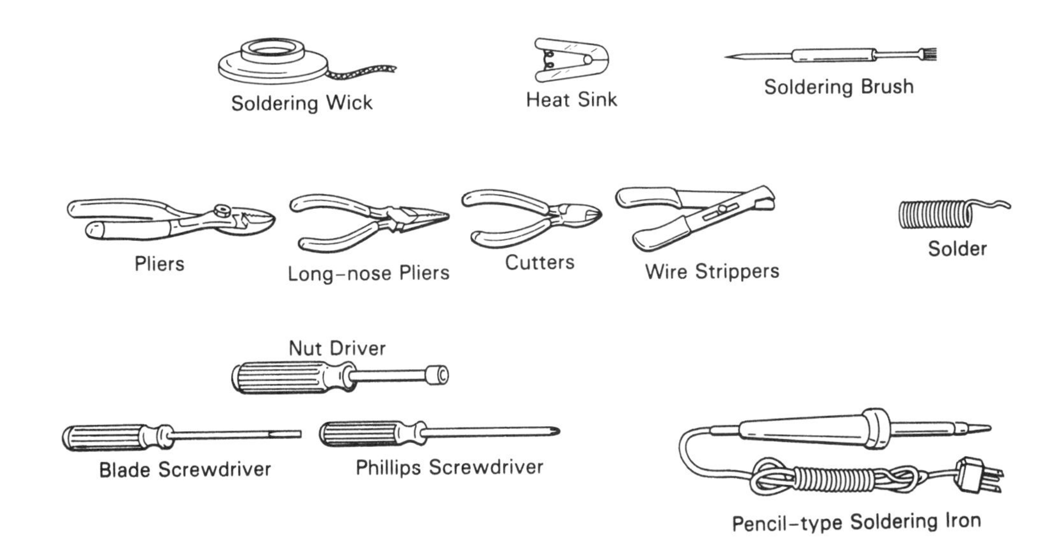

QUICK REFERENCE SUMMARY SHEET

$$W = Q \times V$$

W = **Energy Stored in Joules (J)**
Q = **Coulombs of Charge**
V = **Voltage in Volts (V)**

$$P = \frac{W}{t}$$

P = **Power in Watts**
W = **Energy in Joules**
t = **Time in Seconds**

$$P = I \times V$$

P = **Power in Watts (W),** I = **Current in Amps (A),** V = **Voltage in Volts (V)**

$$V = \frac{P}{I} \qquad I = \frac{P}{V}$$

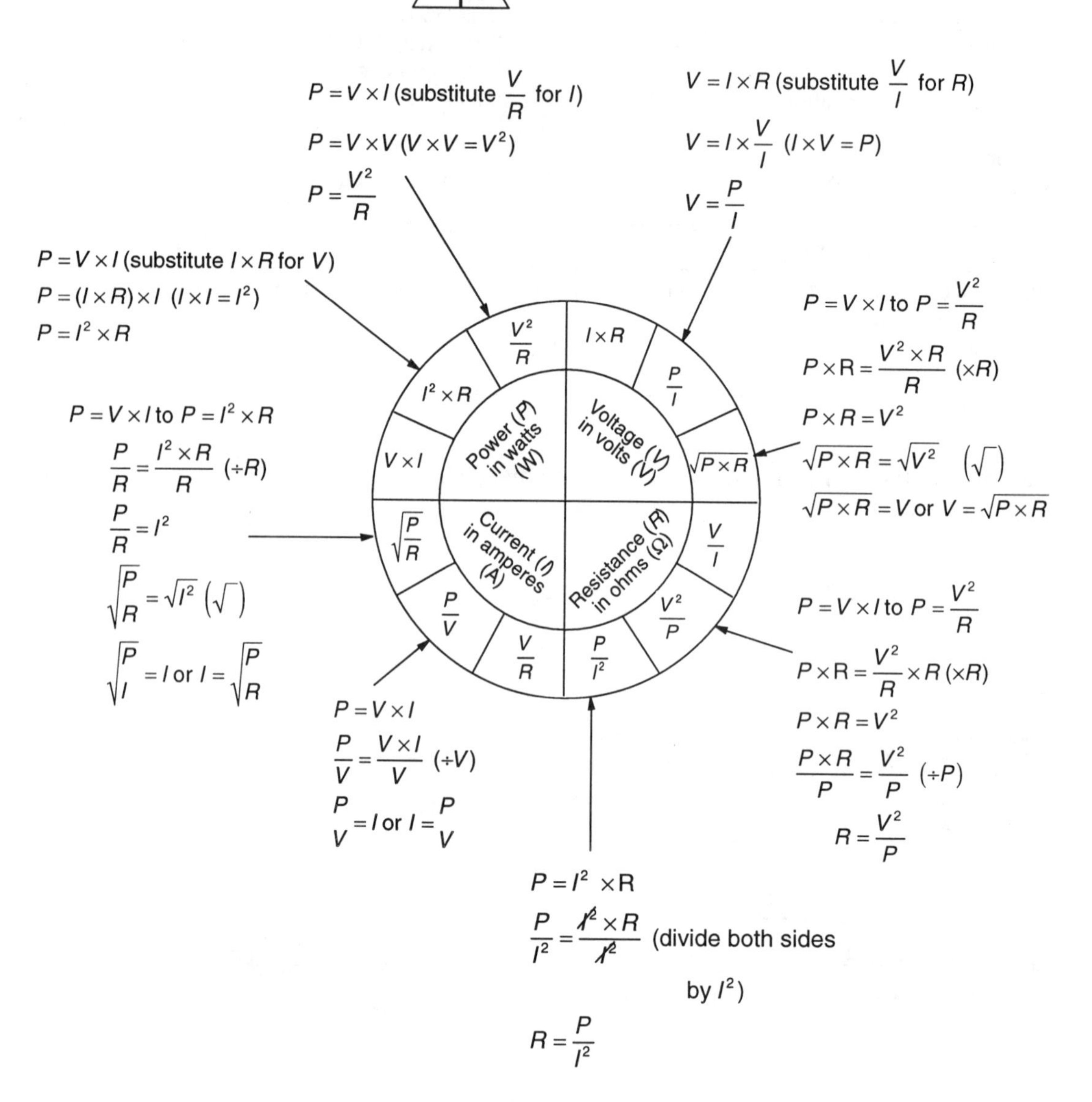

Energy Consumed (kWh) = Power (kW) × time (hours)

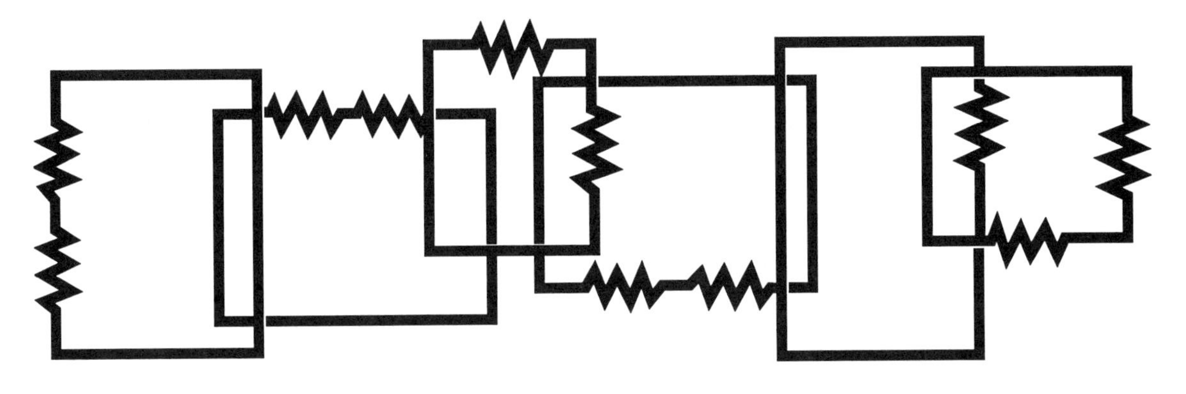

1

The Prototyping Board—Buses

Objectives

After completing this experiment, you will be able to:

1. Measure circuit continuity with the ohmmeter.
2. Determine the patterns of conductive paths in the prototype board.
3. Insert test points into the prototype board for measurements.
4. Accurately confirm the low resistance characteristic of conductors.
5. Confirm the nonconductive characteristic of insulating material.

Introduction

The prototyping board, or "proto" board, will be used extensively as you proceed through the experiments in this manual. The prototyping board will allow you to quickly assemble electrical circuits and easily perform tests and measurements upon the circuits. The purpose of this experiment is to become familiar with the prototyping board and to learn its electrical characteristics. To do this experiment you will need the following materials:

A prototyping board
An Ohmmeter with test leads (Analog or Digital or both)
Two test leads with alligator clips
Six 1.5 inch lengths of #24 hookup wire
A wire cutter/stripper tool.

The prototyping board, as shown in figures 1–1 and 1–2, is a rectangular plastic block with many small holes in the top. The holes are arranged in rows and their purpose is to enable you to insert wires into them to make electrical contact. Within the plastic material are metallic conductors underneath the holes. The metallic conductors interconnect the holes in patterns to make groups of connection points around which you can build electrical circuits. The holes serve as electrical connection points which are able to receive wires and gently clip onto them to hold them in place.

Along the edges of the two long sides of the prototyping board are the Bus Lines. There are two pairs of bus lines at each side of the board which extend down the length of the board. The bus lines are a part of the connection scheme on the board which are usually wired to a power source when using the board to test electrical circuits. The long buses conveniently distribute electrical power down the length of the board making it easy to get power to circuits at any location on the board. When you get into more advanced uses of the board you will appreciate this feature a lot.

FIGURE 1–1 Prototyping board

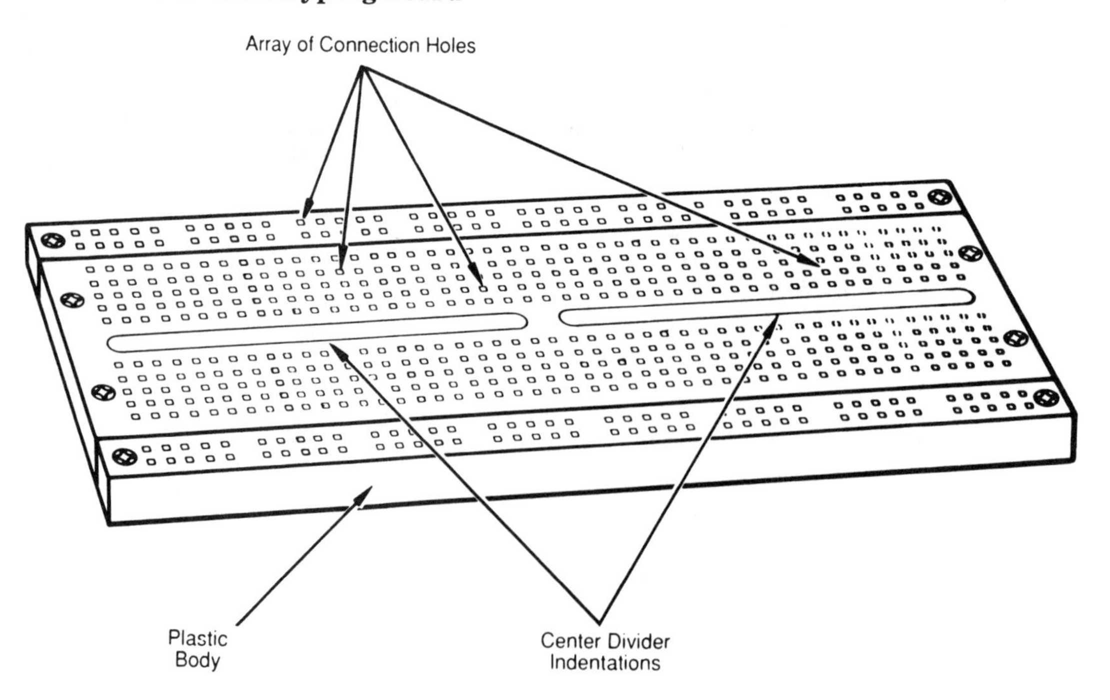

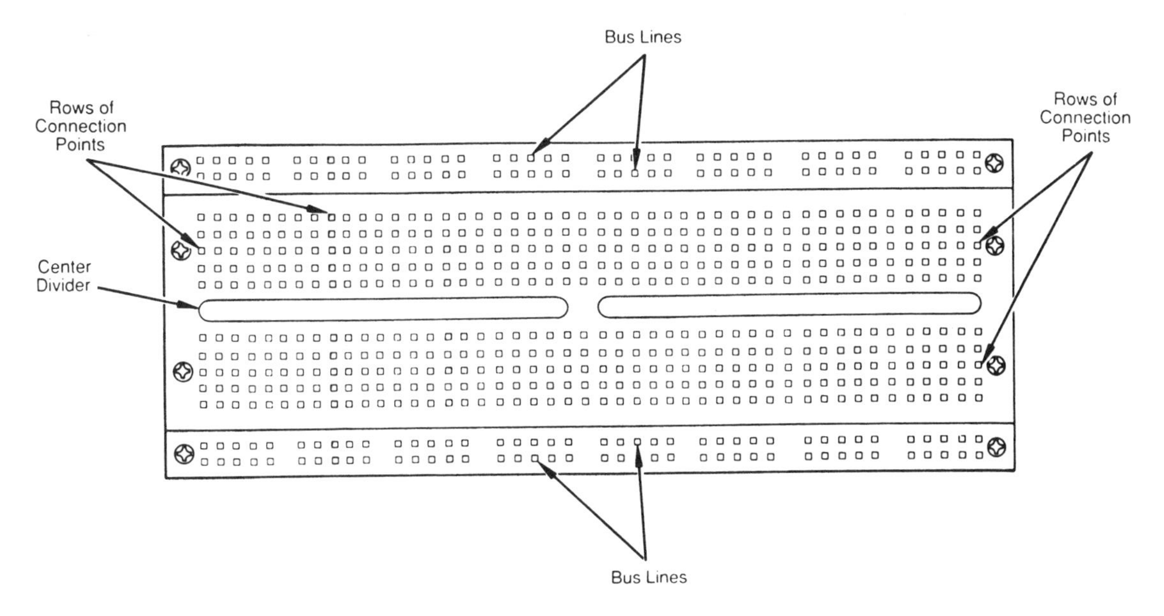

FIGURE 1–2 Top view prototyping board

Figure 1–3 shows only the bus line pairs of the prototyping board. The other areas of the board have been omitted for clarity. The bus line pairs are labelled A and B at the top side, C and D at the bottom side. Notice that all holes in each of the bus lines are connected together in the respective buses. All holes in bus line A are connected together as are all holes in bus lines B, C and D. Notice also that each of the bus lines is isolated from each of the others. Each of the bus lines is electrically independent of all of the others. Although the conductor is shown in Figure 1–3, when you actually look at the prototyping board the conductor is hidden by the plastic body and cannot be seen. That is why it is very important to know exactly how

FIGURE 1–3 Bus lines of a prototyping board

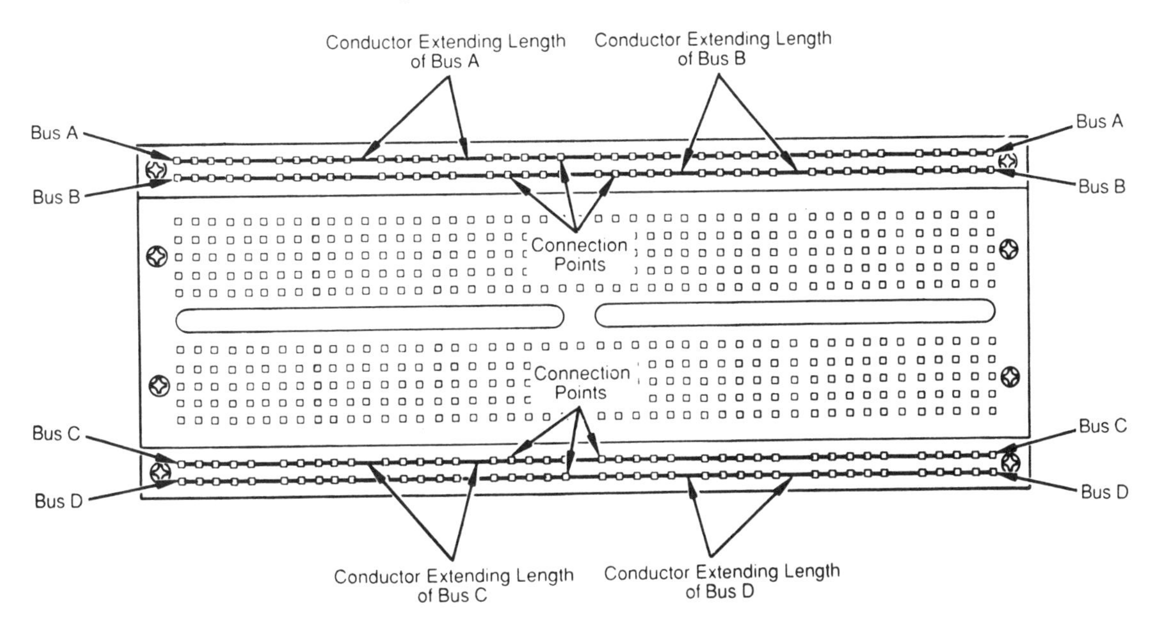

the buses are connected electrically because the internal metallic conductor is not visible.

Procedure

1. Gather the necessary materials. Prepare the six 1 inch lengths of hookup wire as shown in Figure 1–4 by cutting the wires to length and stripping about a quarter inch of insulation from each end.
2. Insert two of the test wires into bus line A. Place one of the wires at the end of the bus. Insert the second wire into the bus several holes away for the first wire. Check to assure that both wires are installed in bus line A and that neither has been accidentally placed into bus line B. The two bus lines are very close together.
3. Prepare the Ohmmeter for use to verify connection, or continuity, between the two test wires in bus line A. If you are using an analog ohmmeter select the RX1 scale, plug the test leads into the meter, and calibrate the meter. Calibration is done by shorting the test leads (simulating a zero ohm resistor) and adjusting the Zero Ohms adjustment until the pointer on the meter scale is positioned exactly over the zero ohms mark. If you are using a digital ohmmeter simply select the lowest resis-

FIGURE 1–4 Closeup of 1.5" length of hookup wire inserted into bus A

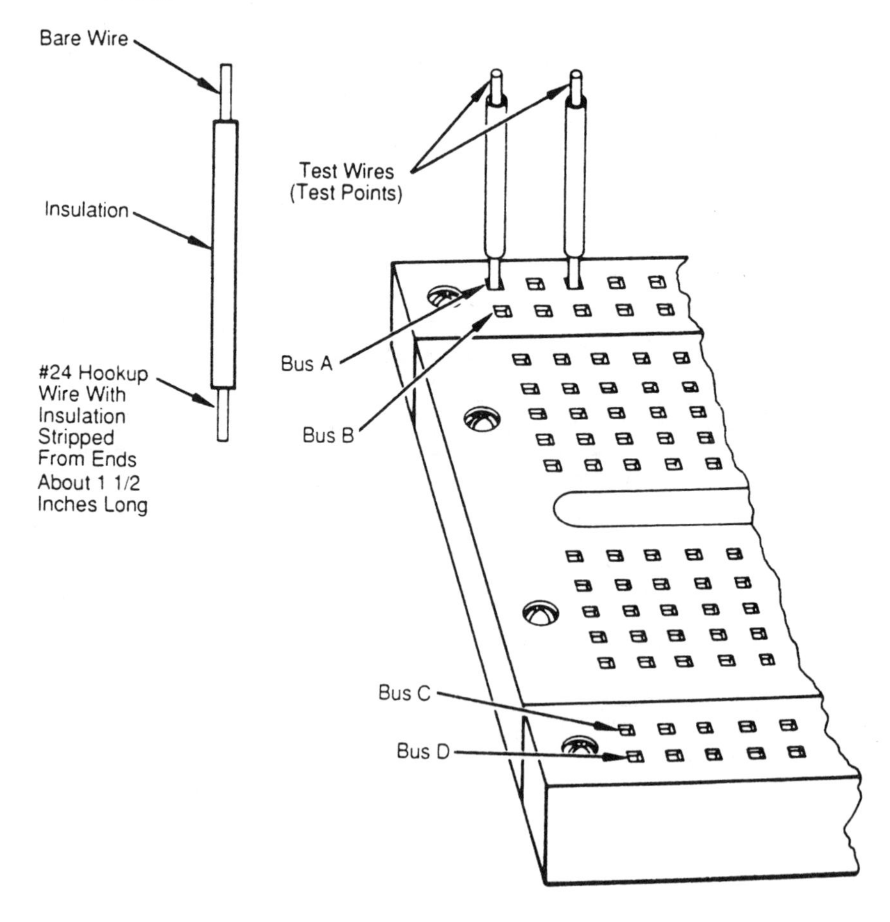

tance scale (usually 10 ohms or 200 ohms). With the digital ohmmeter it is not necessary to do any calibration by shorting the leads together before use. It is a good idea to short the leads together with the digital meter turned on to assure that it correctly indicates zero ohms. You will then be sure that it is working properly. If possible use both meters for the following tests, one after the other, to get experience with their different display styles.

4. Connect the ohmmeter test leads to the test wires which are installed in bus line A. Connect one lead to each test wire. It would be a good idea to use the alligator lead test wires for these connections so that you won't have to hold the meter test leads against the test wires in bus A. That way your hands will be free to do more important work. Look at the ohmmeter indication. Is the ohmmeter showing a value of zero ohms?
5. Insert a third test wire into bus line A several holes down the line from the second test wire. Look carefully to see that the third test wire has been inserted into bus A and not inadvertently into bus B. Connect the ohmmeter to test wires 1 and 3. What does the ohmmeter indicate? Now connect the ohmmeter leads to test wires 2 and 3. What does the ohmmeter indicate? Does the ohmmeter indicate zero ohms in both cases?
6. Now insert a fourth wire into bus line A at the opposite end of the bus away from test wire 1 as seen in figure 1–5 . Use the ohmmeter to measure continuity between all four of the test wires (1 to 2, 1 to 3 and 1 to 4). Does the ohmmeter show zero ohms for each of the tests?
7. Insert the fifth test wire into bus B at the same end of the bus as test wire 1 as seen in figure 1–6. Test wire 5 and test wire 1 will now be alongside each other but in their separate buses. If bus line A and bus line B are really isolated electrically from each other there should be no continuity, or electrical connection, between any test wires in bus A and any test wires in bus B.
8. Connect the ohmmeter test leads to test wires 1 and 5. What does the ohmmeter indicate?
9. Change the ohmmeter scale to its highest range, which is RX10,000 for the analog ohmmeter and for the digital ohmmeter 20 Megohms (20 MΩ) or higher if there is a higher position. Before using the analog ohmmeter on this new scale it will be necessary to re-calibrate it. Do it as before by shorting the ohmmeter test leads together and adjusting the Ohms Adjust for

FIGURE 1–5 Simplified schematic representation of bus A

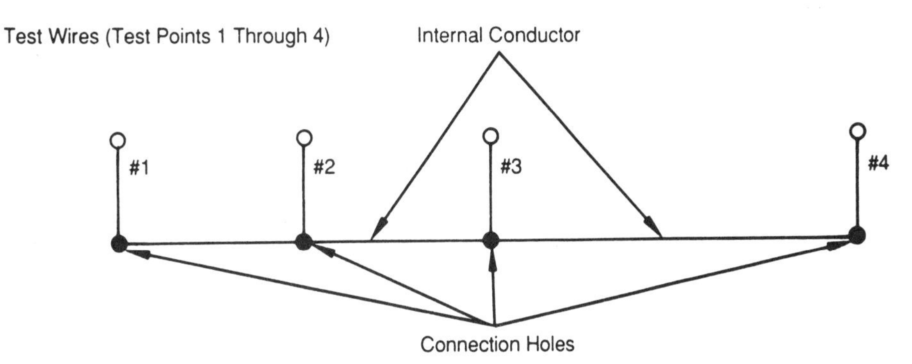

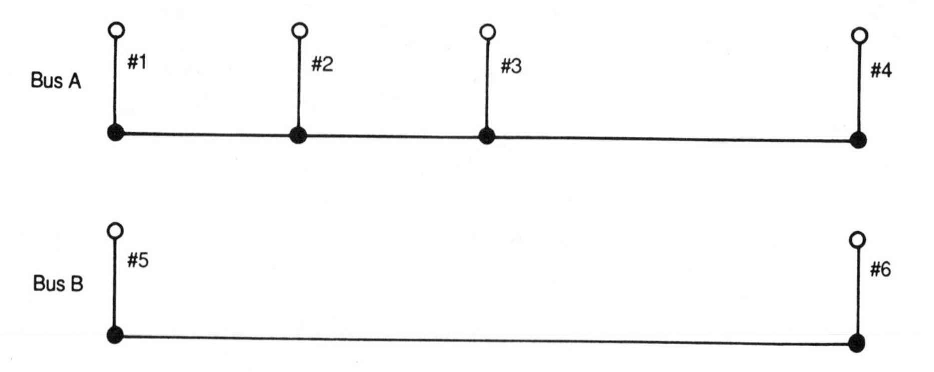

FIGURE 1–6 Simplified schematic representation of buses A and B

exactly zero ohms on the meter dial. After calibration it will give accurate measurements. Remember that it will be necessary to re-calibrate the analog ohmmeter each time you change its measurement scale. If the re-calibration is not done then any measurements made with the ohmmeter will not be accurately indicated on the ohmmeter scale. Connect the ohmmeter leads again to test wires 1 and 5 on the prototyping board. What does the ohmmeter indicate?

10. Insert test wire 6 into bus B at the far end of the bus away from test wire 5. Test wires 4 and 6 will now be adjacent to each other but in their separate buses. Test for continuity with the ohmmeter between all test wires on bus A and the test wires on bus B, checking to see if there is any connection between bus A and bus B. What does the ohmmeter tell you?

11. Set the ohmmeters back to their low resistance modes (don't forget to recalibrate the analog meter) and test for continuity between test wires 5 and 6. These are the test wires inserted into opposite ends of bus B. What does the ohmmeter indicate?

Review Break

What have you done so far? You inserted test wires into buses A and B in order to make electrical measurements with the ohmmeter. The test wires installed into the buses served as "test points" for the measurements that you performed. Initially you used the ohmmeter on its low resistance scale to verify continuity between several test points on bus A. The ohmmeter proved that the resistance between the test points on bus A was very low, essentially zero ohms.

You then tested for isolation between bus A and bus B. The ohmmeter proved that buses A and B are indeed electrically isolated because it showed infinite resistance between them. And finally you verified that there is continuity between test points in bus B, proving that bus B is electrically similar to bus A. You have learned the electrical characteristics of buses A and B. Perhaps even more importantly you learned something about the ohmmeter, that it has certain limitations in its ability to measure low resistances and very high resistances. You discovered that in order to measure continuity between test points it is necessary to use the lowest resistance scale of the meter to obtain the most accurate measurement. On the other hand, when attempting to verify isolation between test points it is necessary to use the highest resistance scale of the ohmmeter. Electrical isolation, or no connection, between test points is ideally an infinitely high resistance. To most accurately verify that condition it is essential to use the highest resistance scale of the meter.

You also learned that you have a choice. It is possible to verify continuity, and isolation, with either an analog ohmmeter or a digital ohmmeter. Later as you learn more about the ohmmeter you will become aware of the differences between the two kinds of meters.

12. Repeat the same measurements that you performed on buses A and B on the buses at the opposite edge of the prototype board, buses C and D. Are they the same electrically?

13. When you have completed the tests, return all of your materials to their proper places. Work on developing a reputation for neatness and orderliness. A truly professional electronics technician or electronics engineer treats all materials and instruments with great care.

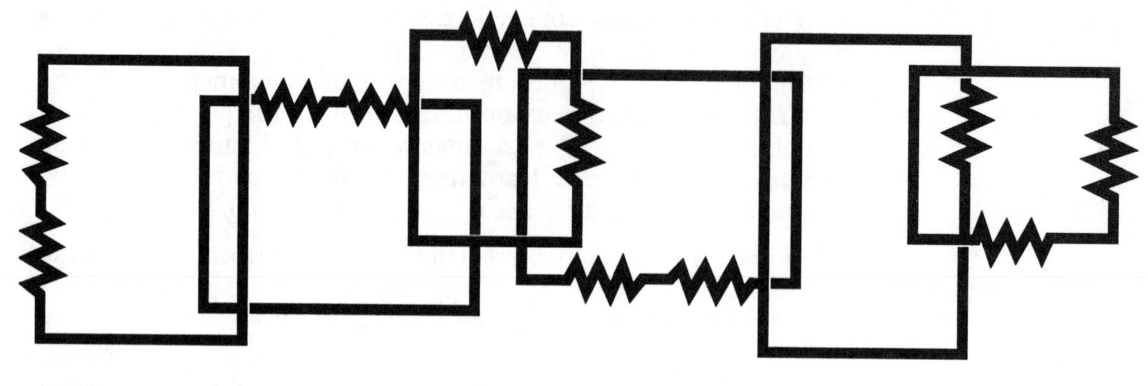

2

The Prototyping Board— Connections

Objectives

After completing this experiment, you will be able to:

1. Determine the patterns of conductive paths in the prototype board.
2. Appropriately select the resistance range of the ohmmeter.
3. Accurately read resistance values from the ohmmeter.
4. Utilize test wires for prototype board continuity measurements.

Introduction

You will now continue your examination of the prototyping board to discover how the remaining connection holes are laid out electrically. The same measurement techniques that you used to investigate the prototyping board buses will be used to check out the rest of the board. To perform this experiment you will need the following materials:

A prototyping board
An Ohmmeter with test leads (Analog or Digital or both)
Two test wires with alligator clips
Six 1.5 inch lengths of #24 hookup wire.

Figure 2–1 is a simplified illustration of the prototyping board showing the rows of connection holes which will be tested to verify either continuity between them, or isolation (no connection) between them. The connection holes extend outwardly from the recessed center dividers. Each small row contains five connection holes which are electrically connected together by a metallic conductor as seen in figure 2–2. The metallic conductor connecting the holes together is imbedded in the plastic of the board so it is not visible to the eye. Even though it cannot be seen visually, its presence can easily be verified electrically with the ohmmeter. Each row of five connection holes is electrically isolated from all of the others. This will also be verified by using the ohmmeter.

Procedure

1. Gather the necessary materials. Use the same 1.5 inch lengths of hookup wire that you used in experiment 1.
2. Set the ohmmeter to measure electrical continuity. If you are using an analog meter, select the RX1 scale, short the test leads together and adjust the Ohms Adjust to cause the meter to show exactly zero ohms. If you are using a digital meter, then simply select the lowest resistance scale on the meter. It is a good idea to short the test leads with the digital meter as well, to assure that it is working properly. With the test leads shorted together it should show exactly zero ohms.

FIGURE 2–1 Prototyping board top view

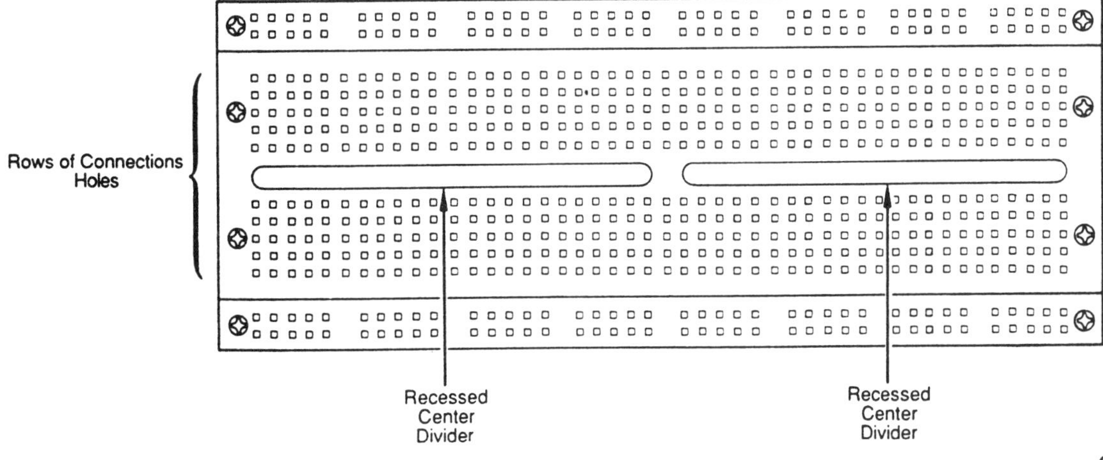

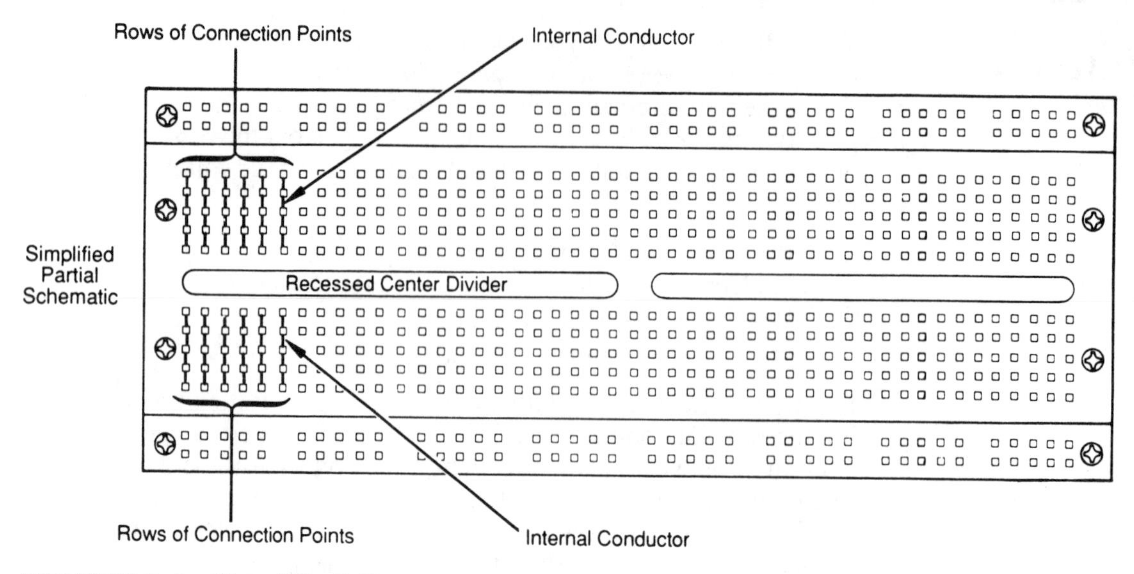

FIGURE 2–2 Simplified diagram of the rows of connection holes

3. Insert the six test wires into the prototyping board at the locations shown in figure 2–3. The test wires will serve as "test points" for the electrical tests which follow.
4. Connect the ohmmeter to test wires 1 and 2. It would be a good idea to again use the alligator clip leads to make the connection. This will allow your hands to be freed for other important tasks.

FIGURE 2–3 Proto board with test wires inserted into selected holes

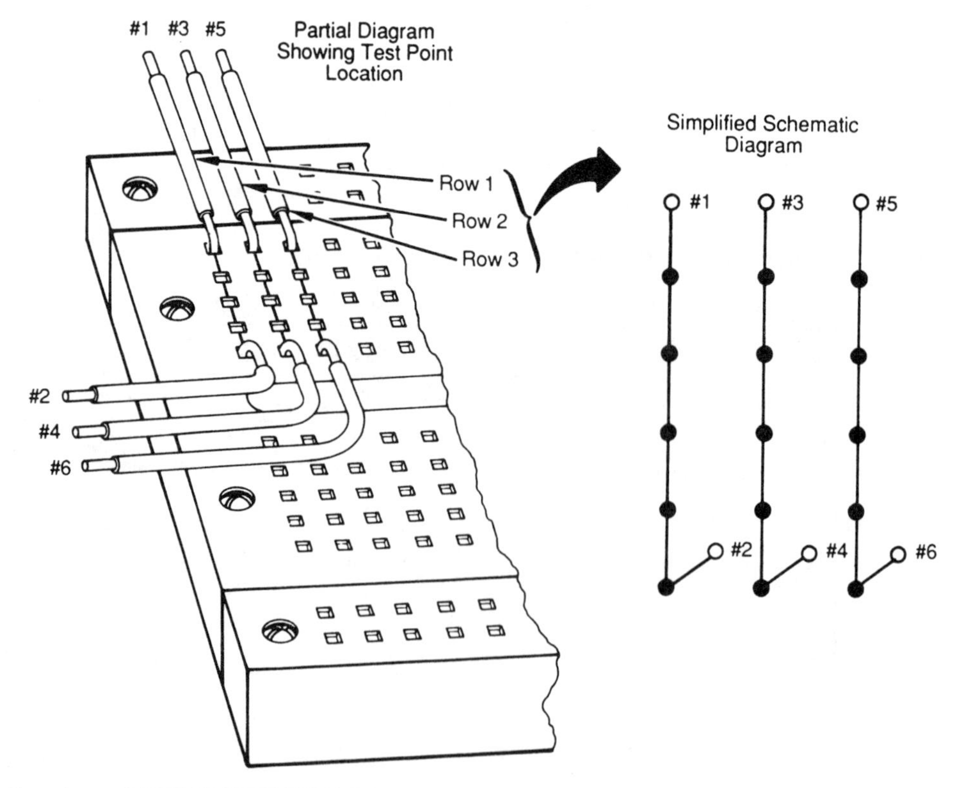

5. Since test wires 1 and 2 are inserted into the same row there should be electrical continuity between them. Does the ohmmeter indicate zero ohms?
6. Connect the ohmmeter to test wires 1 and 3. Would you expect to find continuity between them? What does the ohmmeter tell you?
7. Now prepare the ohmmeter to measure high resistance in order to verify conditions of electrical isolation or no connection. If you are using the analog ohmmeter, select the RX10,000 scale and calibrate the meter for accurate measurements. If you are using the digital ohmmeter select the highest resistance scale.
8. Connect the ohmmeter to test wires 1 and 4. These test wires are adjacent to one another but they are located in different rows. There should be no continuity between these test points. What does the ohmmeter show?
9. Connect the ohmmeter to test wires 4 and 5. These test wires are also adjacent to one another but are in different rows. What does the ohmmeter indicate?
10. Finally, connect the ohmmeter to test wires 5 and 6. These wires are installed in the same row, so electrical continuity between them would be expected. What does the ohmmeter tell you?
11. Look at the ohmmeter to see whether you are set up to measure high resistance or low resistance. The ohmmeter should still be in the high resistance mode. Although it indicated zero ohms for step 10, it is best to be in the low resistance mode when verifying continuity. Change the ohmmeter back to its low resistance setting (don't forget to re-calibrate if it's an analog meter) and connect the ohmmeter again to test wires 5 and 6. What does the ohmmeter indicate?

Review Break

You've done basically the same tests that you did in experiment 1. The tests were done in a different area of the prototyping board though. You inserted test wires into select holes on the prototyping board and verified continuity between wires in the same row. While verifying continuity you used the ohmmeter in its low resistance mode to obtain the most accurate measurement. You then shifted the ohmmeter into the high resistance mode to verify no connection between test wires inserted into different rows. You used the high resistance scale to obtain the most accurate measurement. You then verified continuity between test wires 5 and 6 with the ohmmeter in its high resistance mode. Although, technically, it is a sign of inexperience to use the ohmmeter in such a fashion you did indeed read zero ohms between the test points. After putting the ohmmeter into its proper mode for continuity tests you again measured the condition between test wires 5 and 6. This time the indicated zero ohms on the meter was more reliable than the previous measurement. As you gain more experience with the use of the ohmmeter, and learn more about how it works, you will understand why it is important to have the ohmmeter set up correctly to monitor low resistance conditions and high resistance conditions. The purpose of experiments 1 and 2 was to give you experience in measuring continuity, measuring no connection (electrical isolation), and reveal to you how the connection holes on the prototyping board relate to each other. If there is still a little confusion on any of those points, feel free to improvise and do some additional experimentation on your own. Creativity is a very important personal attribute in the technical professions and the sooner you begin to develop yours the better.

12. When you are satisfied that you have accomplished the objectives of the experiment, return all materials to their proper places.

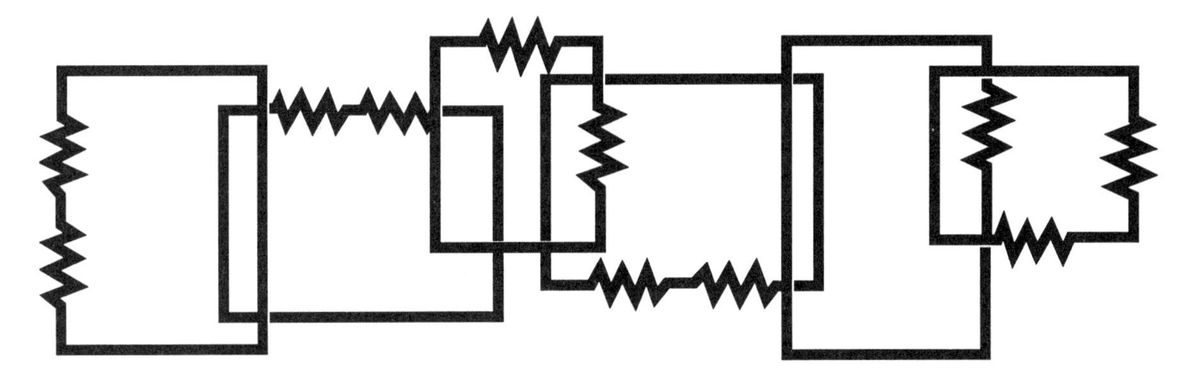

3

Circuit Wiring—Continuity

Objectives

After completing this experiment, you will be able to:

1. Prepare the variable dc power supply for use as a circuit source.
2. Construct a simple series circuit.
3. Determine the electrical characteristic of the conductor.
4. Use the incandescent lamp as an indicator of current flow.
5. Determine the resistance of the incandescent lamp.

Introduction

Now that you know how the prototyping board operates, it is time to use it to construct a simple electrical circuit. With the circuit you are about to build you will perform some basic, but very important, electrical tests. In order to do the experiment which follows you will need these materials:

A prototyping board
An adjustable dc power supply
An Ohmmeter with test leads (Analog, Digital, or both)
Ten 1.5 inch lengths of test wire such as used in experiments 1 and 2
Two test wires with alligator clips
One T $1\frac{3}{4}$ B1-PIN light bulb (#7382 14V 80 mA)
A wire cutting/stripping tool.

The purpose of this experiment is to construct an electrical circuit using several small pieces of wire connected to a light bulb and an electrical power source. Several of the 1.5 inch test wires will be used as circuit test points and others will be used as circuit wiring. You will test the electrical continuity of the circuit using the ohmmeter, and you will observe how the light bulb in the circuit affects continuity measurements. After fully testing the circuit with the ohmmeter you will then connect power to the circuit to see whether it works. Although there is little personal danger associated with this experiment, you will be working with electrical power. Exercise care in following the procedures to protect the instruments you will be using from being damaged.

Procedure

1. Gather the necessary materials, and using figure 3–1 as a guide, build the electrical circuit. Note that no electrical power is yet connected to the circuit. Power will be applied after all preliminary tests are performed. It may be necessary to prepare additional 1.5 inch test wires, with the insulation stripped from the ends, so that you will have enough both for circuit wiring and circuit test points. Prepare the wires now if necessary.
2. Set the ohmmeter up to measure circuit continuity. In the case of the analog meter, select the RX1 scale, short the test leads together, and adjust the Ohms Adjust until the meter indicates exactly zero ohms. With the digital ohmmeter, select the lowest resistance scale. Test the meter to assure that it indicates exactly zero ohms with the test leads shorted.
3. Connect the alligator clip leads to the ends of the ohmmeter test leads, then connect the alligator clips to test points 1 and 2. You are using the alligator clip leads as an extension of the ohmmeter test leads in order to free your hands for more important work. What does the ohmmeter indicate?
4. Now move the ohmmeter test lead clipped to test point 2 to test point 3. Leave the ohmmeter lead connected to test point 1 so that you are measuring for continuity between test points 1 and 3. What does the ohmmeter now indicate? Why do you suppose the ohmmeter reading has changed?

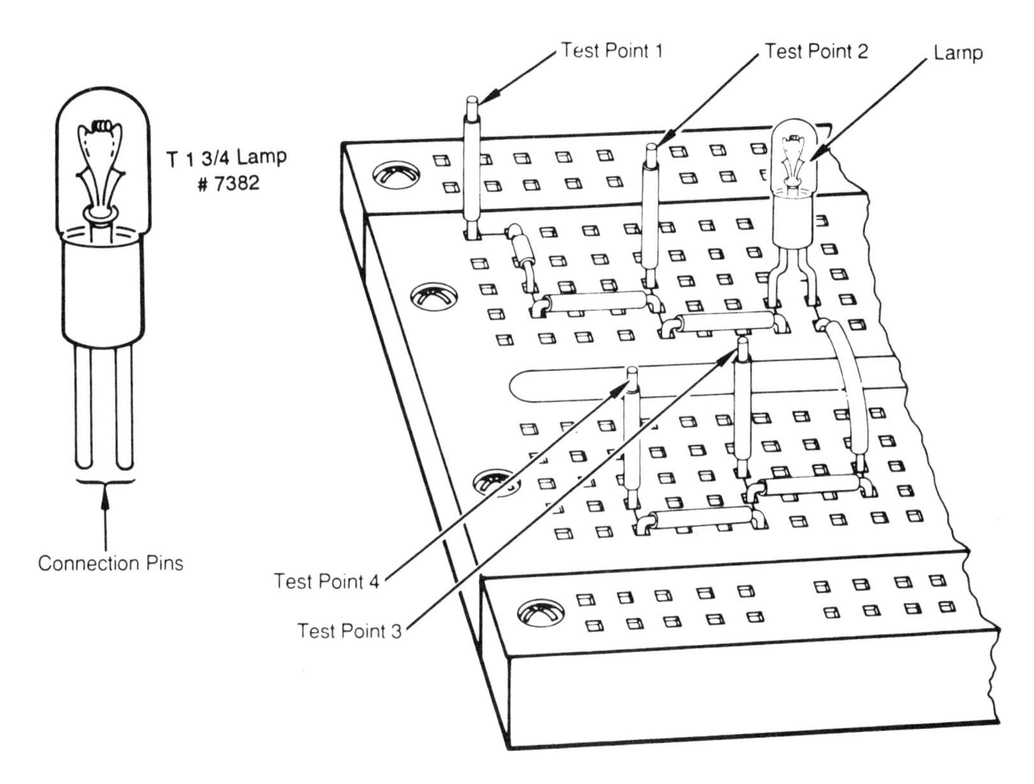

FIGURE 3–1 Proto board showing position of all circuit parts and test points

5. Leave the ohmmeter lead connected to test point 1 still connected, but now move the ohmmeter lead connected to test point 3 on to test point 4. The ohmmeter leads should then be connected between test points 1 and 4. What does the ohmmeter indicate? Is the reading the same as you obtained in step 4?

Review Break

What have you done so far? You have built an electrical circuit consisting of several short lengths of wire with several other short lengths of wire strategically placed to serve as test points. You then proceeded to evaluate the electrical continuity of the circuit by using the ohmmeter in its low resistance mode to measure circuit resistance from one test point to each of the others in turn. If the electrical circuit had been constructed wholly of wire, and nothing else, its resistance would have been essentially zero ohms.

But since the circuit also has a light bulb, the total circuit resistance will be somewhat greater that zero ohms. You verified that the continuity from test point 1 to test point 3 was slightly higher than zero ohms. This tells you that the light bulb is not the same as a piece of hookup wire. You further found that the continuity from test point 1 to test point 4 was the same as from test point 1 to test point 3. This showed you that the resistance of the remaining wire in the circuit had good continuity since it didn't cause the circuit resistance to increase more. What is the moral of this little story? Copper wire has very low resistance in electrical circuits, but things like light bulbs have a higher resistance. This is quite normal and your understanding of this important point will serve you well in the weeks to come. Continue now with the procedure to shed some additional "light" on the subject.

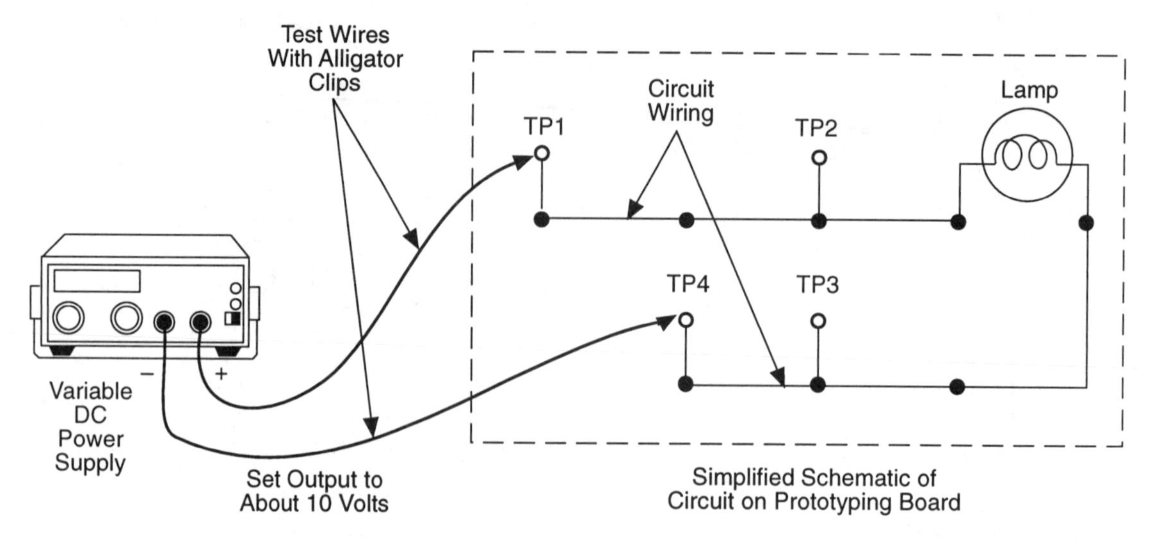

FIGURE 3–2 Simplified schematic of circuit

6. Turn on the variable output dc power supply and set the output to approximately ten volts as seen in figure 3–2. Using the alligator clip leads connect the output terminals of the power supply to test points 1 and 4. This will complete the electrical circuit and allow current flow. It isn't important which terminal of the power supply connects to which test point since the light bulb will work equally well with current flow in either direction. You have just created a working (although very simple) electrical circuit. If for some reason the light does not shine, double check everything and try again. You will more than likely find a very minor mistake.
7. When you are satisfied that everything is in working order, disassemble the circuit and return all materials to their proper places.

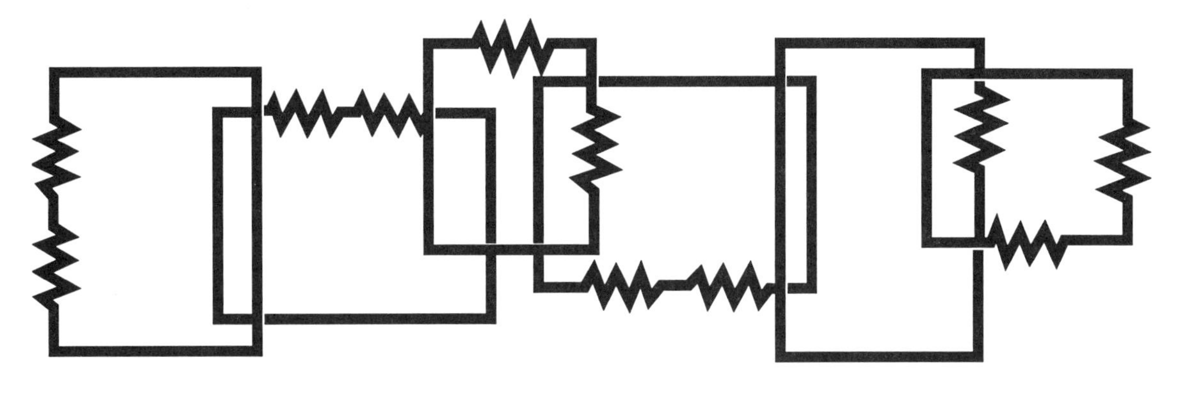

4

Circuit Wiring—Measurements

Objectives

After completing this experiment, you will be able to:

1. Evaluate the electrical properties of a simple circuit.
2. Set the variable dc power supply output accurately.
3. Measure the voltage across a circuit load.
4. Measure the current flow in an electrical circuit.
5. Measure voltage and current flow simultaneously.

Introduction

In the previous experiment you constructed a simple electrical circuit to demonstrate how wires can be used to carry electrical energy to a "load," which in this case was an incandescent lamp. When power was applied to the circuit the lamp illuminated in the same way that lights in your home burn brightly when switched on. The only measurements that you made were to establish continuity in the circuit before power was applied.

In this experiment you will make a similar circuit and use it as a basis to measure electrical characteristics. You will measure the resistance of the circuit, the source voltage, the current flow through the circuit, and the voltage drop across the load. As in the prior experiment, the load will be the incandescent lamp. The following materials will be needed to perform this experiment:

A T $1\frac{3}{4}$ incandescent lamp #7382 (14V 80 mA)
A Volt-Ohm-Milliammeter (VOM), either Analog, Digital, or both, with test leads
Six 1.5 inch #24 wires
A prototyping board
Four test wires with alligator clips

The purposes of this experiment are to use the Volt-Ohm-Milliammeter to measure the electrical quantities of resistance, voltage and current flow. You will also be able to calculate a resistance value which cannot be measured directly, based on measured voltage and current flow. Perhaps the most important purpose of the experiment is to learn to use the VOM to measure the desired electrical characteristic, (resistance, voltage or current flow) with the best precision or maximum accuracy. You have no doubt noticed that the multimeter (VOM) has a variety of functions, and scales within the functions. It is the trademark of an experienced technician or engineer to know which scale will produce the most accurate, precise measurement.

Procedure

1. Collect the materials necessary to do this experiment. Using figure 4–1 as an example, wire the circuit together. Note that the power supply is not connected to the circuit. Power will be applied after circuit resistance is measured.
2. Remove the activity sheet. It will be necessary for you to write down the measurements you will be making during the course of the experiment. It will also be necessary to make the measurements with the best accuracy you are capable of, because the data will be used to calculate a value of resistance that cannot be measured directly.
3. Set the VOM to operate as an ohmmeter, select the scale which will give you the most accurate measurement, then measure circuit resistance between test points 1 and 4 as seen in figure 4–2. Record the measured resistance in the appropriate space on the experiment activity sheet. Remember to use the alligator clip leads when making measurements to free your hands for more important tasks. The resistance you have just measured repre-

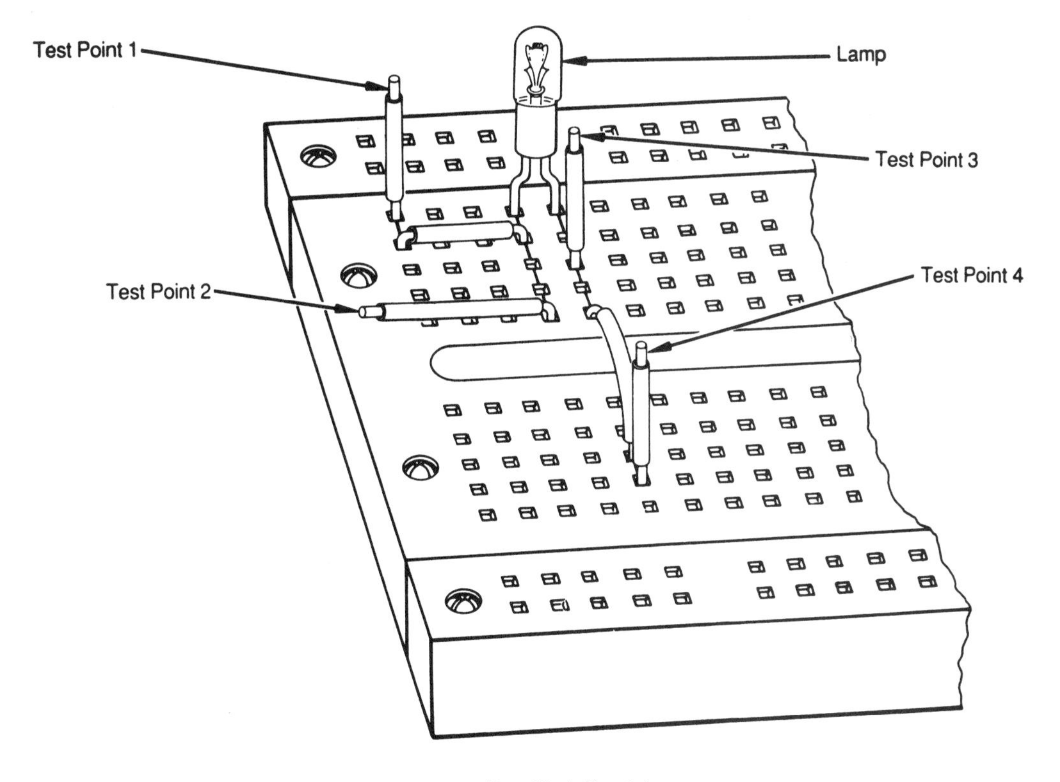

FIGURE 4–1 Proto board with circuit wiring, test points and circuit load installed

sents total circuit resistance. Disconnect the ohmmeter from the circuit.

4. Be sure that the power supply is turned off, then connect the power supply terminals to the circuit at test points 1 and 4 using a set of test wires with alligator clips. You are about to measure voltage with the VOM and it will be necessary to connect the power supply properly with respect to the polarity of its output voltage. Connect the positive side of the supply to test point 1, and connect the negative side of the supply to test point 4.

FIGURE 4–2 Simplified schematic showing ohmmeter connecting to test points 1 and 4

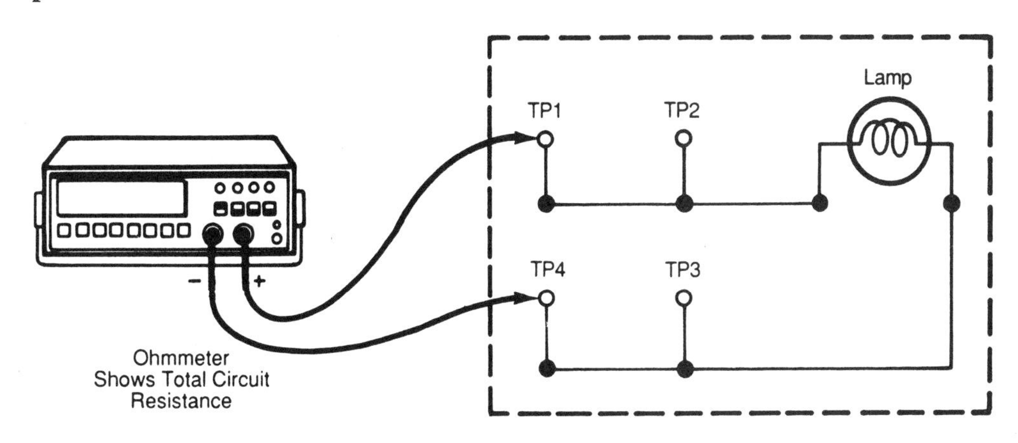

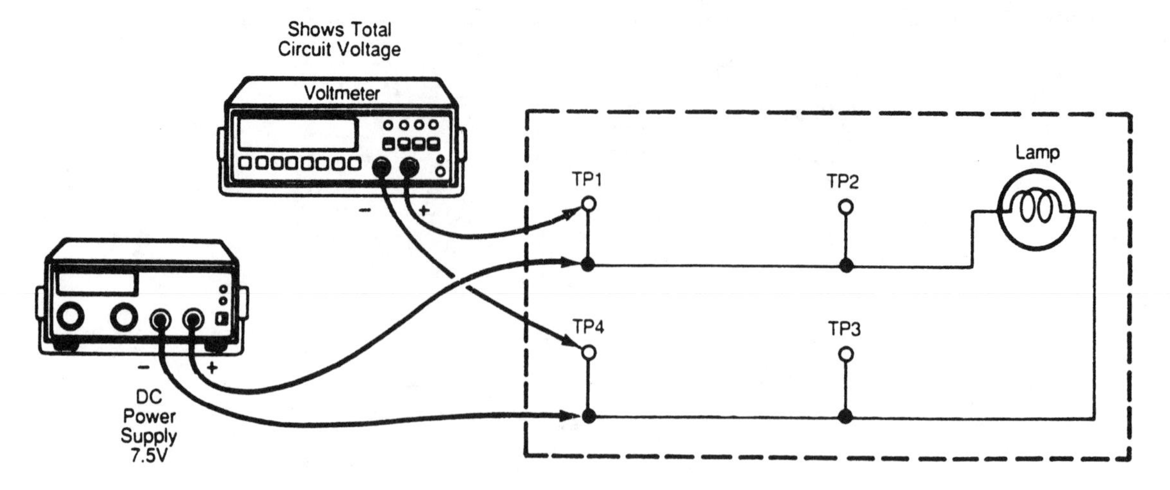

FIGURE 4–3 Simplified schematic showing voltmeter connected to test points 1 and 4

5. Set the VOM to operate as a voltmeter, select the scale which will give you the most accurate measurement, then connect the voltmeter leads to test points 1 and 4 as seen in figure 4–3. Use test wires with alligator clips, connect the positive lead of the voltmeter to test point 1, and connect the negative lead of the voltmeter to test point 4. You are now ready to measure the total voltage which will be applied to the circuit.
6. Assure that the output voltage control on the power supply is set for minimum output. You may now safely turn on the power supply and slowly begin to increase its output voltage. While watching the voltmeter, increase the output of the voltmeter to exactly 7.5 volts. Record this voltage in the appropriate blank on the experiment activity sheet. Incidentally, the lamp should now be illuminated.
7. Disconnect the positive lead of the voltmeter from test point 1 and connect it to test point 2. Read the voltmeter indication. Has it changed?
8. Disconnect the negative lead of the voltmeter from test point 4 and connect it to test point 3 as seen in figure 4–4. Look again at the voltmeter indication. Has it changed? You are now measuring the voltage across the load. Write this voltage down in the appropriate blank on the experiment activity sheet.
9. The next part of the experiment, current flow measurement, is a little more complicated. Pay extremely careful attention each step of the way to avoid damage to the meter.
10. Disconnect the meter from the circuit and turn the power supply off. Be careful not to disturb the output voltage setting at the voltage adjust knob. Remove the wire between test points 3 and 4. Set the VOM to operate as a milliammeter, select the scale which will safely result in the most accurate measurement, and connect the milliammeter to test points 3 and 4 as seen in figure 4–5. It is very important to observe proper polarity so connect the positive lead of the milliammeter to test point 3 and connect the negative lead of the milliammeter to test point 4. Use the alligator clip leads to free your hands for more important work.

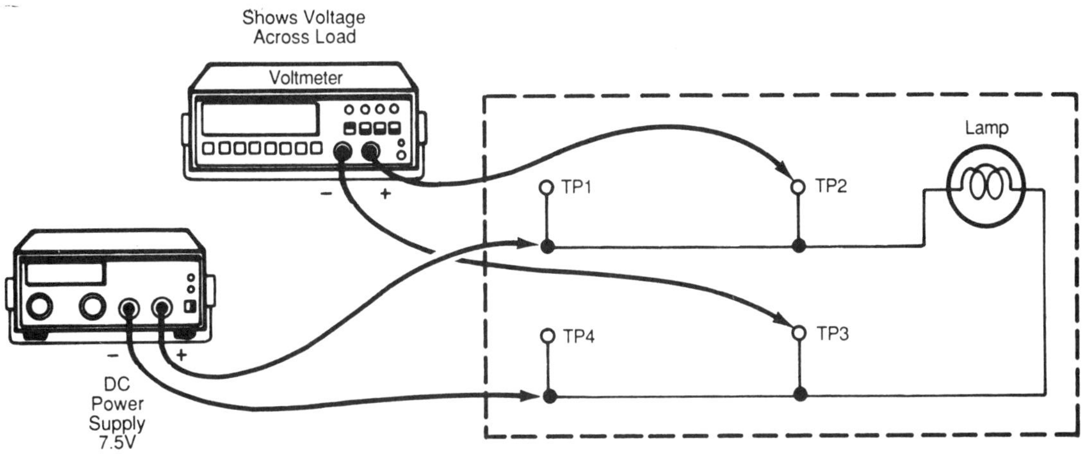

FIGURE 4–4 Simplified schematic diagram showing voltmeter connection to test points 2 and 3

11. Now turn on the power supply. The lamp should again illuminate, telling you that the circuit is operating normally. Look at the milliammeter. Does it indicate any current flow?
12. Now remove the circuit wire which is connecting test point 1 to test point 2 as seen in figure 4–6. This will allow all circuit current to flow through the milliammeter and the milliammeter should now be correctly indicating the value of circuit current. You are now monitoring the amount of total current flowing through the circuit and the load. Record this current value in the appropriate blank on the experiment activity sheet. You may now turn the power supply off. You have completed all electrical measurements for this experiment.

Review Break

You have made several measurements in rapid succession as seen in figure 4–7, and what you have done may seem somewhat confusing.

FIGURE 4–5 Simplified schematic diagram showing milliammeter position to measure circuit current flow

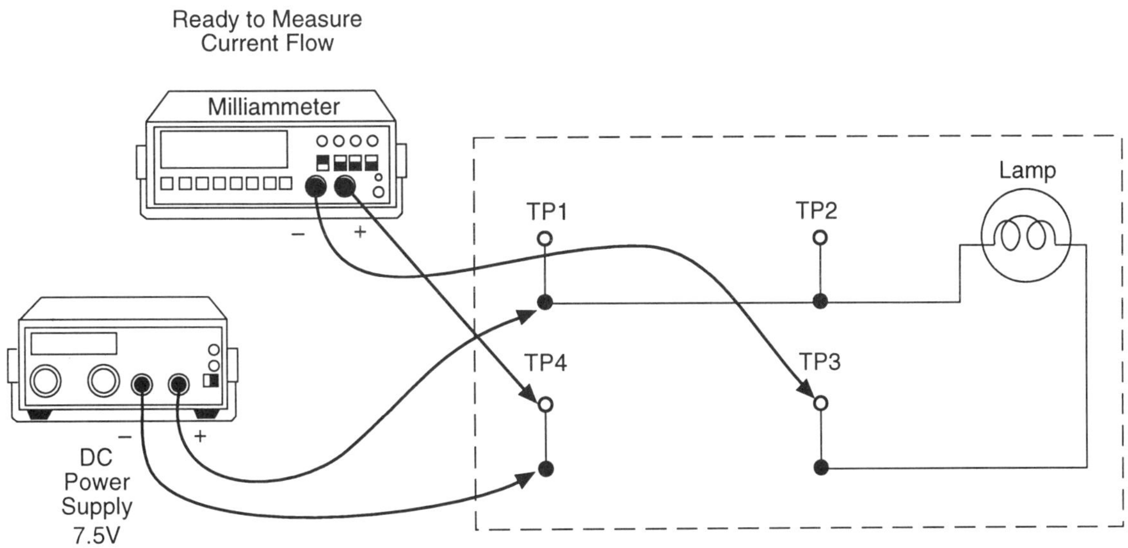

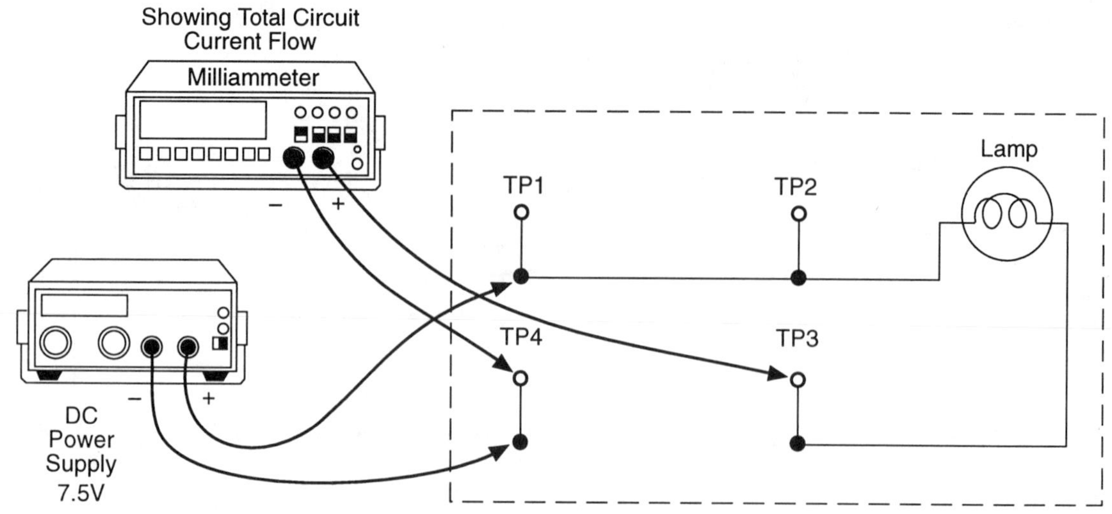

FIGURE 4–6 Simplified schematic showing milliammeter now in circuit and measuring total circuit current flow

Let's review each step of the procedure individually. You constructed the circuit and measured its total resistance. This was done with no power applied to the circuit, which is a most important point. Resistance must always be measured when the circuit under test is removed from any electrical power. This is necessary for three reasons: (1) To prevent any danger to you the measurer. (2) To

FIGURE 4–7 Simplified schematic diagram showing circuit test connections to measure voltage and current

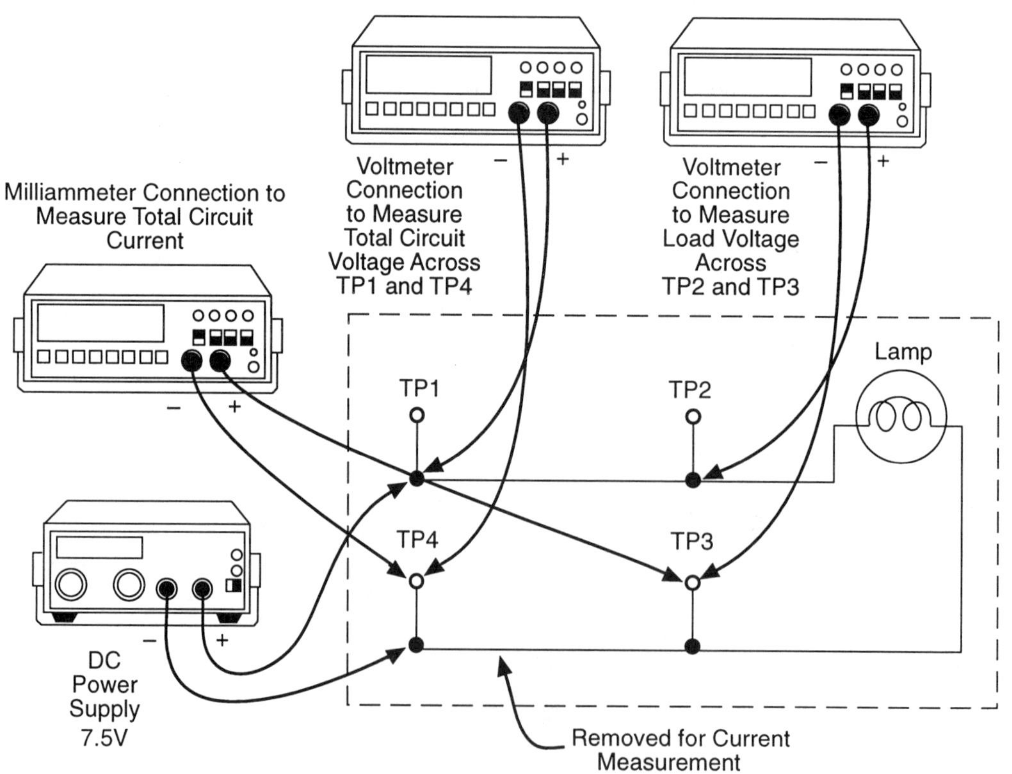

prevent any damage to the ohmmeter. (3) To assure that the measured resistance is accurate. When you get to advanced study of the ohmmeter you will understand why this is so.

Next you connected the power supply to the circuit, observing proper polarity, and also connected the voltmeter to the circuit to match the polarity of the power supply. This allowed you to measure the total voltage applied to the circuit.

You then shifted the voltmeter leads directly across the load to measure the load voltage and found that the total circuit voltage and the load voltage are exactly the same. Why? Because the circuit wiring has such low resistance that it doesn't drop any appreciable voltage. This is normal, and in fact, desirable. We want all of the electrical power to reach the load.

Finally you measured circuit current flow. You initially connected the milliammeter across the circuit wire connecting test points 1 and 2. While connected in this way the milliammeter showed zero current flow because circuit current was still flowing through the circuit wire. Then you removed the circuit wire from between test points 1 and 2 which caused the milliammeter to become part of the circuit. All circuit current had to flow through the milliammeter and it was able to accurately show you how much current was flowing in the circuit. This is probably one of the most important points in all of your studies of basic electronics. In order to measure current flow, the milliammeter must be part of the circuit. All circuit current must flow through the current measurement meter.

While performing each of the measurements you recorded the values of resistance, voltage, and circuit current on the experiment activity sheet. Now for the last phase of the experiment.

13. Use the measured load voltage and the measured load current to calculate circuit resistance. Does the calculated value agree with the value measured at the beginning of the experiment? Why not? It has to do with the light bulb. The incandescent lamp is a special kind of resistor. It has what is called a cold resistance characteristic which can be measured, as you did. While it is in operation it has a "hot" resistance characteristic, which cannot be measured directly because the lamp has current flow through it. But if you know the value of voltage across the lamp and the value of current flow through the lamp, then Ohm's law will enable you to calculate the resistance that can't be measured. You have learned one of the great secrets of electronics. The power of knowledge is awesome!

14. Disassemble the test circuit and return all materials to their proper places. Complete the required computation on the experiment activity sheet.

ACTIVITY SHEET EXPERIMENT 4

NAME ______________________________

DATE ______________________________

Step 3 **A.** Measured total circuit resistance between test points 1 and 4.

____

Step 6 **B.** Measured total circuit voltage between test points 1 and 4.

____

Step 8 **C.** Measured load voltage between test points 2 and 3.

____

Step 12 **D.** Measured total circuit current flow with milliammeter completing current path between test points 1 and 2. ____________

Step 13 **E.** Using Ohm's Law to calculate circuit resistance with power applied (lamp illuminated).

$$R_{\text{Circuit}} = \frac{V_{\text{Load}}}{I_{\text{Load}}}$$ ____________________

F. Is there a difference between the measured resistance in part A and the calculated resistance in Part E? Why? ______________

____

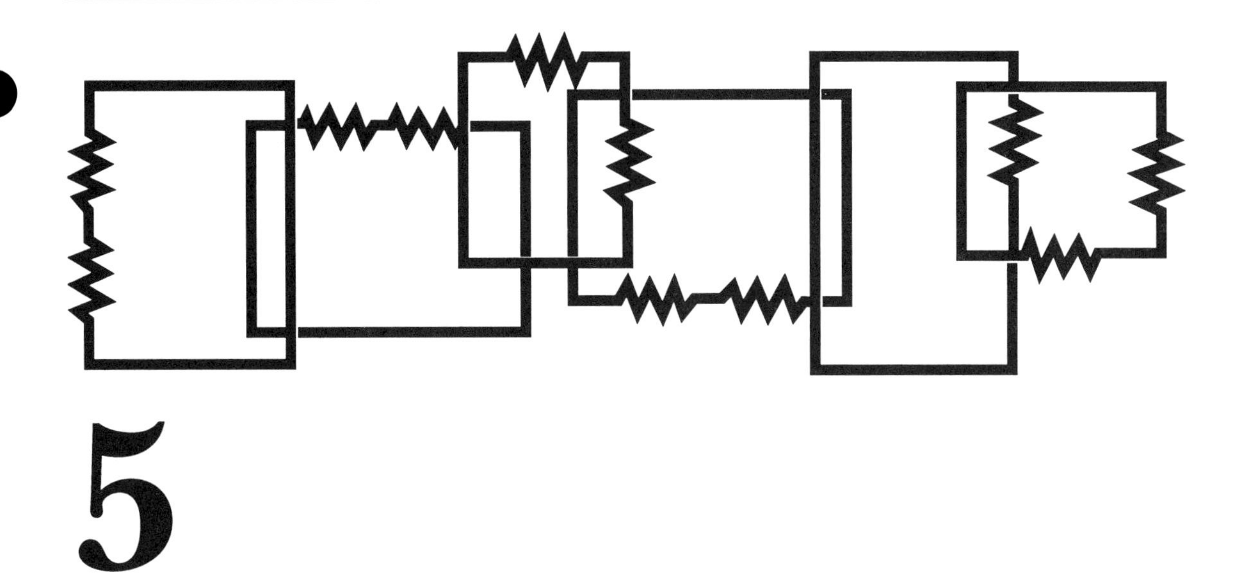

5

Using Ohm's Law

Objectives

After completing this experiment, you will be able to:

1. Determine the excess voltage when comparing a source to load mismatch.
2. Calculate the effectiveness of a series dropping resistor in use.
3. Calculate and select the value of resistance needed as a series drop.
4. Measure load voltage and current flow.

Introduction

There are occasions when it is necessary to operate some electrical device from a source which provides too much voltage for the load. In this experiment we will learn how to safely operate a 14 Volt device from a 25 Volt source. You will need the following materials:

- A variable dc power supply
- A VOM (Analog, Digital or both)
- One #7382 Bi-pin lamp (14V 80 mA)
- Two 8 to 12 inch lengths of #24 hookup wire
- Five $1\frac{1}{2}$ inch lengths of #24 hookup wire
- An assortment of resistors to select from
- A prototyping board

The key to operating an electrical load from a source voltage that is too high is to use a series dropping resistor to drop the excess voltage. In order to determine how much resistance is appropriate it is necessary to know the load characteristics. We must know its voltage and current requirements as well as the source voltage. Then we can proceed to make it work.

Procedure

1. Remove the experiment Activity Sheet to use as you work through the following steps.
2. Analyze the situation. Here is the basic problem—the source voltage is 25 V and the load requires only 14V @ 80 mA. How much voltage must the series dropping resistor drop? Figure 5–1 shows the location of the yet to be determined dropping resistance in the circuit you will construct.

FIGURE 5–1 Schematic and pictorial of test circuit

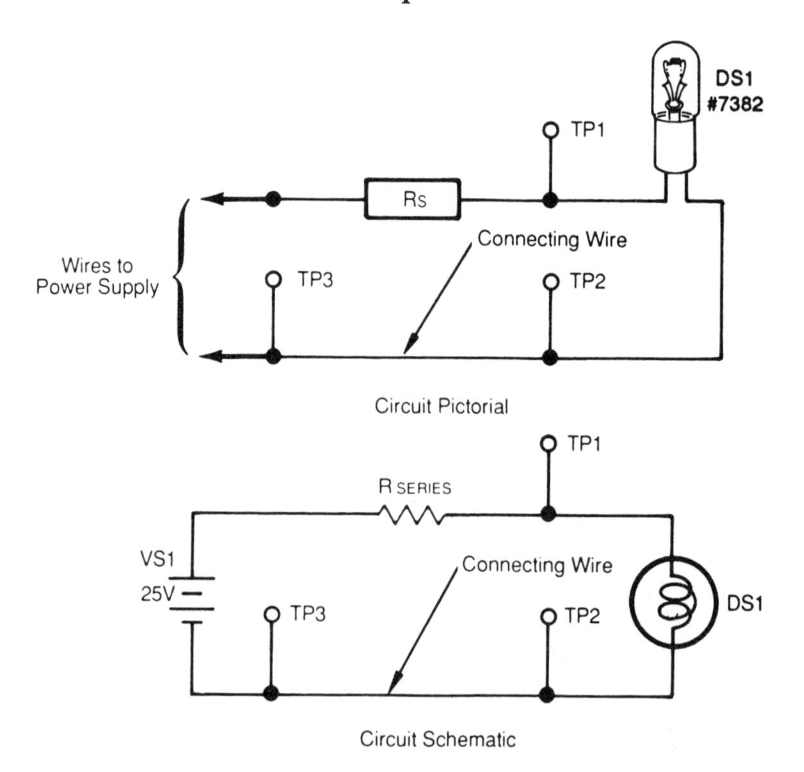

3. Construct the circuit, minus the series dropping resistor, on the prototype board.
4. Perform the calculations on the Activity Sheet to determine the necessary value of dropping resistor. Then find the value needed in your resistor assortment, or as close as you can match, and insert it into the circuit.
5. Adjust the variable dc power supply to exactly 25 volts using the voltmeter function of the VOM to verify accuracy.
6. Connect the power supply to the circuit and observe the results of your handiwork. Does the lamp glow with normal brilliance?
7. Using the voltmeter function of your VOM measure the voltage between test points 1 and 2, across the lamp DS1. Is the voltage correct for the load?

Review Break

You've methodically and scientifically calculated the amount of resistance necessary to operate a load from a higher than needed voltage. The resistance when placed into the circuit in series with the load drops the excess voltage, assuring that the load receives the proper amount of voltage and therefore the correct current flow. If the lamp is lit at its normal brilliance level then you can be assured that you've succeeded. But just to be sure. . . .

8. Now remove the circuit wire between test points 2 and 3. Set the VOM up to measure current flow and connect it to test points 2 and 3. Be sure to observe proper polarity if you're using the analog meter. This causes the milliammeter to now complete the circuit and also indicate the amount of current flow. Is the measured current flow correct for the load?

ACTIVITY SHEET EXPERIMENT 5

NAME ______________________________

DATE ______________________________

As you proceed through the experiment put the required information into the appropriate spaces.

Step 2

A. Situation analysis

Source voltage ______________________ V_S

Load voltage ______________________ $-V_L$

Excess voltage ______________________ V_{excess}

Load current ______________________

Step 4

B. Series dropping resistance (Step 4)

$$R_{series} = \frac{\text{Excess Voltage}}{\text{Load (circuit) Current}} = \text{\_\_\_\_\_\_\_\_\_\_\_\_\_\_\_\_}$$

R_{series} = ____________________ Ω

Steps 5–6

C. After connecting the series dropping resistor into the circuit and attaching the 25V source, does the lamp illuminate to normal intensity? Does it seem too dim or too bright?

____

Step 7

D. Load Voltage: Measured load voltage ____________________

Step 8

E. Load current: Measured load current ____________________

F. Did the completed circuit perform as you expected it to? ________

____

G. Were any circuit components damaged? ____________________

Why? ____

____

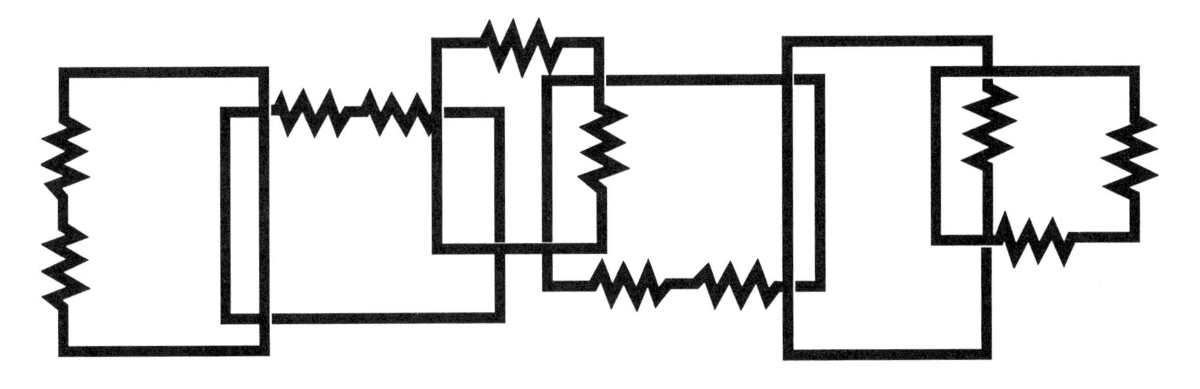

6

The Series Circuit—Measurements

Objectives

After completing this experiment, you will be able to:

1. Evaluate voltage distribution in a series circuit.
2. Determine actual resistances of individual resistors.
3. Confirm that the sum of series resistive drops equals source voltage.
4. Confirm that series resistive drops are proportional to resistance.
5. Verify that circuit current flow is predictable and measurable.

Introduction

The experiments that you have done so far have introduced you to the prototyping board, basic circuit construction techniques, the value of test points in the electrical circuit, and basic electrical measurements using the VOM. In experiment 6 you will explore the characteristics of the series circuit. The circuit upon which you will perform the tests will be made with fixed resistors. Fixed resistors are one of the most common parts in all electronic circuits and your ability to understand their physical as well as electrical traits is very important.

The purpose of this experiment is to show you how to use fixed resistors to verify the laws of electricity. You will make the same kinds of measurements that you have already done, but you will be verifying both Ohm's law and Kirchhoff's voltage law.

The parts needed to perform this experiment are:

A prototyping board
A VOM with test leads (Analog or Digital or both)
One 560 ohm resistor, 5% tolerance
One 330 ohm resistor, 5% tolerance
Seven 1.5 inch #24 hookup test wires
Four test wires with alligator clips
A variable output dc power supply

Procedure

1. Remove the experiment 6 Activity Sheet. On it you will find several locations to record measured values of resistance, voltage, and current flow. You will also see that computations will be done to verify the electrical measurements that you have done. Make all electrical measurements with the best precision and accuracy that you are capable of.
2. Gather together all of the necessary materials and construct the test circuit as it is shown in figure 6–1. Notice that the power supply is not connected to the circuit.
3. Set the VOM to function as an ohmmeter, select the scale which will give you the most accurate measurements, and measure the actual values of the two resistors. Consider the 560Ω resistor to be R1 and the 330Ω resistor to be R2. Connect the ohmmeter to test points 1 and 2, which are across R1, and measure its resistance. Record the measured value, as accurately as you are able, in the appropriate space on the experiment activity sheet.
4. Now move the test leads to points 3 and 4, which are across R2, and measure its resistance as accurately as you are able. Record the measured resistance in the appropriate space on the activity sheet.
5. Now connect the ohmmeter test leads to test points 1 and 4, across both resistors, and measure the total resistance as accurately as possible. Record the measured value in the appropriate space on the activity sheet. Does it appear that the total resistance value is equal to the sum of R1 and R2? Perform the calculation on the activity sheet to verify that this is so.
6. Assure that the power supply is turned off and that the output voltage adjust knob is set for minimum output. Connect the terminals of the power supply to test points 1 and 4, positive

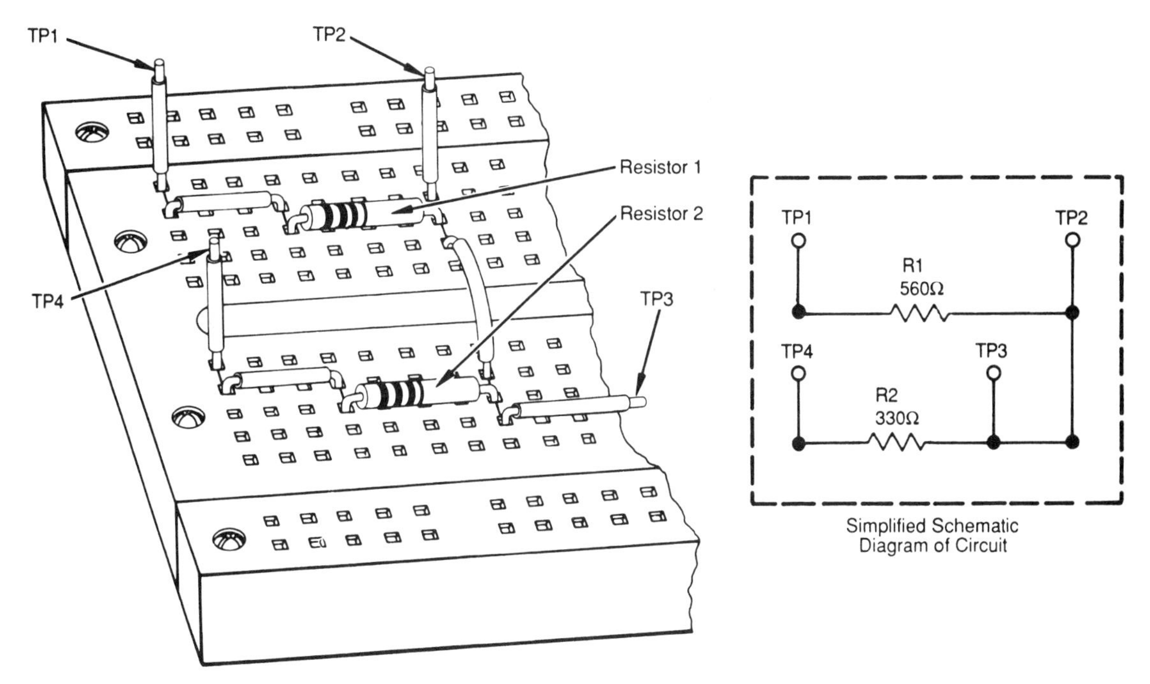

FIGURE 6–1 Pictorial and schematic of a series circuit

terminal to test point 1 and negative terminal to test point 4. Set the VOM to function as a voltmeter, select the appropriate scale to provide the most accurate measurement, and connect the voltmeter to test points 1 and 4, observing the same polarity as the power supply output.

7. Turn on the power supply and slowly increase its output to exactly 10.0 volts. Record this voltage on the activity sheet in the appropriate space. You have just measured the total voltage applied to the circuit.
8. Now shift the voltmeter lead attached to test point 4 to test point 2. This will allow you to measure the voltage dropped across R1. Measure the voltage as accurately as possible and record the value on the activity sheet.
9. Move the negative lead of the voltmeter back to test point 4, and shift the positive lead of the voltmeter to test point 3. This will allow you to measure the voltage dropped across R2. Record the measured voltage, with maximum accuracy, on the activity sheet. Does it appear that the sum of the two voltages is equal to the source voltage? Turn off the power supply. Perform the required calculation on the activity sheet.
10. The next thing you need to do is measure the circuit current flow. Set the VOM to operate as a milliammeter, select the appropriate scale to give you the most accurate measurement, and connect the milliammeter leads to test points 2 and 3. Notice that a circuit wire is in position between these test points. Remove the circuit wire to allow the milliammeter to measure current flow. Turn on the power supply. Record the measured current flow, as accurately as possible, in the appropriate space on the activity sheet.

Review Break

You have accomplished several things up to this point in the procedure. You have measured the actual resistances of the two circuit resistors individually. You then measured the resistance of the series connected combination. Your calculations should have proven that the sum of the two individual resistors is equal to the total resistance of the circuit.

You then set the power supply output to 10.0 volts as it was applied to the circuit, the total circuit voltage.

Next you measured the voltage drops across resistors R1 and R2, then you added them to verify that the sum of the individual drops is equal to the total applied voltage from the source.

The last measurement made was total circuit current. This is always the most difficult measurement because it necessitates a slight modification to the circuit. The milliammeter is inserted into the circuit in place of one of the circuit wires. This forces the current to flow through the milliammeter where it can be accurately measured. Each of the measurements done, though basic, is important. Next you will perform some calculations to test the accuracy of your measurements.

11. You have made all necessary circuit measurements so you may disassemble the circuit and return all parts to their proper places if you desire. If there are any lingering doubts, leave the circuit intact so that you can redo as many of the measurements as necessary to clear up any confusion. Then you can disassemble the circuit and put things away.

12. Perform the required computations on the activity sheet to prove that the measured electrical parameters are extremely close to the calculated electrical parameters. The closeness is dependent upon the accuracy of your measurements. When you have completed all activities on the sheet, turn it in to your instructor.

ACTIVITY SHEET EXPERIMENT 6

NAME ______________________

DATE ______________________

Step 3 **A.** Measured resistance of R1 (560Ω) between test points 1 and 2.

____

Step 4 **B.** Measured resistance of R2 (330Ω) between test points 3 and 4.

____

Step 5 **C.** Total circuit resistance measured between test points 1 and 4.

____

D. Add resistance values measures in Parts A and B. Write sum here.

____

Step 7 **E.** Measured total circuit voltage between test points 1 and 4.

(Should be 10.0 volts) ______________

Step 8 **F.** Measured voltage drop across R1 between test points 1 and 2.

____

Step 9 **G.** Measured voltage drop across R2 between test points 3 and 4.

____

H. Add voltages from Part F and Part G. Is the sum equal to the voltage measured in Part E? ______________

Step 10 **I.** Measured total circuit current flow between test points 2 and 3.

____

Step 12 **J.** Calculate the total circuit resistance by use of Ohm's Law,

$$R_{Circuit} = \frac{V_{Circuit}}{I_{Circuit}}$$

Does the calculated resistance equal the measured resistance from Part C? ______________

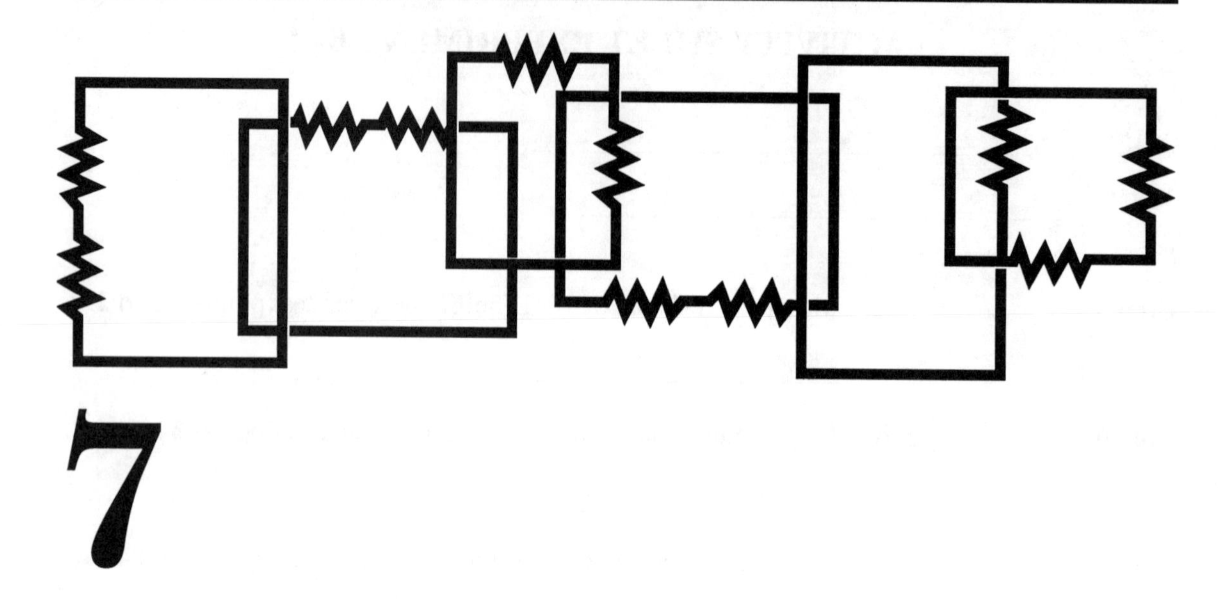

7

Series Circuit Application—Illumination

Objectives

After completing this experiment, you will be able to:

1. Accurately set the output of the dc power supply to a desired value.
2. Observe the effect of eliminating series resistance in a circuit.
3. Evaluate the effect of decreased series resistance upon load operation.
4. Utilize the variable resistor as a load current control.

Introduction

The purpose of this experiment is to learn how it is possible to vary the intensity of a light bulb. You will be using a fixed voltage source, and by changing the amount of resistance connected in series with the lamp you will see that you can control its brightness. Let's get started!

You will need the following materials:

A prototyping board
A variable dc power supply
One #7382 T $1\frac{3}{4}$ Bi-Pin lamp (14V 80mA)
Two 8 to 12 inch lengths of #24 hookup wire
Four $1\frac{1}{2}$ inch lengths of #24 hookup wire
Two test leads with alligator clips
One 1000 ohm variable resistor (Potentiometer)
Four fixed resistors (one 15Ω, two 120Ω, one 510Ω)
A VOM (Analog or Digital or both)

Procedure

1. Assemble the circuit shown in figure 7–1 on the prototyping board. Use the illustration as a guide to position the circuit components on the board. Remove activity sheet for this experiment and use it as you proceed.
2. Adjust the output voltage of the variable power supply to exactly 14 volts. Use the voltmeter to verify the accuracy of the power supply output.
3. Connect the circuit on the prototyping board to the power supply using the two long lengths of hookup wire. Now that you have applied power to the circuit, do you see any evidence of the circuit doing anything? Is the lamp lit or not?
4. Now take one of the test leads with alligator clips and use it to connect test points 1 and 2, effectively shorting R1 and remov-

FIGURE 7–1 Simplified schematic of first test circuit with pictorial layout

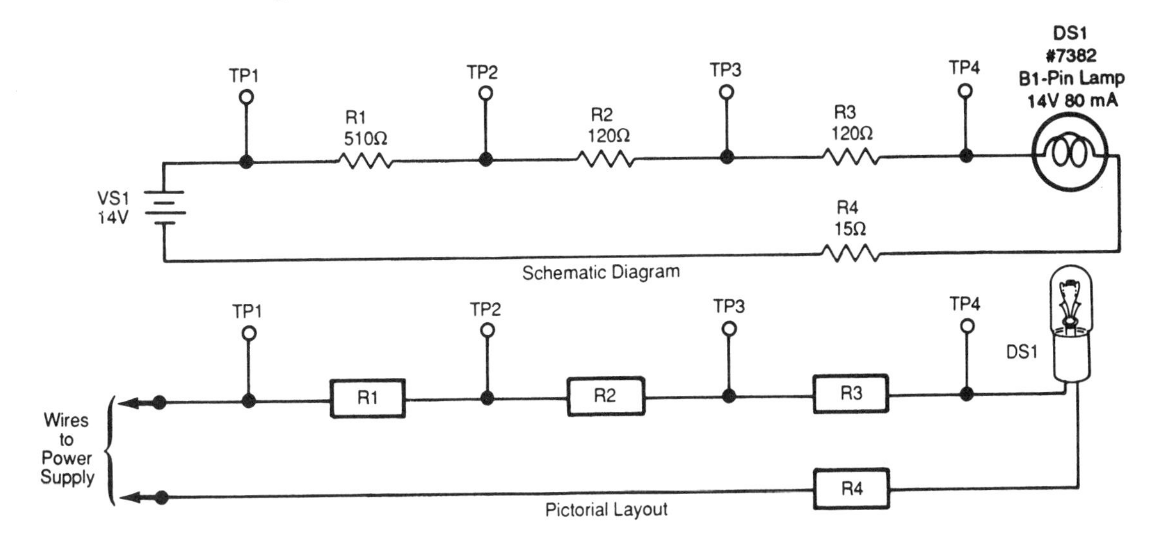

ing it from the circuit. What effect does this have on circuit operation? Do you see anything happening?

5. Move the alligator clip connected to test point 2 to test point 3, leaving the end connected to test point 1 intact. This connects test points 1 and 3, effectively shorting and removing from the circuit resistors R1 and R2. What do you observe now? Is there a difference in the brightness of the light?
6. Now move the test lead connected to test point 3 to test point 4. This results in resistors R1, R2 and R3 being shorted and effectively out of the circuit. What is the result at the light?

Review Break

You have constructed a somewhat more complex series circuit consisting of 4 resistors and a light bulb. Resistor R4, 15Ω, is simply a safety resistor to protect the lamp from burnout in case the applied voltage is higher than it should be. It isn't 100% effective but is sometimes included in electric circuits. The other resistors have considerably more resistance and cause the greatest change in the operation of the circuit as they are shorted out.

When you first applied power to the circuit you noticed that the lamp was not illuminated. The relatively large amount of series resistance did not permit enough current flow to cause the lamp to light. Then you shorted R1 and noticed that light, though dim, was now visible. As you shorted the other resistors, R2 and R3 with R1 in succession, you observed that the lamp became progressively brighter, being brightest when all three resistors were shorted. You have demonstrated one method of varying the brightness of a lamp when it is powered by a fixed voltage source.

7. Remove resistors R1, R2 and R3 from the circuit as well as test points 2 and 3. Using both test leads with alligator clips, carefully clip one end of each lead to the center and one end terminal on the potentiometer, and connect the opposite ends of the test leads to test points 1 and 4. You are replacing the fixed resistors with a variable resistor, as seen in figure 7–2.
8. Once you are satisfied that your connections are correct (use figure 7–2 as a guide) rotate the shaft of the potentiometer slowly fully clockwise and fully counterclockwise several times. What did you notice about how turning the variable resistance affects circuit operation?

Review Break

After replacing the fixed resistors with the variable resistor you discovered that you could vary the brightness of the lamp smoothly from off to maximum brightness. This is a technique that is used often in automobiles to allow the driver to vary the brightness of the dash panel background illumination. The final circuit was very simple but effective as a means of adjusting the brightness of a lamp. As you continue in your studies you will find that there are several other ways to accomplish this electronically with semiconductors.

9. Return all materials to their proper places.

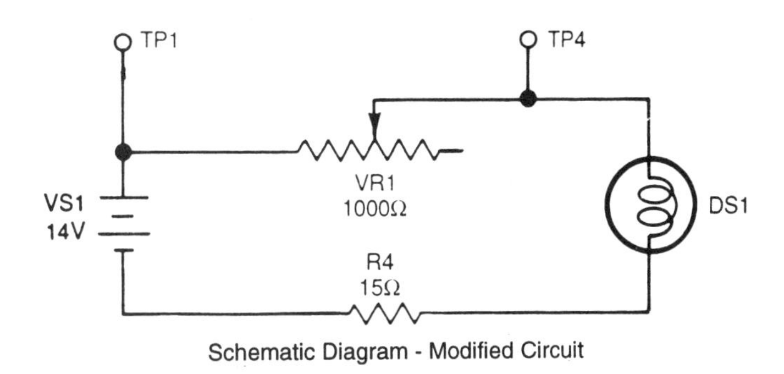

Schematic Diagram - Modified Circuit

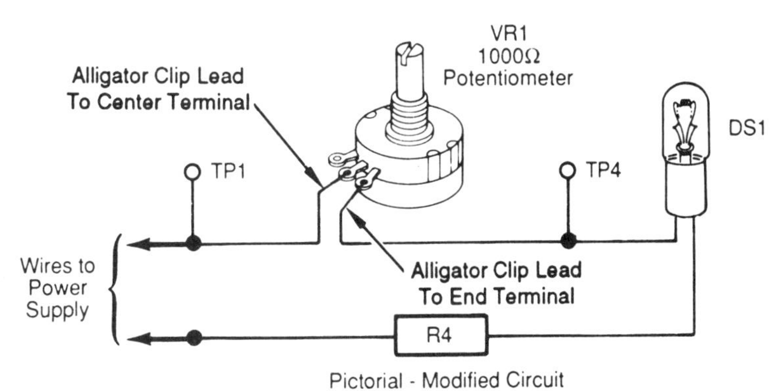

Pictorial - Modified Circuit

FIGURE 7–2 Modified circuit with a potentiometer

FIGURE 7–3

ACTIVITY SHEET EXPERIMENT 7

NAME ______________________________

DATE ______________________________

As you proceed through the steps of the experiment fill in the blank spaces with your answers to the questions:

Step 3 **A.** Is the lamp illuminated? ____________________

Step 4 **B.** Is the lamp now glowing? ____________________

Step 5 **C.** Is the lamp glowing more brightly? ____________________

Step 6 **D.** Is the lamp brightest now? ____________________

E. Read the review break. ____________________

Step 8 **F.** Are you able to vary the lamp's brightness smoothly from off to full on? ____________________

G. In the first part of the experiment why did the lamp glow more brightly as more circuit resistance was eliminated? ____________________

____

____

H. In the second part of the experiment how do you explain the smooth, continuous control of the lamp's intensity? ____________________

____

____

I. What have you learned about the use and effect of a variable resistor in an electrical circuit? ____________________

____

____

____

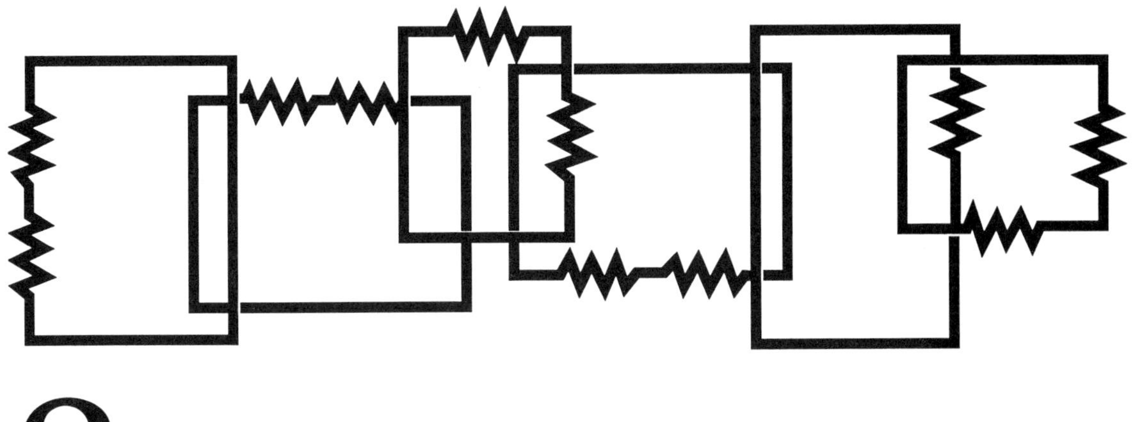

8

Resistors—Values and Tolerance

Objectives

After completing this experiment, you will be able to:

1. Accurately measure the actual resistance of a resistor.
2. Compare the coded and actual values of resistors.
3. Determine whether a resistor is within tolerance.

Introduction

The resistor ranks high in the field of electronics, not because of any unusual abilities, but in terms of sheer numbers. There are probably more resistors in all of the electronic equipment on earth, than any other singular component. There must be something important about what the resistor does that makes it so popular and indeed there is. As you continue your studies you will develop a great appreciation for its purposes and its applications.

All resistors are manufactured to have certain precise values of resistance. The value of the resistor, in ohms, is placed on the outside of the resistor usually as color coded bands. Included in the code is the resistor's tolerance characteristic. Common tolerances are 1%, 2%, 5%, and 10%. The tolerance figure of the resistor tells us how close its actual resistance is to its coded resistance. The smaller numerical value of the tolerance percentage, the greater the accuracy of the resistor.

In this experiment you will measure the actual value of several resistors with a digital ohmmeter to compare their real values with their coded values. An analog meter could also be used, but it would be much more difficult to accurately determine the measured values of the resistors. You will also determine whether the resistors are within their tolerance ratings. You will need the following materials:

A Digital VOM
One 15Ω resistor
One 120Ω resistor
One 330Ω resistor
One 4700Ω resistor
One 5600Ω resistor
Test leads with alligator clips

Procedure

1. Remove the activity sheet for this experiment and record the measured data and calculations.
2. In the appropriate spaces on the activity sheet write down the color coded values for each of the resistors. Record the resistance values in ascending order.
3. In the appropriate spaces on the sheet write down the resistor's tolerance ratings as read from the resistor.
4. Now calculate the permissible variation of resistance for each resistor by multiplying the resistors' coded value by its tolerance. Record this value for each resistor in the appropriate spaces on the sheet.
5. Now determine the upper and lower tolerance limits for each resistor by first adding the permissible variation to the coded value. This will give you the upper limit. Then subtract the permissible variation from the coded value to obtain the lower tolerance limit. Do this for each of the resistors and record the data in the appropriate spaces on the sheet.
6. Now set up the digital VOM to measure resistance and measure the actual value of each of the resistors. Be sure to change scales on the ohmmeter when necessary to obtain the most accurate

measurements. Write the actual measured value of each of the resistors in the appropriate spaces on the sheet.

7. Now analyze your data. Do the resistors' actual values fall within the upper and lower limits of its tolerance range? If so, then you have some reasonably accurate resistors and can use them with confidence. Occasionally the actual measured value will not be within the expected tolerance limits. Does this mean the resistor is bad? Not necessarily. It does mean however that it is out of tolerance and shouldn't be used in a circuit that demands high accuracy. It can still be used in less demanding applications though.
8. Return all materials to their proper places.

ACTIVITY SHEET EXPERIMENT 8

NAME ______________________________

DATE ______________________________

Step 2 **A.** Enter each resistor's coded value into the table below.

Step 3 **B.** Enter the resistor's tolerance characteristic for each.

Step 4 **C.** Calculate and enter into the appropriate spaces the permissible variation.

Step 5 **D.** Calculate and enter the tolerance limits.

Step 6 **E.** Enter actual measured resistance values.

	Resistor Color Coded Values	Tolerance	Permissible Variation	Upper Limit	Lower Limit	Measured Resistance	Error
1.	Ω	%	± Ω	Ω	Ω	Ω	Ω
2.	Ω	%	± Ω	Ω	Ω	Ω	Ω
3.	Ω	%	± Ω	Ω	Ω	Ω	Ω
4.	Ω	%	± Ω	Ω	Ω	Ω	Ω
5.	Ω	%	± Ω	Ω	Ω	Ω	Ω

F. Calculate the error for each resistor by comparing the coded value with the actual value. If the resistor's actual value is greater than its coded value, assign a positive polarity to the error. If the actual value is less than the code value assign a negative polarity. If done with a calculator use the formula (Error = Actual Value – Code Value) and use the indicated sign.

G. Were all of your resistors within tolerance? ____________________

____

H. What have you learned about resistors? ____________________

____

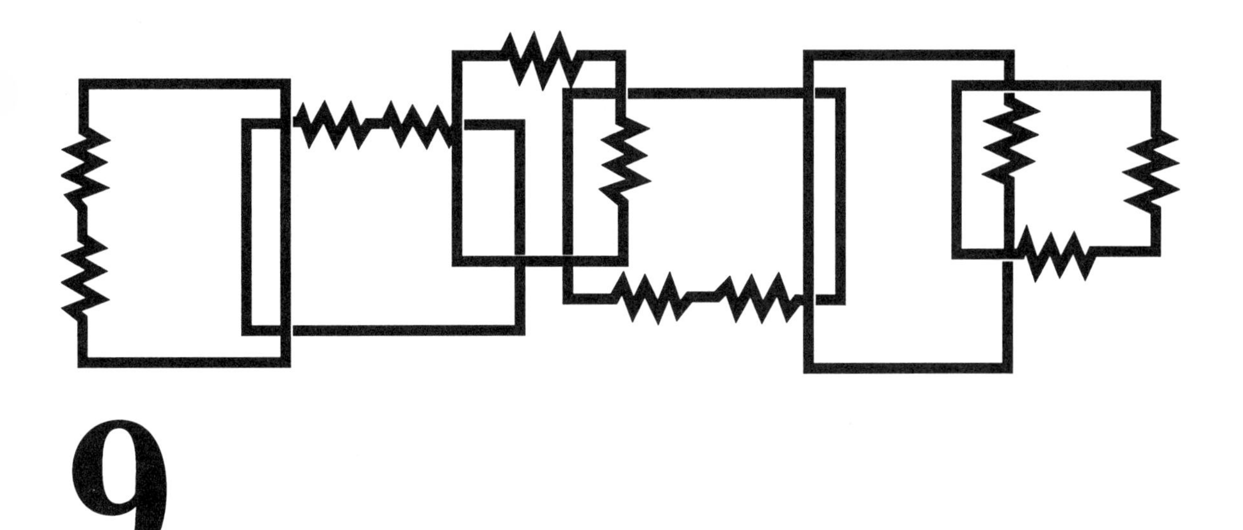

9

The Smoke Theory

Objectives

After completing this experiment, you will be able to:

1. Experience the destruction of a resistor.
2. Purposely subject a resistor to excessive electrical power.
3. Evaluate the condition of a resistor which has been stressed.
4. View the changes in resistor appearance caused by stress.
5. Experience the smell of a resistor under stress.

Introduction

Some of you have undoubtedly already experienced what this experiment is about to demonstrate, either accidentally or by design. But that's OK. Doing it one more time with purpose will tend to reinforce the lesson.

For those of you who have yet to experience one of life's greatest joys, being destructive without any fear of punishment, you are about to prove (or disprove) the infamous Smoke Theory. This theory states that all electronic and electrical devices, including resistors of course, depend upon smoke to make them work. If you cause the smoke to escape, then the device is ruined. Let us see if this theory is truth or myth. You will need:

- A variable dc power supply
- Two 8 to 12 inch lengths of #24 hookup wire
- A VOM (Analog, Digital or both)
- One resistor, 120Ω $\frac{1}{4}$ to $\frac{1}{2}$ watt
- Two $\frac{1}{2}$ inch lengths of #24 hookup wire
- A prototyping board

Note to Instructor: You may opt to demonstrate this exercise if class size or room ventilation pose a problem.

CAUTION

There is some hazard associated with this experiment. It is possible that the resistor could ignite into flame briefly. Assure that the resistor is inserted into the prototyping board as shown in figure 9–1, at least 1 inch above the surface of the board. Should the resistor burst into flame it will burn for only a brief time and should pose no threat to you providing you are careful not to touch it or place your face or hands close to it. Please proceed carefully and you will not be caught by surprise.

Procedure

1. Pull the activity sheet for this experiment to record your observations as you follow the procedure.
2. Wire the circuit shown in figure 9–1 on the prototyping board.
3. Set the output of the dc power supply to minimum voltage. Set the VOM to measure voltage and connect it to test points 1 and 2, across R1.
4. Slowly increase the output voltage of the power supply until you see the first wisp of smoke curl up from the body of the resistor. Record the measured voltage in the appropriate space on the activity sheet.
5. Without increasing the output of the power supply, allow the resistor to "cook" for several minutes, savoring the peculiar aroma.
6. When the resistor is fully cooked, decrease the power supply output voltage to minimum and disconnect the supply from the circuit.

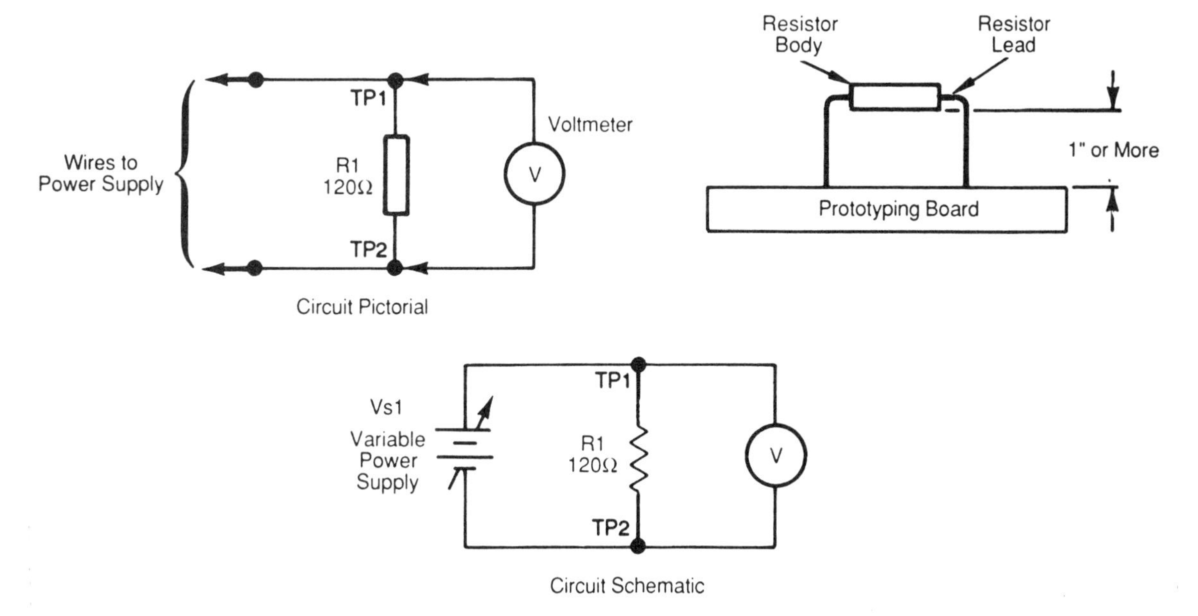

FIGURE 9–1 Simplified schematic of the test circuit

7. With the power supply disconnected from the circuit, remove the voltmeter from test points 1 and 2.
8. Set the VOM to measure resistance, re-connect to test points 1 and 2 and measure the resistor's resistance. Record the measured resistance in the appropriate space on the activity sheet. Has the resistor's resistance changed?
9. Remove the ohmmeter from test points 1 and 2 and reconnect the power supply to the circuit once again. Increase the power supply's output voltage to maximum. What do you see?
10. After about 15 seconds turn the output voltage of the supply to minimum and disconnect the power supply from the circuit.
11. With the power supply disconnected, measure the resistance of the resistor once again. Record the reading in the appropriate space on the sheet.

Review Break

You have just pushed a resistor to its limits and beyond. It looks like a mere shell of its former self, burnt almost beyond recognition. You have in effect proven the smoke theory, once you let the smoke out of a resistor it is indeed a "has been." But the important point of this whole experience is to observe the effects of excessive current flow which causes excessive power dissipation, which results in destruction. Another important point is that you have experienced the smell of a resistor which has gone bad. You will encounter this same smell many times throughout your career as an electronics technician or engineer, and your remembering this can serve as a valuable troubleshooting aid. After allowing all hot spots to cool, continue. . . .

12. Return all materials to their proper places, do whatever you wish with the remains of the resistor. Complete the activity sheet for this experiment and pat yourself on the back. Now you know.

ACTIVITY SHEET EXPERIMENT 9

NAME ______________________________

DATE ______________________________

As you conduct the experiment fill in all required information.

Step 4 **A.** At what voltage setting did you first observe visually the presence of smoke? ____________________

Step 8 **B.** What is the measured resistance of the "cooked" resistor?

____

Step 10 **C.** Describe the appearance of the resistor after applying maximum voltage.

____

Step 11 **D.** What is the measured resistance of the resistor at this point?

____

E. Describe the smell of a resistor as it is being destroyed. _______

____

F. What have you learned about resistors as a result of this experiment?

____

____

____

____

____

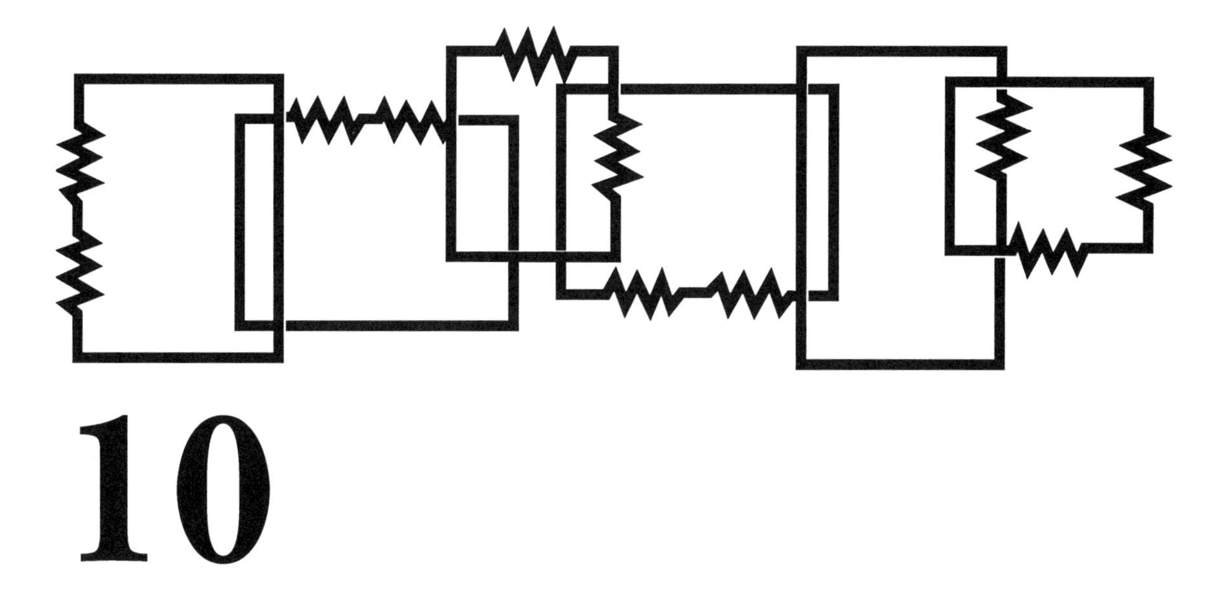

10

The Electric Battery

Objectives

After completing this experiment, you will be able to:

1. Create an electrochemical power source.
2. Evaluate the electrical properties of an electrochemical source.
3. Determine the electrical characteristics of a battery under load.
4. Use the light emitting diode as an indicator of relative current flow.

Introduction

You are about to duplicate a variation of an experiment which was first performed over 200 years ago by Alessandro Volta and others exploring the phenomenon of electricity in the 1700's. The materials that you will use won't be the same as those used by Volta, but the end product will be very similar. You may have to do some scrounging or scavenging to come up with some of the materials. The effort will be well worth it though. You will have to do some preparation for the experiment at home, such as the next paragraph explains, and you will have to bring some materials from home. The rest of the experiment can be done in the laboratory.

Get a couple of used flashlight cells of any size (AA, A, C or D, but not the alkaline kind) which are completely dead, and with a large pair of pliers carefully squeeze or squash them to loosen up the carbon rods which run through the center and gently remove them. Avoid breaking the rods or removing the metallic cap at the top end. The chemicals in the flashlight cells, although not deadly, are slightly corrosive so wear gloves while removing the carbon rods. Wash the rods thoroughly with detergent and water after you've gotten them out to remove the residual chemicals and graphite coating, then rinse thoroughly in clean water. Use some old rags or paper towels while doing this. This whole operation should be performed outside or over several thicknesses of old newspapers if done inside. Discard what is left of the flashlight cells after removing the carbon rods. The other materials you will need are:

Two pieces of aluminum foil about 1 inch wide and 3 inches long
Two plastic or glass containers such as glasses or cups
A partially full bottle of household chlorine bleach (approx. 1 cup)
Three test leads with alligator clips
One Light Emitting Diode (Red, yellow or green)
A VOM (analog, digital, or both)
A plastic tray about 4" by 6" or larger
Two paper clips
Two 4 inch lengths of #24 hookup wire
A small bottle of tap water

Procedure

1. After organizing all necessary materials, put the cells together as shown in figure 10-1. Use the plastic tray for protection in case of spillage. Clip the aluminum foil strips to one side of each cup with the paper clips. Wrap two turns of the 4 inch lengths of hookup wire around the tops of the carbon rods and with the remainder hook the rods to the sides of the cups opposite the aluminum foil. The carbon rods and the aluminum foil strips cannot touch each other or an internal short circuit within the cell will result.

CAUTION

Chlorine bleach is corrosive and harmful to eyes and skin. Handle very carefully to avoid skin or eye contact. Flush hands or eyes immediately with fresh water in case of contact.

2. When you've completed assembling the two cells pull the activity sheet to record your measurements and observations.
3. Fill each of the cell cups about $\frac{1}{3}$ full with ordinary tap water.
4. Then add to the water slowly the household bleach until the cups are about $\frac{2}{3}$ full.

FIGURE 10–1 Pictorial representation of cell structure showing layout and hookup

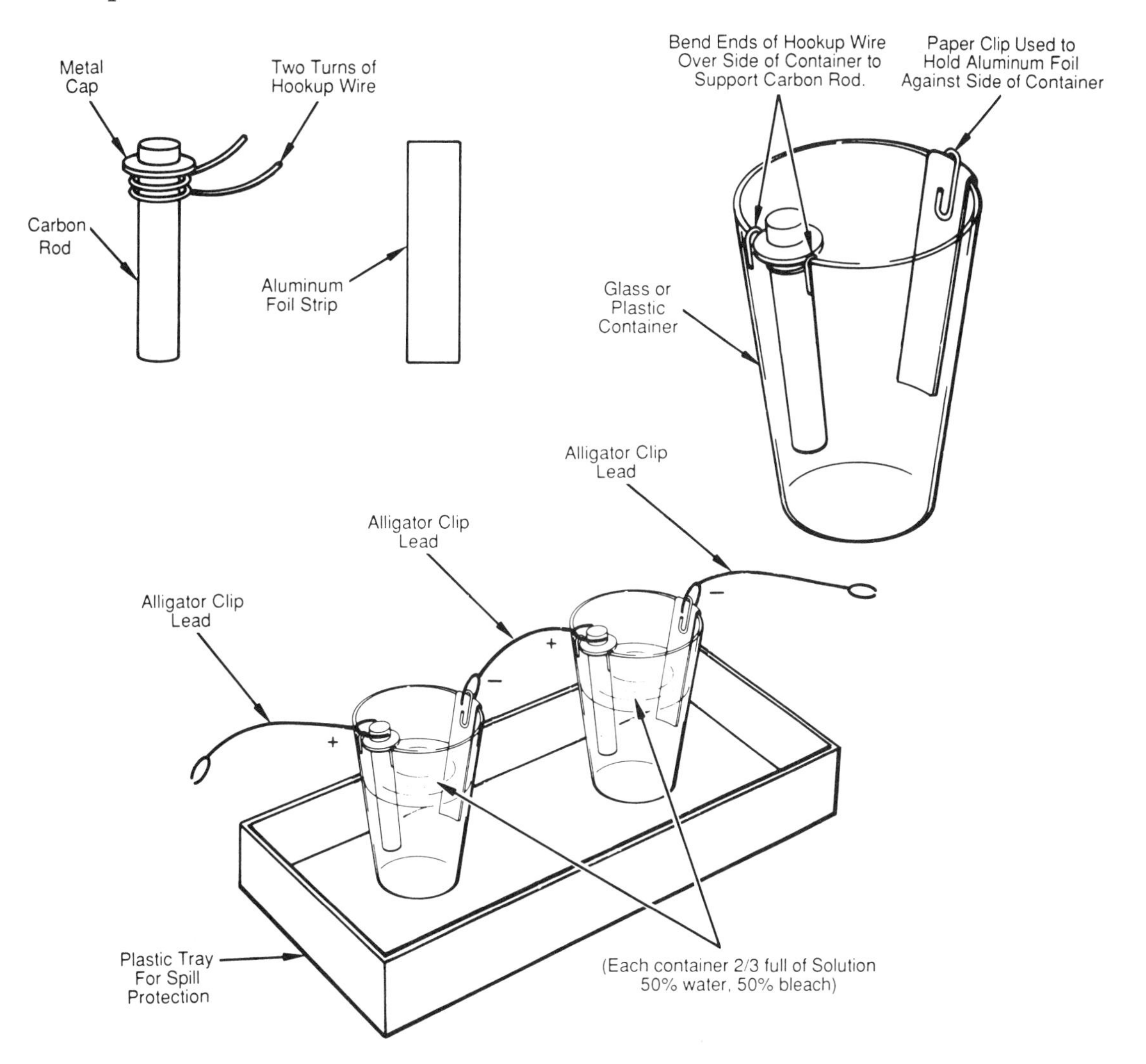

CAUTION

Use caution while handling the bleach, and if you should spill any on your hands wash them immediately for several minutes in running water at a sink. The household bleach contains caustic chemicals which though not deadly, should be handled with respect and extreme caution.

5. Set the VOM to function as a voltmeter and measure the voltages across the terminals of the cells. The output of the cells will be dc with the carbon rod the positive terminal, and the aluminum foil the negative terminal. Record the measurements in the appropriate spaces on the activity sheet.
6. Now measure the voltage of the series combination of the two cells and record the value in the appropriate space on the sheet.
7. Set the VOM to function as a milliammeter. Select the highest scale and connect the milliammeter to the battery (both cells in series) to measure its short circuit output current. The milliammeter will effectively be a short circuit for this test. It will be necessary to shift downscale in order to obtain the most precise measurement since the current capacity of the battery will not be great. Once you have found the most accurate scale, record the measured current on the activity sheet.
8. With the alligator clip test leads connect the LED to the battery as shown in figure 10–2. The LED is polarity sensitive so it will only work when connected correctly. What do you see?
9. Set the VOM to measure voltage and measure the voltage dropped across the LED. Select the voltage scale which gives you the most accurate result and record the measured value on the sheet. After making this measurement leave the LED connected to the battery to see how long it will continue to work.

Review Break

You've constructed a battery made up of two cells and some readily available materials. You've verified that the cells are dc sources and that when connected in series their voltages do add. You measured a characteristic of batteries known as its short circuit current charac-

FIGURE 10–2 Outline drawing of the LED showing polarity of the leads—and circuit diagram.

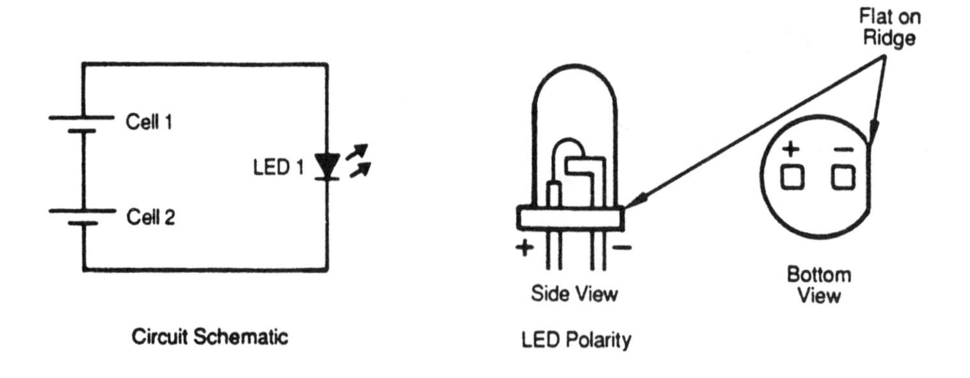

teristic. Although the battery is definitely not a powerhouse it does produce enough current to perform some useful work. When you connected the LED to the battery you found that it did emit visible light, evidence that work is being done and energy is being consumed. An even larger battery made of bigger cells and more of them in series could perform even more work, such as operating a transistor radio. The LED was selected for this experiment because it is capable of producing light at low voltage and current levels and effectively matched the characteristics of the battery which you made. It is a special semiconductor device which you will study about in more advanced sections of the course.

10. Set the VOM to measure current once again and insert it in series with the LED to measure the current consumed by the LED. Record this measurement on the activity sheet. You may allow the LED to operate for the remainder of the experiment to see if the battery runs down.

11. With the data you've gathered it is now possible to calculate the internal resistance characteristic of the battery. Follow the procedures on the activity sheet and you will have completed this experiment.

12. Disconnect all circuit wiring from the battery and carefully carry your tray with the cells to a sink. Dump the bleach solution down the drain, rinse the cell cups out thoroughly, then remove the aluminum foil and the carbon rods. You may keep the carbon rods for future experiments or dispose of them as you see fit. Discard the aluminum foil and generally clean up your materials. Put all equipments back in their proper places and give some thought to what you've accomplished.

ACTIVITY SHEET EXPERIMENT 10

NAME ______________________________

DATE ______________________________

As you perform the experiment record all required measurements, observations, and calculations.

Step 5 **A.** Measure each individual cell voltage.

Cell 1____________________ Cell 2____________________

Step 6 **B.** Measure the total battery voltage.

Battery voltage ____________________

Step 7 **C.** Measure the short circuit current of the battery.

$I_{short\ circuit}$ ____________________

Step 8 **D.** With the LED properly connected to the battery what do you observe?

____

____

Step 9 **E.** Measure the voltage dropped by the LED.

V_{LED} ____________________

Step 10 **F.** Measure current flow through the LED.

I_{LED} ____________________

After several minutes of continuous operation has the brightness

of the LED diminished? ______________________________

____

Step 11

G. Calculate the internal resistance of the battery by use of this formula:

$$R_{INT} = \frac{V_{BATTERY} - V_{LOAD}}{I_{LOAD}}$$

$$R_{internal} = \frac{\underset{V_{BATTERY}}{\_\_\_\_\_\_\_\_} - \underset{V_{Led}}{\_\_\_\_\_\_\_\_}}{\underset{I_{LED}}{\_\_\_\_\_\_\_\_\_\_\_\_\_\_\_\_}} = \underset{R_{internal}}{\_\_\_\_\_\_\_\_\_\_}$$

Enter data needed into spaces above and solve. Enter internal resistance value into space at right. ____________________

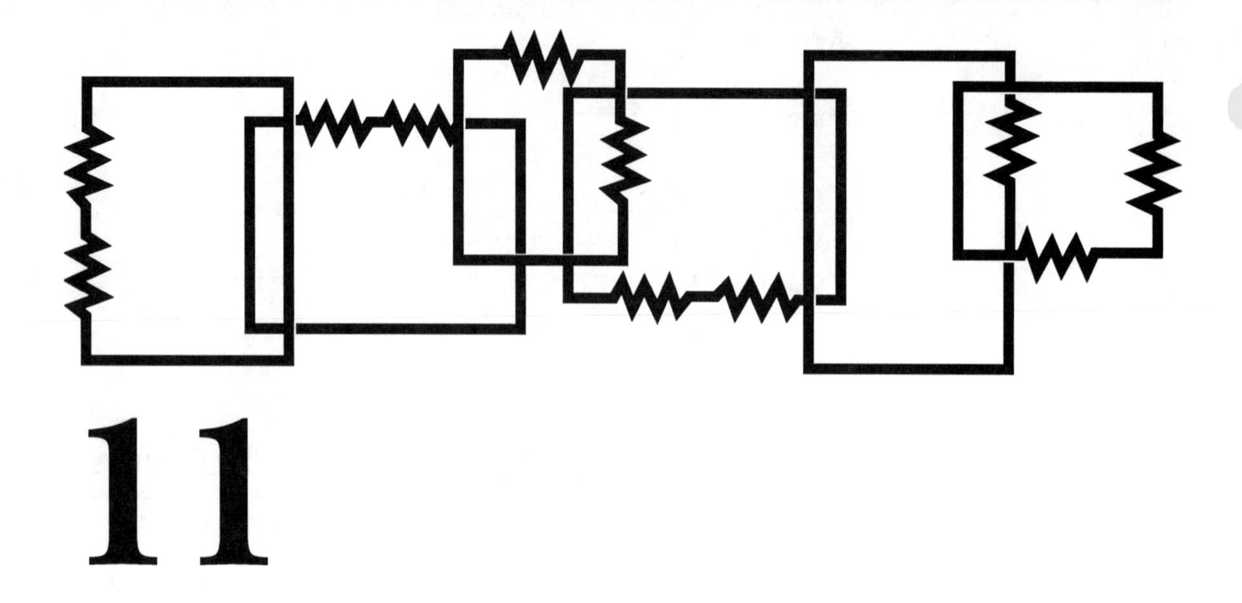

11

Fuses for Protection

Objectives

After completing this experiment, you will be able to:

1. Actively monitor increasing current flow in an electrical circuit.
2. Incorporate a fuse into an electrical circuit.
3. Evaluate the response of a fuse to potentially damaging current flow.
4. Observe the protective ability of the fuse.

Introduction

In a prior experiment you observed the effects of excessive current flow through a resistor: heat, smoke and ultimately destruction. All electronic components are susceptible to damage by excess current flow, although not all produce smoke as they draw their last electrical gasp.

In this experiment you will use light bulbs to observe how they respond to excessive current flow. Then you will insert a protective device into the circuit to prevent excess current flow. To perform the experiment you will need:

Eight $1\frac{1}{2}$ inch lengths of #24 hookup wire
Two 8 to 12 inch lengths of #24 hookup wire
Two #7382 bi-pin lamps (14V 80 mA)
A VOM (analog, digital, or both) with leads
One $\frac{1}{8}$ Amp fuse
Four test leads with alligator clips
A variable dc power supply
A prototyping board

Procedure

1. Pull the activity sheet for this experiment then assemble the circuit as shown in figure 11–1. Before connecting the power supply set its output to zero volts. Set VOM to operate as milliammeter and insert into the circuit between test points 3 and 4.
2. Slowly increase the power supply output voltage while watching both the lamps and the milliammeter.
3. As you approach 160 milliamperes output proceed with caution so as not to destroy the lamps. Very slowly increase the power supply output to 180 milliamperes. Quickly note and record the appearance of the lamps and the current flow, then decrease the output to zero volts.
4. Remove the circuit wire between test points 1 and 2. Use a set of test leads with alligator clips to connect the fuse into the circuit across the same test points.

FIGURE 11–1 Schematic diagram of circuit with shorting wire between fuse connection points

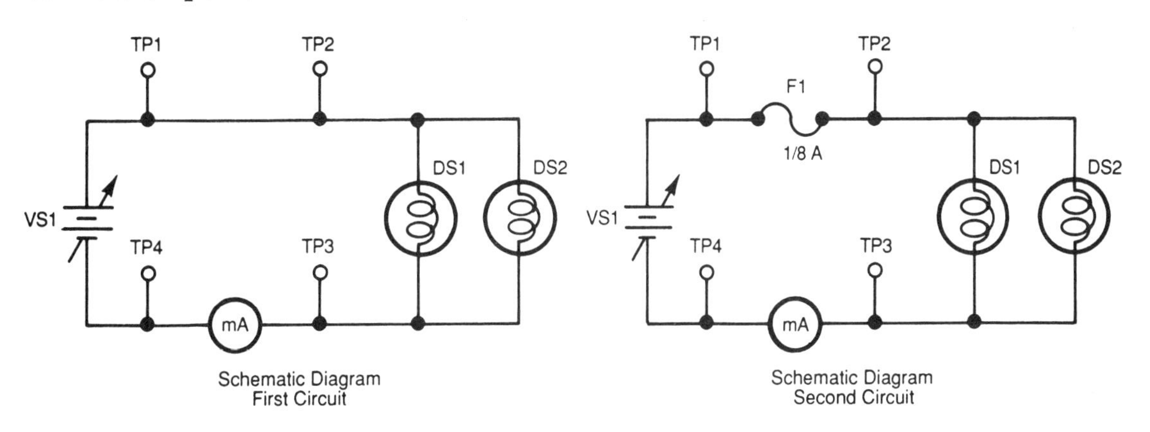

5. Slowly increase the power supply output while observing the milliammeter and the lamps. As you approach 160 milliamperes what happens? Can you increase the current to a dangerously high level (high enough to destroy the lamps)? Increase the power supply output to maximum, then to zero.

Review Break

After constructing the circuit you applied power from the variable power supply as a slowly increasing current. You purposely stopped the current increase at 180 milliamperes in order to avoid damage to the lamps. If you had continued to increase current flow the lamps would have burned out. Next you modified the circuit to include a fuse. Again you slowly increased circuit current and observed that the fuse prevented you from damaging the lamps.

6. You have demonstrated the value of fuses with this very simple circuit and may now conclude the experiment and put all materials back to their proper places.

7. Return all materials to their proper places, you have collected sufficient empirical data for the time being.

ACTIVITY SHEET EXPERIMENT 11

NAME ______________________________

DATE ______________________________

As you perform the experiment, record all necessary data in the appropriate spaces.

Step 3

A. Measured current flow ______________________ maximum.

Is the lamp brighter than it should be? ____________________

____

Step 5

B. At what current level does the fuse protect the lamps by opening the circuit? ____________________

C. What caused the fuse to respond as it did to the increasing circuit current?

____

____

D. What would happen if you were to replace the fuse with a piece of hookup wire and increase the power supply output to maximum? ____

____

E. Would it be a good idea to replace a fuse with a piece of wire in a real electrical circuit? ____________________

Explain why. ______________________________________

____

____

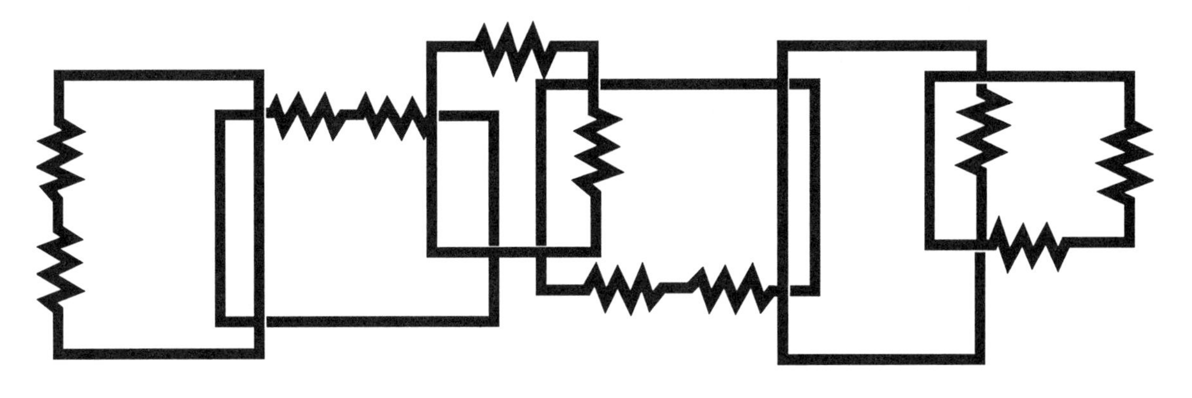

12

Troubleshooting Series Circuits

Objectives

After completing this experiment, you will be able to:

1. Evaluate the effects of an open circuit defect on circuit operation.
2. Evaluate the effects of a short circuit defect on circuit operation.
3. Isolate an open circuit defect in a series circuit.
4. Isolate a short circuit defect in a series circuit.

Introduction

Now that you've had quite a bit of experience working with series circuits it is time to examine two of the most common problems in electric circuits, and to learn how to find the problems. Even though the circuit will be a simple one, the techniques that you learn will apply to the more advanced circuits which you will encounter in the weeks to follow. Learn these techniques well and soon troubleshooting will be a simple process. You will need the following materials:

A VOM (Analog, Digital or both)
A variable dc power supply
Two 8 to 12 inch lengths of #24 hookup wire
Six $\frac{1}{2}$ inch lengths of #24 hookup wire
One T 1 $\frac{3}{4}$ Bi-pin lamp (#7382) 14V 80mA
Two fixed resistors (120Ω each)
Two test leads with alligator clips
A prototyping board

Procedure

1. Construct the test circuit on the prototyping board using figure 12–1 as a guide.
2. Adjust the output voltage of the variable power supply to exactly 18 volts. Use the voltmeter function of the VOM to verify the accuracy of the power supply. Connect the circuit on the prototyping board to the power supply. The lamp should be illuminated. This level of brightness will be considered "normal" for the purposes of this experiment.
3. Get the experiment Activity Sheet and use it to record observations as you proceed through the experiment.
4. With one of the alligator clip test leads connect test points 1 and 2, shorting resistor R1. How does this affect the brightness of the lamp?

CAUTION

The next step in the procedure can be hazardous to the resistors in the circuit and therefore must be done very quickly and only very briefly.

5. Remove the short from test points 1 and 2, and connect the test lead to test points 2 and 3, shorting lamp DS1. Be sure to connect the test lead for only a brief moment to avoid damage to the circuit. You may make the connection as many times as you need to observe the effect, just be sure each time you short test points 2 and 3 that you do so only momentarily. How does this affect the brightness of the lamp?
6. Now move the alligator clip test lead to test points 3 and 4 from test points 2 and 3. What effect do you observe with respect to the brightness of the lamp?

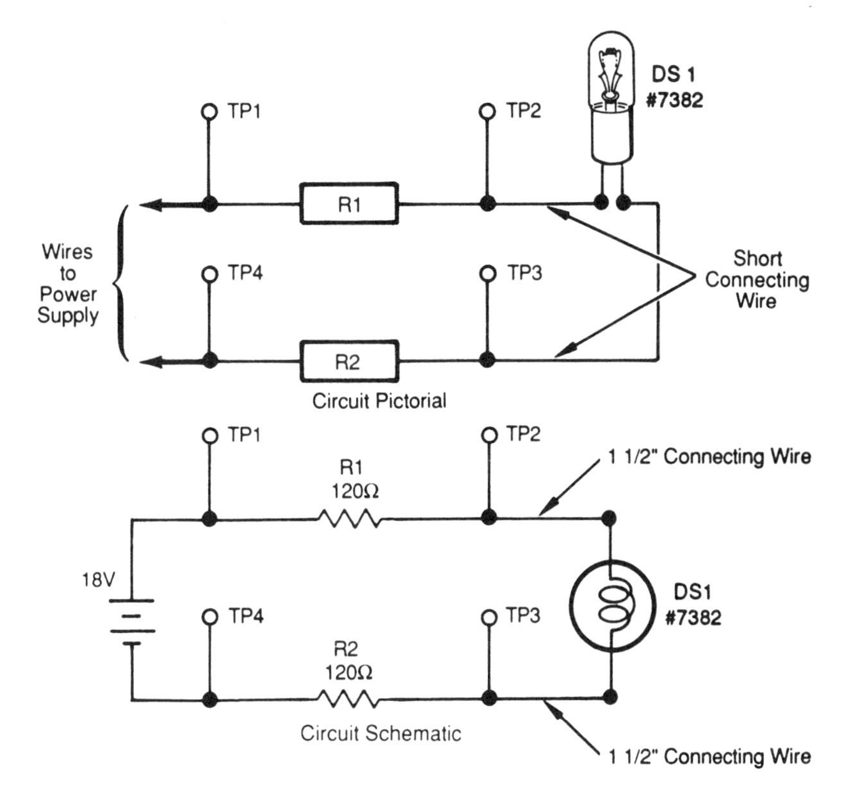

FIGURE 12–1 Simplified schematic of a test circuit with pictorial of layout alongside

Review Break

You've used the circuit you made to simulate the effects of short circuits in various parts of the circuit. We've used an incandescent lamp to give a visual indication of the problem. The visual indications can be used as clues to the nature of the problem as you troubleshoot the circuit defect. You should have observed that when you shorted R1 the lamp became brighter than normal. This means that the resistance of the circuit is less than normal which allows more current to flow than should. The brighter than normal light is a definite clue.

When you shorted DS1 briefly, to avoid damage to the circuit, you saw that the lamp went out. The short circuit across the lamp is the path of least resistance for current flow and acts as a current "hog," not allowing any current flow through the lamp and hence no light. The temptation was strong to leave the short across DS1 in place for a long time just to see what the nature of the damage would be. If you succumbed to the temptation you have learned another lesson. If you did not succumb and followed instructions, then you deserve a compliment. In any case the cause of the hazard will be covered in a later experiment, just be patient! Shorting R2 had exactly the same effect as shorting R1 since their values are the same.

Since the short circuits were simulated with a test lead, you were able to see their physical location. In real circuits the short may not be so obvious, but later you will learn techniques to isolate the faulty part by use of test equipment. For now it is important to be aware that short circuits almost always produce a visual and smelly (more on that later) indication of their existence. Sometimes the short circuit can even do damage. The damaging aspects of short circuits will be explored in the future.

7. Now remove the circuit connecting wire between test point 2 and the lamp. What effect does this have upon the lamp's brightness?
8. Re-connect the connection wire. Does this restore the circuit to normal operation?
9. Remove the circuit connection wire between the lamp and test point 3. How does this affect circuit operation?
10. Re-connect the wire. Does the circuit now work properly?

Review Break

You have just examined the effect on circuit operation of an "open" circuit. In each case the open resulted in the lamp being extinguished, no matter where in the circuit the open was located. The main difference between the open circuit and the short circuit is that when the circuit is open, there is no current flow. Did you also notice that the visual indication can sometimes be the same? Since a short circuit and an open circuit can sometimes give the same visual indication, we must learn how to distinguish one form the other by electrical measurements. You will learn those secrets in the experiments to follow.

11. Disconnect your circuit and return all materials to their proper places. And remember that troubleshooting is mostly a mental process. Observe, then think.

ACTIVITY SHEET EXPERIMENT 12

NAME ______________________________

DATE ______________________________

As you proceed through the steps of procedure in performing the experiment, fill in measurements, observations, or answers to questions.

Step 4 **A.** What effect does shorting R1 have upon the brightness of DS1?

____

____

Step 5 **B.** What happens when you momentarily short test points 2 and 3?

____

Step 6 **C.** What do you observe when test points 3 and 4 are shorted?

____

D. In all cases, what does the short circuit do? ________________

____

How does the short circuit affect normal circuit current flow?

____

Step 7 **E.** What does the open circuit cause to happen to the light bulb?

____

Step 8 **F.** After repairing the open circuit, does the circuit work correctly?

____

Step 9 **G.** Does the open circuit cause the same effect as observed in step 7? ______________________________

Step 10 **H.** Why is the circuit now working properly again? ____________

____

I. Can a short circuit cause damage to any part of an electrical circuit? ______________________________

Why? ____

____

J. Can an open circuit cause damage to any part of an electrical circuit? ____

____

____

Why? ____

____

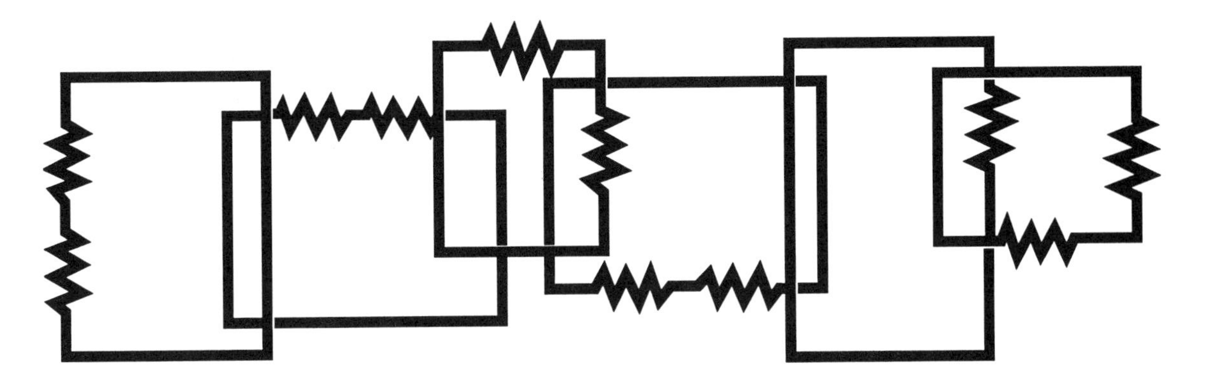

13

Series Circuit Voltage Division

Objectives

After completing this experiment, you will be able to:

1. Evaluate the voltage distribution in a series circuit.
2. Calculate resistance values needed for desired drops in a series string.
3. Select resistance values for series circuit to meet design goals.
4. Evaluate effectiveness of selected resistances in a series circuit.

Introduction

This experiment will demonstrate that a series circuit is a voltage divider. You will construct a series circuit made up of three resistors in order to divide a source voltage into three smaller, predetermined voltages. The calculations are simple and require some understanding of Ohm's Law and Kirchhoff's Voltage Law. The parts necessary to perform the experiment are:

- A variable dc power supply
- Six $1\frac{1}{2}$ inch lengths of #24 hookup wire
- Two 8 to 12 inch lengths of #24 hookup wire
- A prototyping board
- A VOM (analog, digital, or both)
- One 120Ω resistor
- Assorted values of resistors to select from

This is the problem you must solve: A source voltage of 25V needs to be divided into three smaller voltages. They are 2V, 8V, and 15V. The 2V drop will be developed across the only known resistor value in the series string, the 120Ω resistor. With your knowledge of Ohm's Law and Kirchhoff's Voltage Law calculate the values of the two other resistors. Refer to figure 13–1 for the locations of the resistors and test points in the circuit you will construct to verify your calculations.

Procedure

1. Pull the activity sheet for this experiment and use the table provided as a guide as you calculate the unknown resistor values. Then select from your assortment of resistors two whose values are as close as possible to your calculated values. Because your choices will be limited to standard values of resistance it will not be possible to attain perfection. This is the reality of circuit design—frequently a compromise must be made because resistor values which are calculated may not be readily available. As long as you are able to come reasonably close to your design objective you will have succeeded.
2. Construct the circuit as shown in figure 13–1. Set the power supply output to 25V.
3. Measure and record the voltage drops across each of the three resistors from test points 1 to 2, 2 to 3, and 3 to 4. Are the measured values close to the expected voltages?

FIGURE 13–1 Schematic of circuit showing all resistors, test points and desired drops

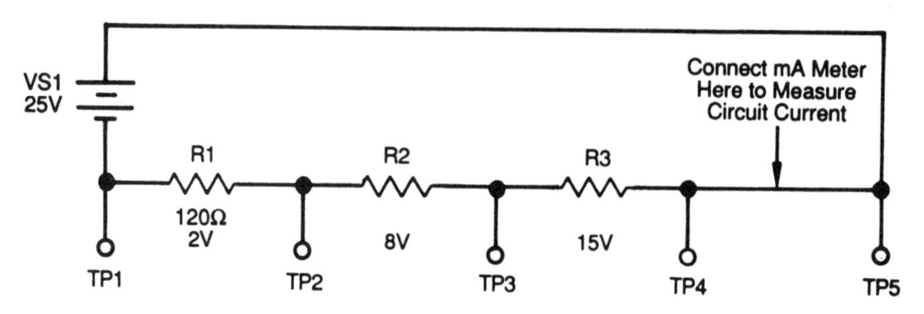

4. Verify Kirchhoff's Voltage Law by adding the three voltage drops. Is the sum equal to the source voltage?
5. Measure the voltage from test point 1 to test point 3. Is the measured voltage equal to the sum of the two individual voltages?
6. Measure the voltage from test point 2 to test point 4. Is the measured voltage equal to the sum of the two individual voltages?
7. Set the VOM to measure current and connect it to test points 4 and 5. Remove the hookup wire connecting the test points to allow current flow through the milliammeter. Measure and record circuit current. Remove the VOM from the circuit and restore the connecting wire between test points 4 and 5.
8. Calculate total circuit resistance by use of source voltage and circuit current.
9. Disconnect the power supply, set the VOM to measure resistance and measure the total resistance of the circuit. Is the measured resistance the same as the calculated resistance?

Review Break

You have designed a circuit to divide a known voltage into three smaller known parts. You constructed a circuit based on your design values and verified by measurement that your circuit did pretty much what you expected it to do. The resistors you used may have varied from the calculated values by more than a hundred ohms, but the result was in a practical sense acceptable.

You then verified Kirchhoff's Voltage Law, proved that voltage drops in series are aiding (additive), calculated total circuit resistance based on measured current and voltage and finally measured the circuit resistance to verify your computation. Do you feel that you could design and implement a voltage divider to develop other combinations of voltages? These techniques will be used again in the future when you study transistor amplifiers and will be of value in designing your own amplifier circuits.

10. **Retain the circuit on the prototyping board for the experiment to follow.** Put all other materials back to their proper places.

ACTIVITY SHEET EXPERIMENT 13

NAME ____________________________

DATE ____________________________

Record calculations and measured data in the appropriate spaces.

A.

	Component →	**R1**	**R2**	**R3**
Step 1	Desired voltage drop	2V	8V	15V
	Desired current flow	mA	mA	mA
	Calculated resistance	120Ω	Ω	Ω
	Standard value used	120Ω	Ω	Ω
Step 3	Measured voltage drops	V	V	V
Step 5	Measured voltage TP1 – TP3		V	
Step 6	Measured voltage TP2 – TP4			V

Step 4 **B.** The sum of the three measured voltage drops.

____________________ V

Step 7 **C.** Measured circuit current flow ____________________ mA

Step 8 **D.** Calculated total circuit resistance ____________________ Ω

Step 9 **E.** Measured total circuit resistance ____________________ Ω

F. Were your measured results close to the design objective? ______

____

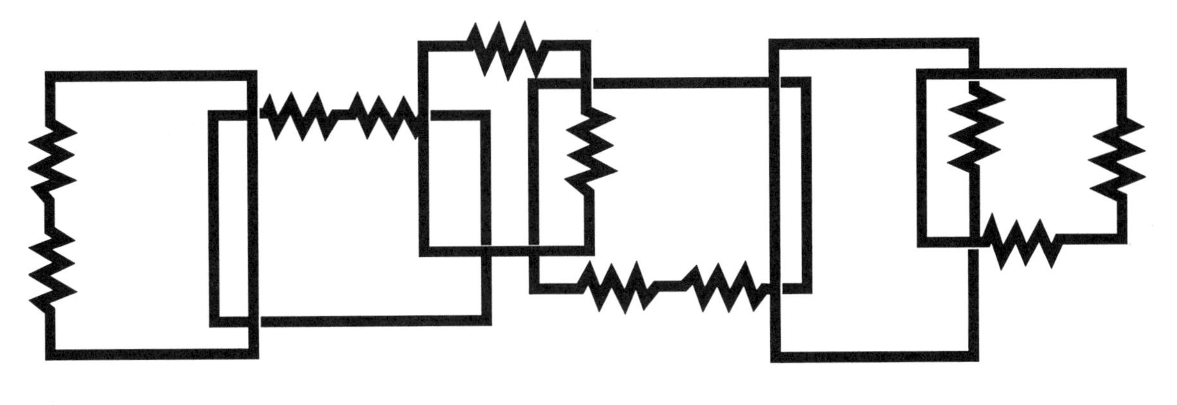

14

Series Circuits—Power Distribution

Objectives

After completing this experiment, you will be able to:

1. Calculate the power dissipated by a resistance when in a circuit.
2. Determine the power distribution in a series circuit.
3. Determine resistor power dissipation by measurement.
4. Confirm the relationship between resistance and power.

Introduction

Resistors are best known for their ability to oppose current flow. It is important to know that as they oppose current flow resistors also dissipate power. You are already aware that when the dissipated power exceeds the power rating of a resistor damage and destruction are likely to result. It is important to know how to determine the power dissipated by all resistors in a series string, and to ascertain whether any of the resistors are being stressed. You should also know intuitively which resistor in the string is dissipating the most power and which resistor is dissipating the least. In order to perform this experiment you will need:

A VOM (digital, analog, or both)
A variable dc power supply
Series circuit constructed in previous exercise
Two 8" to 12" lengths of hookup wire

Procedure

1. Pull the activity sheet for this experiment. Before connecting the power supply to the circuit, measure each of the individual resistors in the circuit and record their actual resistance values.
2. Connect the power supply to the circuit as shown in figure 14–1 and set its output to 25V. Measure and record each of the resistor voltage drops.
3. Connect the VOM as a milliammeter between test points 4 and 5. Remove the circuit wire to place the meter in series with the resistors. Measure and record circuit current.
4. You have now collected all data necessary to complete the activity sheet. You may disconnect the circuit and return all materials to their proper places.
5. Calculate the power dissipated by each of the resistors using the measured voltage, current and resistance values. You will be able to calculate the dissipated power for each resistor by two methods. Do the two calculations agree?
6. Based on your calculations which resistor dissipates the most power? Which resistor dissipates the least power?
7. Determine total power dissipation by three methods. Add the powers dissipated by the three resistors, calculate the total power by multiplying measured voltage and measured current, and calculate total power by multiplying current squared and total measured resistance. Do the three calculations agree?

FIGURE 14–1 Schematic diagram of a series circuit

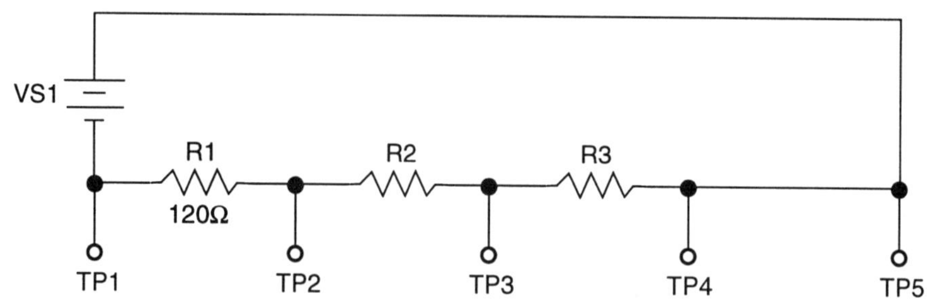

You measured the electrical characteristics of the circuit and based on the measured values of voltage, current and resistance determined the power dissipated by each resistor. You used two methods of computation to verify that either method will give you an accurate result. You then calculated total power dissipation based on measured parameters by three methods. You found that the results were virtually the same. The lesson learned is that when it comes to calculating dissipated power it doesn't matter which method is used providing your measurements are as accurate as possible. You were then able to note a correlation between a resistor's resistance and its power dissipation in a series circuit. Is the power dissipated by series resistors directly or inversely proportional to their resistance?

ACTIVITY SHEET EXPERIMENT 14

NAME ______________________________

DATE ______________________________

Record measured and calculated data in the appropriate spaces.

Step 1 **A.** The measured resistance of each resistor is

R1 ____________________ Ω R2 ____________________ Ω

R3 ____________________ Ω

Step 2 **B.** The measured voltage drops across each resistor are:

V_{R1} ____________________ V V_{R2} ____________________ V

V_{R3} ____________________ V

Step 3 **C.** The measured circuit current is ____________________ mA

Step 4 **D.** Calculated power dissipation for each resistor

	P = VI	P = I²R
P_{R1}	____________________ mW	____________________ mW
P_{R2}	____________________ mW	____________________ mW
P_{R3}	____________________ mW	____________________ mW

Step 6 **E.** The resistor dissipating the most power is ____________________

The resistor dissipating the least power is ____________________

Step 7 **F.** Total power dissipation

$\frac{V^2}{R_T}$ ____________________ mW

$I^2 R_T$ ____________________ mW

$V_T I$ ____________________ mW

The sum of the individual resistor powers is ____________________ mW

G. In the series circuit, how is the power dissipated by a resistor related to the resistance of the resistor? ____________________

____

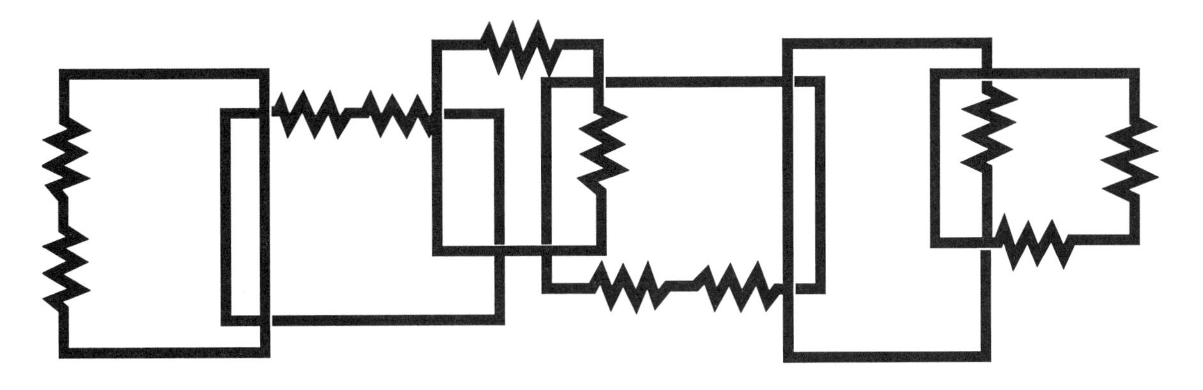

15

Troubleshooting Series Circuits

Objectives

After completing this experiment, you will be able to:

1. Evaluate circuit operation by measurement of voltage and current.
2. Isolate an open circuit by measurement.
3. Isolate a short circuit by measurement.
4. Evaluate voltage and current measurements as indicators of trouble.
5. Measure "in circuit" resistances as indicators of circuit problems.

Introduction

Troubleshooting electrical circuits has been likened to solving a problem or a puzzle. It is mainly a mental skill requiring a lot of thought, digesting all the clues that we can gather to find the circuit component which is faulty. How well we are able as troubleshooters to solve a problem is dependent upon two things: first our understanding of the circuit, and second our ability to find clues which will lead us to the problem. As you approach a circuit which is defective, with the intent of fixing it, there are several steps of clue (or symptom) recognition which should always be followed. These steps are:

1. Visually inspect the circuit for indications of trouble such as broken parts, loose connections, burnt components, corroded or soiled parts, and other obvious things which might be seen. It is a good idea to use your nose while doing the visual inspection because quite often you will be able to smell something which has burned out. The visual inspection is usually done before turning the circuit on but you have a choice. There will be times when the visual inspection will be most fruitful while power is applied. With power applied you may see smoke or other indications of abnormality that might not be apparent while the circuit is de-energized.
2. Energize the circuit and check the power supply voltage or voltages. Never assume that the power source is good. Checking it early in the troubleshooting process can save you much time.
3. Measure circuit voltages, resistance, and current at select areas and compare your measurements to normal circuit characteristics. To make the comparisons you will need technical information about the circuit such as a technical manual.
4. If the circuit is a low voltage circuit, 25 volts or less, then you may carefully touch components in the circuit feeling for parts that are unusually warm. This must be done with extreme caution to avoid injury by getting burned since some parts can become very hot when they are defective.
5. Seek the advice and assistance of a more experienced troubleshooter if you are unable to isolate the problem or get stalled in the process of gathering symptoms. The person you ask will most certainly be eager to assist you, and you will gain valuable experience as a result. There is no shame in asking for help because even the most sagacious technical experts have resorted to this step countless times.
6. Repair the circuit by replacing the faulty part, fixing the bad connection, or correcting whatever you have found to be defective.
7. Test the circuit thoroughly to assure that it works as it should. If there still appear to be problems, go back to step 1.

As you proceed through this experiment you will troubleshoot a simple series circuit and exercise all the options at your disposal in order to find whatever defects may exist. Then you will correct each problem and verify that the circuit is operating normally. You will need:

A VOM (digital, analog or both)
Three 120Ω 5% resistors
One 560Ω 5% resistor

A variable dc power supply
Four $1\frac{1}{2}$" lengths of hookup wire
Two 8 to 12 inch lengths of #24 hookup wire
Two test leads with alligator clips
A prototyping board
One #7382 bi-pin lamp (14V 80 mA)
One 2 inch length of #24 hookup wire

The most common problems you will encounter in simple electric circuits are (1) resistors which have changed value and have more resistance or less resistance than they are supposed to, (2) open circuits and (3) short circuits. In determining the nature of the circuit problem the VOM is a very valuable tool. In addition to making visual and olfactory observations it will probably be necessary to measure electrical parameters of the circuit to find the exact problem. The questions arise: Which measurements are most meaningful? Is it best to measure voltages, resistances or current? To aid you in answering those questions you will perform all three measurements as you search for simulated problems. Then you can decide which method to use when. In the final analysis the choice is yours.

Procedure

1. Pull the activity sheet for this experiment. Construct the circuit as shown in figure 15–1. Before connecting the circuit to the power supply measure and record the resistance of the circuit. Set the output of the power supply to 12.0 volts and connect the circuit to the supply. The lamp will be visibly glowing but at less than its normal intensity. For this circuit application a dim light is desired and is therefore a normal circuit characteristic.
2. Measure all circuit voltages and record them in the appropriate spaces on the sheet. You are establishing for reference purposes the operating characteristics of the circuit.
3. Measure and record normal circuit current for reference purposes.
4. Create an open circuit by lifting the lead of R3 which is connected to R2 and placing it in a hole on the board which has no continuity to R2. See figure 15–2. You should observe that the lamp is no longer illuminated, a clear indication that something is wrong. You of course know where the problem is. Now you will make some measurements to learn what the electrical indications of the problem are.

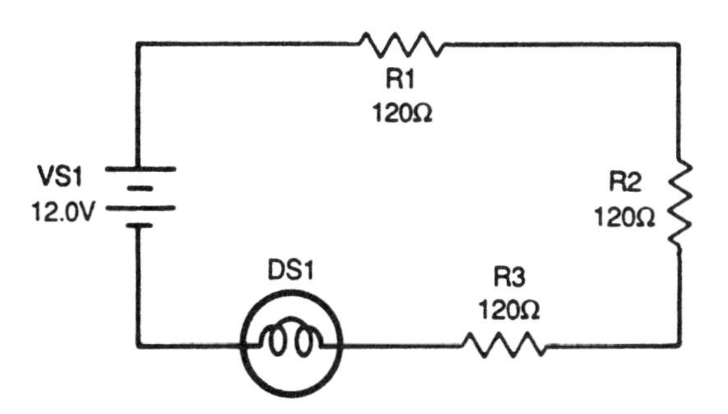

FIGURE 15–1 Schematic diagram of series circuit

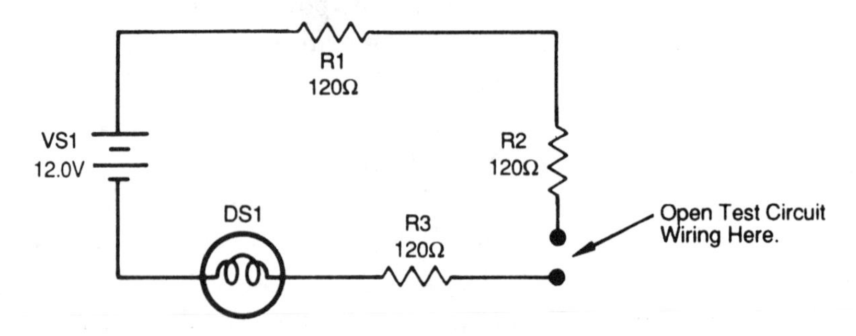

FIGURE 15–2 Schematic diagram - open circuit problem

TROUBLESHOOTING TIP

According to Ohm's Law voltage drops can be developed only when there is current flow through the resistances of a circuit. Now that the test circuit has an open, will there be any voltage drops to measure?

5. Measure the source voltage to ascertain whether it is still correct.
6. Measure each of the resistive voltage drops to ascertain whether they are still correct.
7. Measure circuit current to ascertain whether it is still correct.
8. Now measure and record the voltages between each resistor in the circuit. Ordinarily there would be no measurable voltage at the connection points because the resistance of the connecting wire is negligible. This circuit will be an exception.
9. Review the data you have collected. Have you measured something suspicious? How can this anomaly be explained? Is it a definite indicator of the circuit problem?
10. Restore the connection of R2 to R3 and measure all voltages to assure that the circuit is working properly.
11. Place the 2 inch piece of hookup wire across R2 to short it as shown in figure 15–3. Once the short is in place is there any visual indication of an abnormality in the circuit? What is it?
12. Measure and record the voltages across the circuit components, then measure and record circuit current.

FIGURE 15–3 Schematic diagram - short circuit problem

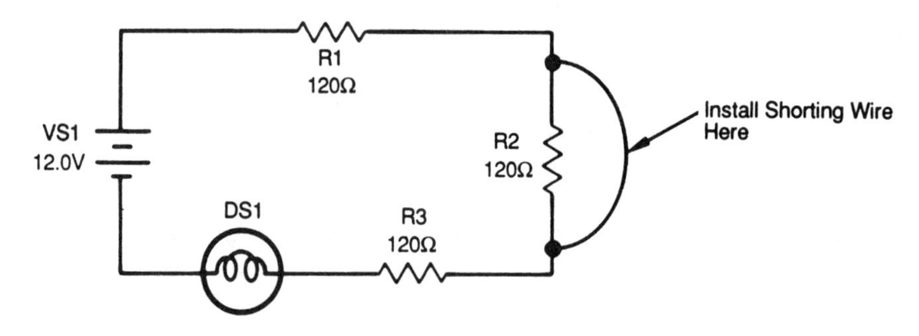

13. Measure the source voltage to assure that it is correct.
14. Review the data that you have collected. Is there anything in your measurements which substantiates the nature of the circuit problem? Which measurements are abnormal when compared to normal circuit characteristics?
15. Using source voltage and circuit current calculate the resistance of the circuit. Then disconnect the circuit from the power supply and measure its resistance. Then compare the measurement with the normal resistance measured earlier. What conclusion can you draw based on the comparison?
16. Remove the short circuit from across R2, reapply 12V dc to the circuit and measure all voltages to assure that the circuit is again working normally.

Review Break

After putting the circuit together and establishing what its normal electrical characteristics were, you simulated an open circuit by moving one of the circuit wires. You observed that the lamp was no longer illuminated, a sign that there may not be any current flow in the circuit. You measured the source and found it to be normal, then you attempted to measure voltage across each of the circuit components and found none. This is a sure sign that there is no current flow. To finally isolate the open circuit you measured for voltage at each of the circuit connections and found a voltage from the lead of R2 to the lead of R3. This condition verified what you had discovered in an earlier experiment. When there is an open circuit the source voltage will be measured across the open. You then shorted one of the resistors and made both visual and electrical observations. You found that the lamp intensity increased indicating an increase in circuit current. You also found that all measured voltages were incorrect and the measured current was incorrect. When you compared the measurements against the normal characteristics the problem was rather obvious. You verified that circuit resistance was incorrect by calculation and by measurement. The open circuit and the short circuit produced different symptoms and by careful consideration of the gathered symptoms you are able to reason out the location of the fault. Now for another problem.

17. Remove R1 and replace it with the 560Ω resistor. The lamp should not be glowing visibly, indicating that something is wrong.
18. Measure and record all circuit voltages then compare them to the normal voltages. Based on observed electrical date is the nature of the problem indicated?
19. Disconnect the circuit from the power supply and measure the resistance of each component separately. Does this reveal the fault with absolute certainty?
20. Return all materials to their proper places; you've finished the experiment.

Review

The final problem simulated a resistance whose value increased substantially. The visual symptom was similar to that of the open circuit so it was necessary to perform measurements in order to determine the real nature of the problem. You measured circuit voltages and when comparing them with the normal circuit voltages you found a likely candidate for the fault. You then measured the individual resistances to remove all doubt. Whether you decide to rely upon voltage, resistance, or current measurements while troubleshooting depends upon the symptoms you observe. Voltage measurements are easy to make and can be very revealing when properly interpreted. Resistance and current measurements can also be of great value.

ACTIVITY SHEET EXPERIMENT 15

NAME ______________________

DATE ______________________

Step 1 **A.** Record the measured total circuit resistance:

______________ Ω

Step 2 **B.** Record all measured circuit voltages:

V_S ______________ V V_{R1} ______________ V

V_{R2} ______________ V V_{R3} ______________ V

V_{DS1} ______________ V

Step 3 **C.** Record measured total circuit current:

______________ mA

Steps 5–6 **D.** Record measured circuit voltages:

V_S ______________ V V_{R1} ______________ V

V_{R2} ______________ V V_{R1} ______________ V

V_{DS1} ______________ V

Step 7 **E.** Record measured circuit current: ______________ mA

Step 8 **F.** Record measured voltages between resistors:

$V_S - V_{R1}$ ______________ V $V_{R1} - V_{R2}$ ______________ V

$V_{R2} - R_{R3}$ ______________ V $V_{R3} - V_{DS1}$ ______________ V

$V_{DS1} - V_S$ ______________ V

Step 12 **G.** Record measured circuit voltages: V_{S1} ______________ V

V_{R1} ______________ V V_{R2} ______________ V

V_{R3} ______________ V V_{DS1} ______________ V

Step 12 **H.** Record measured circuit current: ______________ mA

I. Record calculated total circuit resistance: ______________ Ω

Step 15 Measured total resistance: ________________ Ω

Step 18 **J.** Record all measured voltages: V_{S1} ____________________ V

V_{R1} ________________ V V_{R2} ________________ V

V_{R3} ________________ V V_{DS1} ________________ V

Step 19 **K.** Record measured resistances:

R1 ________________ Ω R2 ________________ Ω

R3 ________________ Ω R_{DS1} ________________ Ω

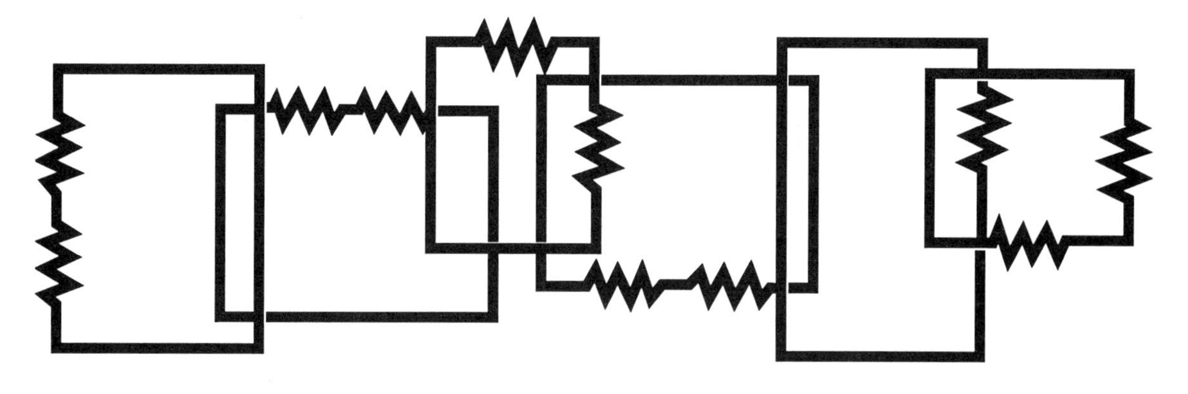

16

Parallel Circuits—Measurements

Objectives

After completing this experiment, you will be able to:

1. Evaluate the operation of a parallel circuit.
2. Measure voltages and currents in the parallel circuit.
3. Evaluate the effect of changing resistances in the parallel circuit.
4. Confirm that total current is the sum of branch currents.
5. Confirm that branch voltages are equal.

Introduction

The purpose of this experiment is to explore the characteristics of parallel circuits. You will use several resistors to determine how their values combine when connected in parallel, and you will verify the current division properties of parallel circuits. To perform this experiment you will need:

A VOM (Digital or Analog)
A prototyping board
Three 120Ω 5% resistors
One 330Ω 5% resistor
One 560Ω 5% resistor
Two 8 to 12 inch lengths of #24 hookup wire
Fifteen $1\frac{1}{2}$ inch lengths of #24 hookup wire
A variable dc power supply

Procedure

1. Pull the activity sheet for this experiment. Before constructing the circuit measure and record the values of the five resistors. Then construct the circuit as shown in figure 16–1.
2. Before connecting the power supply to the circuit measure and record total circuit resistance. Set the power supply output to 3.6V, then connect it to the circuit.
3. Now that the power source is connected to the circuit, measure and record the voltages across the three resistors.
4. Set the VOM to function as a milliammeter. Then measure and record the current flow in the individual circuit branches.
5. Measure and record the total circuit current.
6. Calculate circuit performance characteristics as instructed on the activity sheet.

Review Break

The circuit you have constructed and evaluated is a special kind of parallel circuit because each of the three branches has approximately the same resistance. You verified by measurement that the voltage across each branch is equal to the applied voltage. You verified that in this special case the currents through the individual branches are almost the same. And you verified that the circuit resistance is equal to $\frac{1}{3}$ the resistance of each branch resistance. Now you will modify the circuit and continue your examination of parallel connections.

7. Replace R2 with the 330Ω resistor and R3 with the 560Ω resistor.
8. Measure and record the voltage drops across the three resistors.
9. Measure and record the individual branch currents.
10. Measure and record the total circuit current.
11. Disconnect the power supply from the circuit and measure the total circuit resistance.

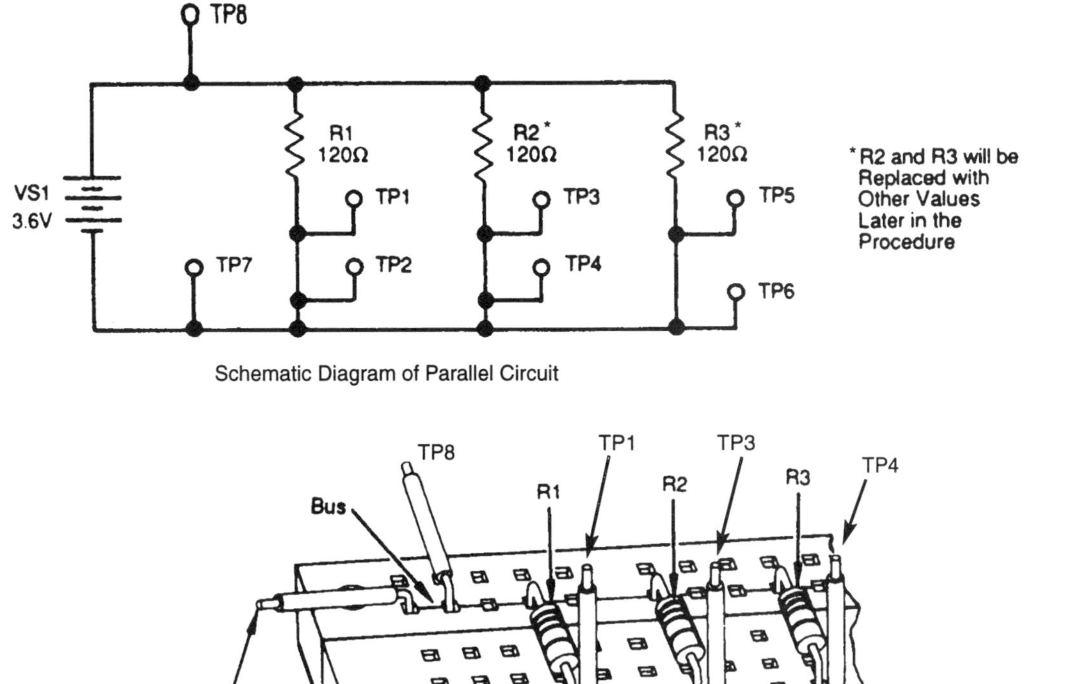

Schematic Diagram of Parallel Circuit

Pictorial of Circuit

FIGURE 16–1 Schematic and pictorial diagram of a parallel circuit

12. Return all materials to their proper places. You have completed the measurements necessary to the experiment.

13. Verify by calculation that the performance of any parallel circuit made up of resistors is predictable. Follow instructions on the activity sheet.

Review

The modified circuit presented some differences as well as some similarities as compared to the original circuit. You found that the voltage across each branch is still the same even though the resistances of the branches are different. You measured the currents in the three branches and verified that the parallel circuit operates as a current divider. The current flow through the individual branches is inversely proportional to the resistance of the branches. Finally, you proved that the electrical parameters of parallel connected circuits can be predicted by calculation.

ACTIVITY SHEET EXPERIMENT 16

NAME ______________________________

DATE ______________________________

Step 1 **A.** Record measured resistances of the five resistors:

R1: 120Ω ____________________ Ω

R2: 120Ω ____________________ Ω

R3: 120Ω ____________________ Ω

(R2′): 330Ω ____________________ Ω

(R3′): 562Ω ____________________ Ω

Step 2 **B.** Record measured circuit resistance: ____________________ Ω

Step 3 **C.** Record measured circuit voltages: V_{S1} ____________________ V

V_{R1} ____________________ V V_{R2} ____________________ V

V_{R3} ____________________ V

Step 4 **D.** Record measured branch currents:

I_{R1} ____________________ mA I_{R2} ____________________ mA

I_{R3} ____________________ mA

Step 5 **E.** Record total circuit current I_T ____________________ mA

F. Using measured resistances and voltages, calculate branch currents:

I_{R1} ____________________ mA I_{R2} ____________________ mA

I_{R3} ____________________ mA

G. Using measured resistances calculate total circuit resistance:

____________________ Ω

Step 8 **H.** Record measured voltages across modified circuit resistors:

V_{R1} ____________________ V V_{R2} ____________________ V

$V_{R3'}$ ____________________ V

Step 9 **I.** Record measured branch currents:

I_{R1} ____________________ mA I_{R2} ____________________ mA

I_{R3} ____________________ mA

Step 10 **J.** Record measured total circuit current:

____________________ mA

Step 11 **K.** Record measured circuit resistance: ____________________ Ω

Step 13 **L.** Using measured resistances and voltages calculate:

I_{R1} ____________________ mA I_{R2} ____________________ mA

I_{R3} ____________________ mA R_T ____________________ Ω

I_T ____________________ mA

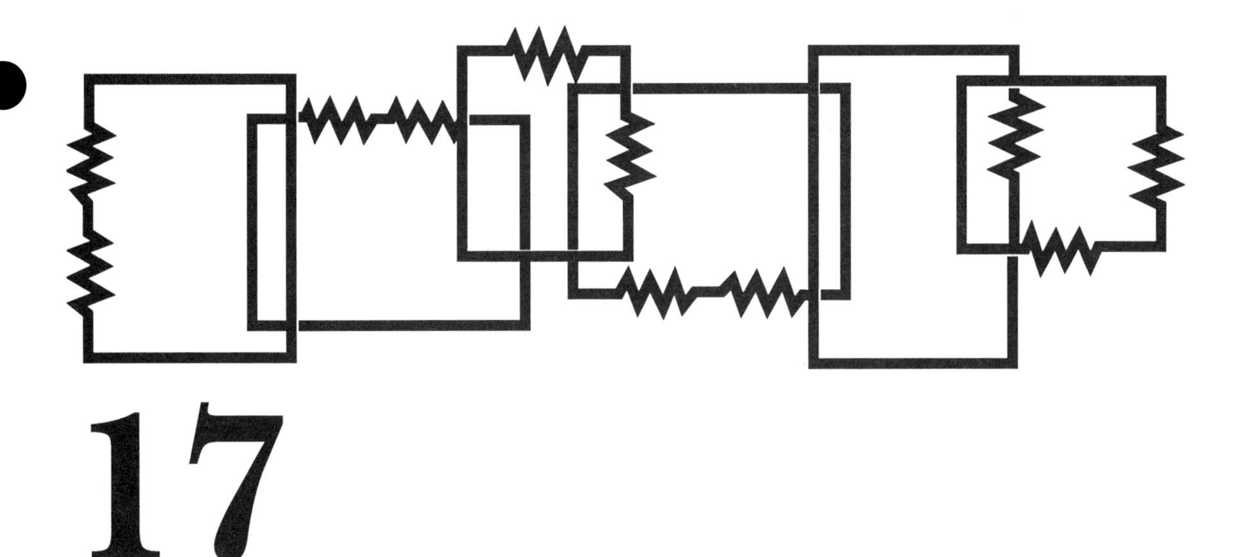

17

Parallel Circuit Applications

Objectives

After completing this experiment, you will be able to:

1. Evaluate the normal operation of parallel connected lamps.
2. Modify a parallel circuit to permit nondestructive short circuits.
3. Observe the characteristics of shorted parallel branches.
4. Confirm that a shorted branch shorts all branches.

Introduction

You have probably wondered "Where can I find parallel circuits in real life? Do they serve a useful purpose?" You may find this difficult to believe, but you are literally surrounded by parallel circuits. There must be some important advantages about parallel circuits to justify their prominence. But there may be some pitfalls to be aware of as well. Let us examine an application. You will need:

A prototyping board
A variable dc power supply
Three #7382 bi-pin lamps (14V 80mA)
Two 8 to 12 inch lengths of #24 hookup wire
Five $1\frac{1}{2}$ inch lengths of #24 hookup wire
One 47 ohm resistor

Procedure

1. Pull the activity sheet for this experiment. Construct the circuit shown in figure 17–1. Before connecting the power supply to the circuit set its output to minimum.
2. Increase the output of the power supply to 14 volts. Vary the supply output between 14 volts and minimum then back to 14 volts several times and record your observations (how do the lamps respond?).
3. Set the power supply output to 14 volts. Remove two lamps from the circuit one at a time and record your observations.
4. Put the lamps back into the circuit one at a time and record your observations.

FIGURE 17–1 Schematic diagram of parallel lamps

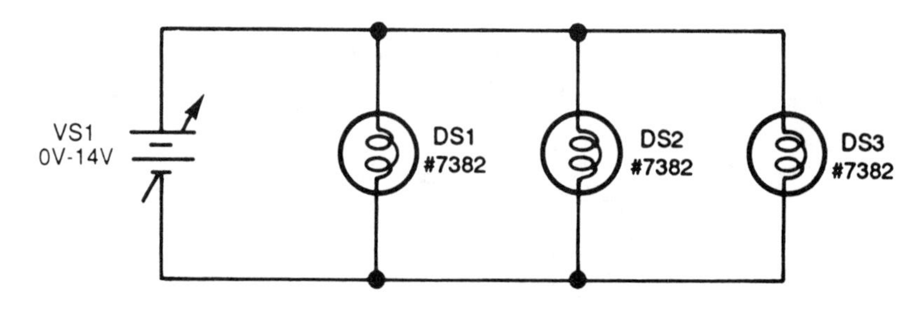

FIGURE 17–2 Schematic of modified circuit showing 47Ω resistor added in series with the source

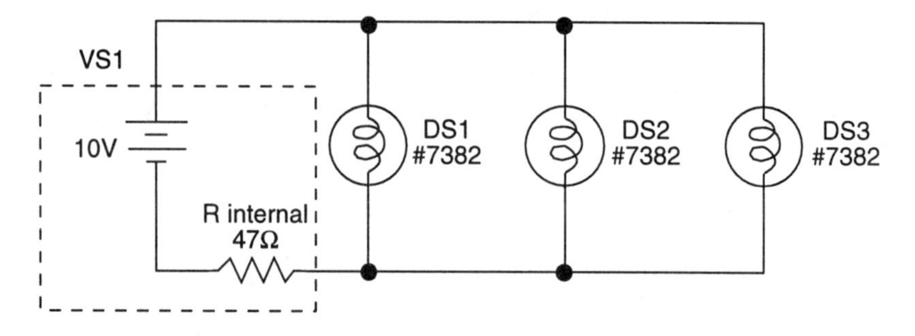

5. Modify your circuit as shown in figure 17–2 and set the power supply output to 10 volts. Consider the added resistor to be the internal resistance of the power supply. Its purpose is to protect the power supply from excessive current demand during the next step.
6. Take a $\frac{1}{2}$ inch length of hookup wire and momentarily short circuit each of the lamps in succession. Record your observations. Do not allow the short circuit across the lamps to remain for any length of time or damage to the 47Ω resistor may result.

Review

You have constructed a simple parallel circuit consisting of three lamps. You found that as you vary the applied voltage the intensity of the lamps will vary in unison. You then disconnected two of the lamps from the circuit one at a time and noted that the intensity of the remaining lamp/s did not vary. You then replaced the lamps one at a time and made the same observation. The circuit was modified to simulate an internal resistance of the power source and the three lamps were short circuited one at a time. What you saw was the disadvantage of parallel circuits. If for some reason one of the parallel branches is shorted it becomes a current hog. What effect did this have on the remaining lamps? Why do you suppose this is so? What could be done with parallel circuits to prevent this disadvantage? (Think of the electrical system in your home and your car.)

7. Return all materials to their proper places. You've passed another milestone in your trek through the mysteries of electric circuits.

ACTIVITY SHEET EXPERIMENT 17

NAME ______________________________

DATE ______________________________

Step 2 **A.** As you vary the output of the power supply, what do you see?

____

Step 3 **B.** As you remove two lamps from the circuit one at a time, how does this affect the intensity of the remaining lamp/s in the circuit?

____

Step 4 **C.** As you re-insert the removed lamps back into the circuit, how is their intensity affected? ______________________________

Step 5 **D.** As each of the lamps is shorted one at a time, how is the intensity of all lamps affected? ______________________________

E. Explain your observations in part A above: ____________________

____

____

F. How do you account for your observations in part B above? _____

____

____

G. Why did your actions in part C above produce the results you observed?

____

____

H. Explain how your observations in part D above can be accounted for electrically. ______________________________

____

____

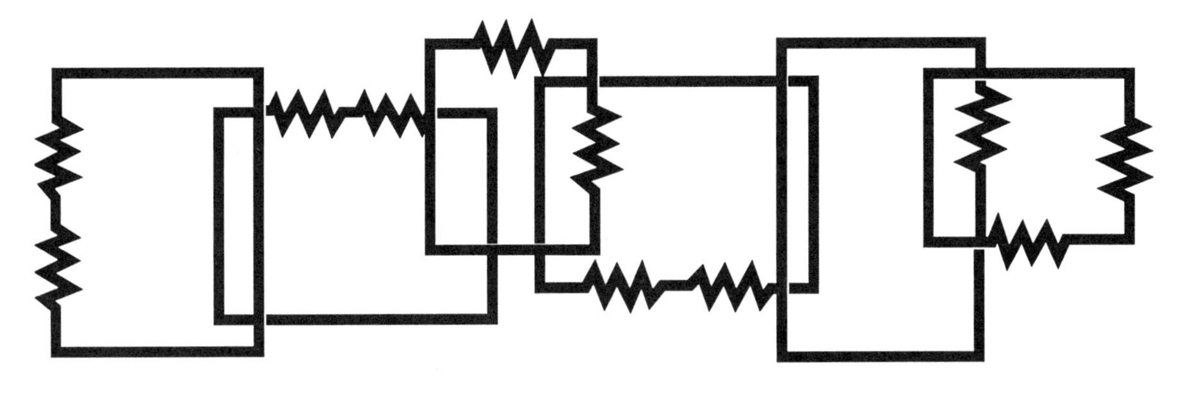

18

Troubleshooting Parallel Circuits

Objectives

After completing this experiment, you will be able to:

1. Investigate the effects of troubles injected into parallel circuits.
2. Determine normal circuit parameters by measurement.
3. Interpret circuit measurements as indicators of problems.
4. Ascertain continuity between wiring and components in parallel circuits.
5. Analyze circuit measurements when troubleshooting parallel circuits.

Introduction

Troubleshooting parallel circuits is not too different from troubleshooting series circuits. The measurement techniques used to gather data about the circuit, as a means of isolating troubles, are virtually the same. The only real difference is the way the circuit is wired. Your knowledge and understanding of parallel circuits will allow you to handle the mental aspect of troubleshooting: reasoning the problem through to solution. But in order not to get confused or stuck, it is of supreme importance that the data you gather by measurement be both accurate and pertinent to the circuit problem. A little practice with a simple circuit will allow you to see for yourself. To perform this experiment you will need:

A variable dc power supply
One 120Ω resistor
One 330Ω resistor
One 560Ω resistor
One #7382 bi-pin lamp (14V 80mA)
Five $1\frac{1}{2}$ inch lengths of #24 hookup wire
Two 8 to 12 inch lengths of #24 hookup wire
A prototyping board
A digital or analog VOM

Procedure

1. Pull the activity sheet for this experiment. Construct the circuit shown in figure 18–1. The circuit doesn't do anything spectacular, it just dissipates power, but it is representative of three devices connected to a common source. The resistors could be transistor radios or lamps or any small electronic device in an actual application. **Do not connect the power supply to the circuit yet.**
2. Before connecting the power supply measure and record the total resistance of the circuit. Then set the power supply output to 10 volts, connect it to the circuit, and measure and record the total circuit current.
3. Using the measured current and the applied voltage, calculate and record the resistance of the circuit. Is the calculated resistance the same as the measured resistance? Does something seem to be wrong? (Remember the characteristics of the light bulb.)

FIGURE 18–1 Schematic diagram of parallel circuit

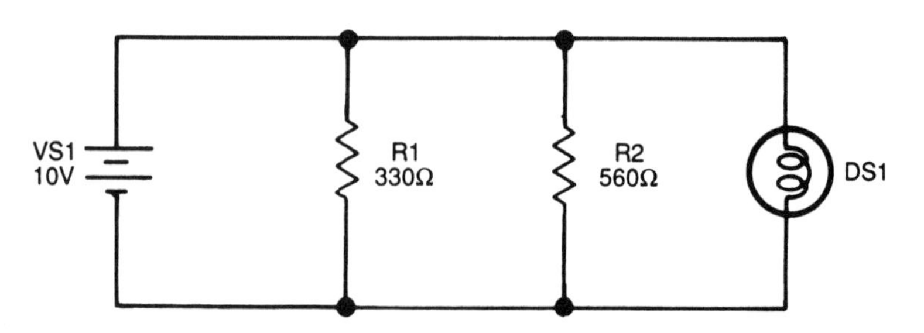

4. To induce a problem into the circuit, disconnect one end of resistor R2 from the circuit. Plug the lead back into the prototyping board in a hole that has no connection with the circuit wiring. This will effectively open the R2 branch of the circuit. Of course you know where the problem is and what it is, but how would you go about locating it by measurements if you didn't know? With this circuit problem are there any visual indications?
5. When a parallel circuit has an open branch, how can you locate the open branch by measurement? What electrical properties of the circuit will be altered by the open? Will a resistance measurement locate the faulty branch, or would a current measurement be better? Or a voltage measurement? Let's perform all three to see which is most meaningful.
6. Measure and record the total circuit current. Compare this measurement to the normal circuit current characteristic you established before putting the problem into the circuit. Does this give you enough information to work with?
7. Disconnect the power supply from the circuit then measure and record the total circuit resistance. Compare this measurement with the normal circuit resistance established earlier. Do you now have enough information to prove which branch is open?
8. With the power supply still disconnected from the circuit, measure and record the resistance of each individual branch by placing the ohmmeter leads directly across each component of the circuit. Study the results carefully. Have you now gathered enough data to prove the nature of the problem?
9. Reconnect the power supply to the circuit so that you can make voltage measurements. Measure and record the voltage at each component by placing the voltmeter leads directly across the three individual components one at a time. These measurements, coupled with all other measurements made to electrically isolate the problem, should prove beyond doubt the nature of the problem. But there is one thing which needs to be established. At which end of the branch is the open? What measurement technique would best illustrate where precisely the open is?
10. Let's try voltage measurements. Observe proper polarity and make two last measurements, from the top end of R2 to the circuit wiring which connects to the positive terminal of the source. Then from the bottom end of R2 to the circuit wiring which connects to the negative terminal of the source.

Review Break

In order to gain some experience in tracking down parallel circuit problems, you induced a connection fault into your test circuit. You then proceeded to see if you could find the defect by measurement of circuit characteristics and comparison of your measurements to known good circuit characteristics. You began by making a circuit current measurement. This indicated that the total current flow was indeed less than it should have been. This measurement alone, and a couple of quick calculations, may have led some of you directly to the faulty branch. The next measurement you made was a total circuit resistance measurement. With a good understanding of parallel circuits, this piece of data, and a couple of more complicated calculations could have pinpointed the fault. To positively locate the fault you then proceeded to measure the resistance of each component in the circuit. Because of the parallel connection, the resistance measured at each component would ordinarily be the same value, the total resistance of the circuit. Because of the open connection in one of the branches you noticed a definite irregularity. Most of you by this time had the problem electrically in the bag. But to be absolutely certain, you then measured voltages across each component in the circuit. This resulted in finding that there was no voltage drop across R2, which in turn led you to correctly deduce that there was no current flow through R2. All that was left was to determine at which end of R2 the loss of continuity was located. A couple of voltage measurements led you quickly to the fault. You measured voltage from one end of the resistor to the circuit wiring to which it was supposed to be connected. Ordinarily you should have found zero volts there. The presence of voltage proved beyond doubt that an open in the circuit had developed. Isn't incredible what can be done with some basic electrical measurements and some serious thought?

11. Reconnect R2 to its proper place in the circuit to restore normal operation, then modify the circuit as shown in figure 18–2. The purpose of the modification is to offer some protection to the power supply. Consider the added resistor to be a part of the power supply; its internal resistance. Reduce the power supply voltage to 7.0 volts to avoid damaging the added resistor as you proceed. At this reduced voltage the lamp will be only dimly illuminated.

12. Now induce another problem into the circuit by placing a short circuit across R1. As soon as the short circuit is in place you should observe a visual indication of circuit problem. What did you see?

13. The shorted branch in a parallel circuit is one of the most difficult problems to locate. You of course again know the cause of the problem and where it is located, but how do you go about proving its location electrically? When you suspect a short circuit the most productive measurements generally are voltage and resistance measurements. Measure and record the voltage across the three components in the circuit. Connect the voltmeter leads directly across the components one at a time. Why do you measure zero volts in each case?

14. Disconnect the power supply from the circuit then measure and record the resistance of each branch (ignoring R_{series}) by connecting the ohmmeter directly across each component one at a time. Why do you get the same reading in each case?

FIGURE 18–2 Schematic diagram of modified circuit

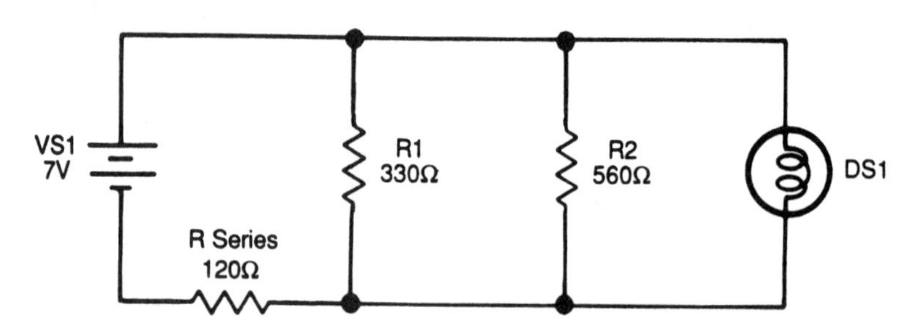

15. The reason that shorted branches are so difficult is that the faulty branch is not easy to isolate electrically without disconnecting things from the circuit. As you have just discovered voltage and resistance measurements do indicate the nature of the problem, but they can't give you a clue as to which branch is the culprit. Place the ohmmeter across R1 and leave it connected.
16. Disconnect DS1 from the circuit. Is the circuit still shorted according to the ohmmeter? Leave DS1 disconnected.
17. Disconnect R2 from the circuit. Has the ohmmeter reading changed? Since you have disconnected all components except R1, it would appear that R1 is the defective branch. To verify, measure the resistance of the lamp DS1 and resistor R1 while they are out of the circuit. Are their resistances normal?

Review

To isolate a shorted branch requires a little more work. Electrical measurements are not very helpful in the case of a shorted branch because the short affects the entire circuit. The only way to positively isolate the problem is to disconnect circuit components one at a time, testing them as you go along, until you find the bad one. Would the inclusion of fuses in the branches be of any help?

18. Return all materials to their proper places. You have just learned that troubleshooting can be a tedious process.

ACTIVITY SHEET EXPERIMENT 18

NAME ____________________

DATE ____________________

Step 2 **A.** Record measured circuit resistance: ____________________ Ω

Step 2 **B.** Record measured total circuit current:

____________________ mA

Step 3 **C.** Using measured voltage and current calculate R_T:

____________________ Ω

Step 4 **D.** After injecting the problem into the circuit, do you notice any visual indications? ____________________ If so, what are they?

____

Step 6 **E.** Record measured circuit current: ____________________ mA

Step 7 **F.** Record measured circuit resistance: ____________________ Ω

Step 8 **G.** Record the measured branch component resistances:

R1 ____________________ Ω R2 ____________________ Ω

DS1 ____________________ Ω

Step 9 **H.** Record the voltage drops-measure directly across each branch resistor:

V_{R1} ____________________ V V_{R2} ____________________ V

V_{DS1} ____________________ V

Step 10 **I.** Record measured voltages from each end of R2 to the circuit wiring it is supposed to be connected to:

V_{R2} (top) to positive circuit wiring: ____________________ V

V_{R2} (bottom) to negative circuit wiring: ____________________ V

Step 13

J. Why do you measure zero volts across each branch resistor in the modified circuit?____

____

Step 14

K. Why are the measured resistances of each branch the same?

____

L. Based on your experiences, what is the best way to isolate a shorted branch in a parallel circuit?____________________________

____

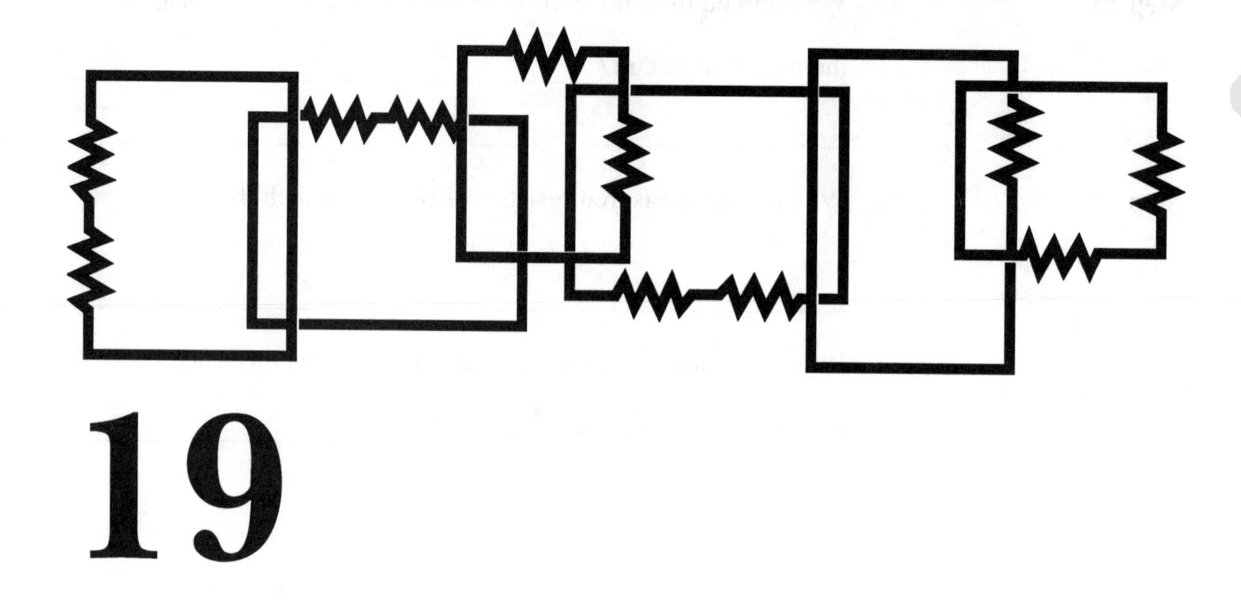

19

Series-Parallel Circuits

Objectives

After completing this experiment, you will be able to:

1. Evaluate the operation of the series-parallel circuit.
2. Measure voltages and currents throughout the series-parallel circuit.
3. Measure the effects of circuit defects in the series-parallel circuit.
4. Confirm that parallel connected components have a common voltage.
5. Confirm that parallel branches are current dividers.

Introduction

You are ready for another advancement in your study of electric circuits. In this experiment you will construct a simple three resistor circuit and make several electrical measurements. The purpose of the measurements is threefold: (1) to continue gaining experience in dealing with electric circuits and measurement instruments, (2) to enhance your theoretical understanding of series-parallel circuits, and (3) to further develop your basic troubleshooting skills. To perform this experiment you will need:

A prototyping board
A variable dc power supply
Two 8 to 12 inch lengths of #24 hookup wire
Six $1\frac{1}{2}$ inch lengths of #24 hookup wire
One 330Ω resistor
One 560Ω resistor
One 1000Ω resistor
An analog or digital VOM

Procedure

1. Pull the activity sheet for this experiment. Before constructing the circuit, measure and record the exact resistance of the three resistors. Then construct the circuit as shown in figure 19–1.
2. Before connecting the power supply to the circuit, measure and record the total circuit resistance.
3. Adjust the power supply output to 9 volts and connect it to the circuit. Measure and record the voltage drops from TP1 to TP2 and from TP2 to TP3. Next measure the voltages across the three resistors individually by connecting the voltmeter leads directly to the resistor leads. Are the voltages across the resistors the same as the voltages across the test points? Why are the voltages across R2 and R3 the same?
4. Measure and record the current flow through each of the resistors. Be sure to place the milliammeter in series with the resistor through which you want to measure current flow. A common mistake is to place the milliammeter across the resistor which will result in erroneous data and possible damage to the milliammeter.

FIGURE 19–1 Schematic diagram of circuit showing resistors and test points

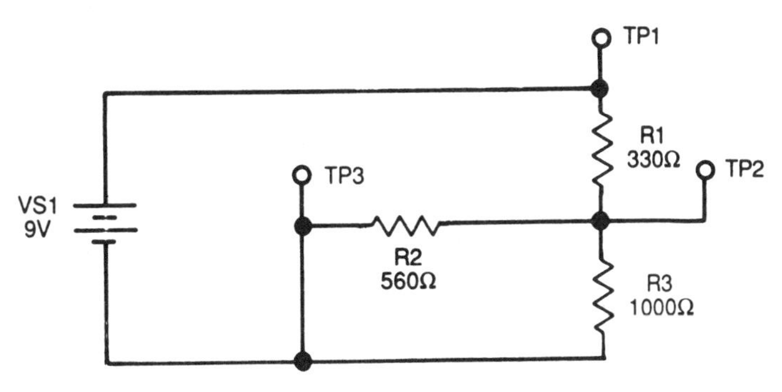

5. Using the measured voltage and current for each of the resistors, use Ohm's law to calculate the resistance of each. Record your calculations in the appropriate space on the activity sheet. Do your calculations compare well with your measurements?
6. Now you will see how faulty components affect voltage and current distribution in the circuit, and the total resistance of the circuit. Simulate R1 opening by lifting the end which connects to TP1 and move it to a hole on the prototyping board which is not in the circuit. Consider the resistor to still be connected between test points 1 and 2. Measure and record the voltages from TP1 to TP2 and from TP2 to TP3. Based on the measurements is it obvious in an electrical sense that R1 is open?
7. Reconnect R1 to its proper place then short circuit R1 with a short length of hookup wire. Repeat and record the voltage measurements from TP1 to TP2 and from TP2 to TP3. Do the measured voltages verify that R1 is a short circuit? Remove the shorting wire from R1.
8. Simulate R2 opening by lifting the end of the resistor which connects to TP2 and moving it to a hole on the prototyping board which is not connected to the circuit. Again measure and record the voltages from TP1 to TP2 and from TP2 to TP3. How do the measured voltages indicate the nature of the problem?
9. Return R2 to its proper place then place a shorting wire across R3. Measure and record the voltages from TP1 to TP2 and from TP2 to TP3. After studying the measurements is it rather clear what the problem is?

Review

You constructed a simple series-parallel circuit then measured its electrical properties. You found that the measured quantities were reasonably accurate when you used them to calculate the resistances of the three resistors. This is one way to verify your measurements. If you had paper analyzed the circuit prior to making any measurements, you would have discovered that all electrical characteristics could have been pre-calculated then measured to verify the accuracy of your analysis. You then simulated several circuit problems by opening and shorting select resistors in the circuit. Voltage measurements made after each problem was induced into the circuit tended to verify the nature of the problem. In short you found that short circuits still develop no voltage drop. You found that opening the series resistor in the circuit resulted in no current flow through the circuit and therefore no voltage drops. You measured the full source voltage from TP1 to TP2. When R1 was shorted you found that it caused the whole source voltage to be dropped from TP2 to TP3. When you opened R2 you found the electrical effect on the circuit was less pronounced. The voltage distribution was changed, demonstrating that the loss of R2 caused more voltage to appear from TP2 to TP3 than was normal, and less voltage from TP1 to TP2. When you shorted R3 you found that it also shorted R2, resulting in the entire source voltage being applied to R1. With continued practice you will soon be able to isolate the area of problem in a series parallel circuit just by measurement of the voltage distribution. To isolate the actual component at fault it may be necessary to disconnect things and perform resistance measurements, especially when the fault is one of the parallel components.

10. Return all materials to their proper places.

ACTIVITY SHEET EXPERIMENT 19

NAME ______________________________

DATE ______________________________

Step 1 **A.** Record measured resistances of the three resistors:

330Ω __________________ Ω 560Ω __________________ Ω

1000Ω __________________ Ω

Step 2 **B.** Record measured circuit total resistance:

__________________ Ω

Step 3 **C.** Record measured voltages: $V_{TP1-TP2}$ __________________ V

$V_{TP2-TP3}$ __________________ V

V_{R1} __________________ V V_{R2} __________________ V

V_{R3} __________________ V

Step 4 **D.** Record measured current flow:

I_{R1} __________________ mA I_{R2} __________________ mA

I_{R3} __________________ mA

Step 5 **E.** Record calculated resistances based on measured voltage and current:

R1 __________________ Ω R2 __________________ Ω

R3 __________________ Ω

Step 6 **F.** Record measured voltages: $V_{TP1-TP2}$ __________________ V

$V_{TP2-TP3}$ __________________ V

Is the problem obvious? __________________

Step 7 **G.** Record measured voltages: $V_{TP1-TP2}$ __________________ V

$V_{TP2-TP3}$ __________________ V

Is the problem clearly visible electrically? __________________

Step 8 **H.** Record measured voltages: $V_{TP1-TP2}$ ____________________ V

$V_{TP2-TP3}$ ____________________ V

Why are these voltages nearer the normal circuit voltages than in the two previous problems? ____________________________

Step 9 **I.** Record measured voltages: $V_{TP1-TP2}$ ____________________ V

$V_{TP2-TP3}$ ____________________ V

With this problem, why is it electrically so obvious? __________

____

____

____

____

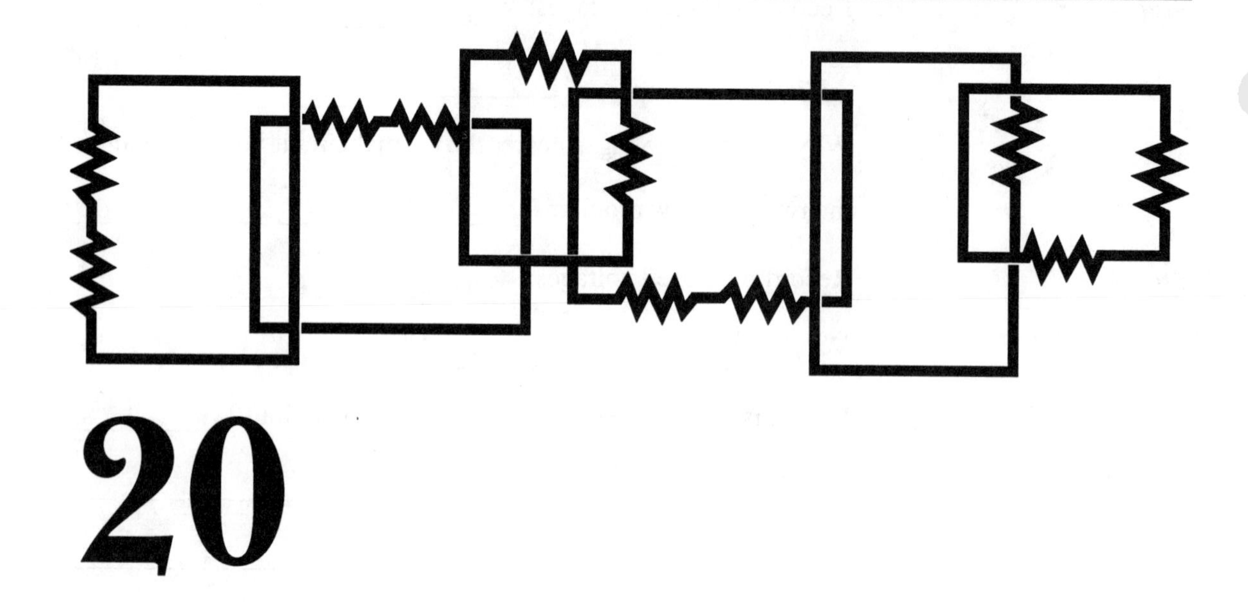

20

Loaded Voltage Dividers

Objectives

After completing this experiment, you will be able to:

1. Evaluate the operation of the loaded voltage divider.
2. Analyze circuit requirements and select appropriate component values.
3. Operate diverse loads from a common power source.
4. Determine the effectiveness of a loaded voltage divider by measurement.

Introduction

It is possible with a voltage divider circuit to provide power to two or more different loads which have differing voltage and current requirements. That will be the object of this experiment, to design and implement a voltage divider capable of properly powering two loads. To accomplish this experiment you will need:

A prototyping board
A variable dc power supply
Two 8 to 12 inch lengths of #24 hookup wire
Four $1\frac{1}{2}$ inch lengths of #24 hookup wire
One 220Ω, 1 watt resistor
One 33Ω, 1 watt resistor
One 18Ω, 1 watt resistor
One 22Ω, 1 watt resistor
One 47Ω, 1 watt resistor
One #7382 bi-pin lamp (14V 80 mA)
One #7328 bi-pin lamp (6V 200 mA)

Procedure

1. Pull the activity sheet for this experiment. Study the circuit shown in figure 20–1 to see how the various parts will be connected. Then follow the procedure on the activity sheet to analyze the circuit requirements (A through D).
2. After calculating the values of the three resistors in the voltage divider network, construct the circuit. (Hint: Refer to the parts list.)
3. Set the power supply output to minimum, connect it to the circuit, then slowly increase its output to about 15 volts. The lamps will not be at their full brilliance yet but should be visibly illuminated. Does their brightness seem approximately the same?
4. Continue to increase the power supply output to 20 volts. Have both lamps reached full brilliance? Measure and record the voltage drops across the lamps.

FIGURE 20–1 Schematic diagram of voltage divider circuit

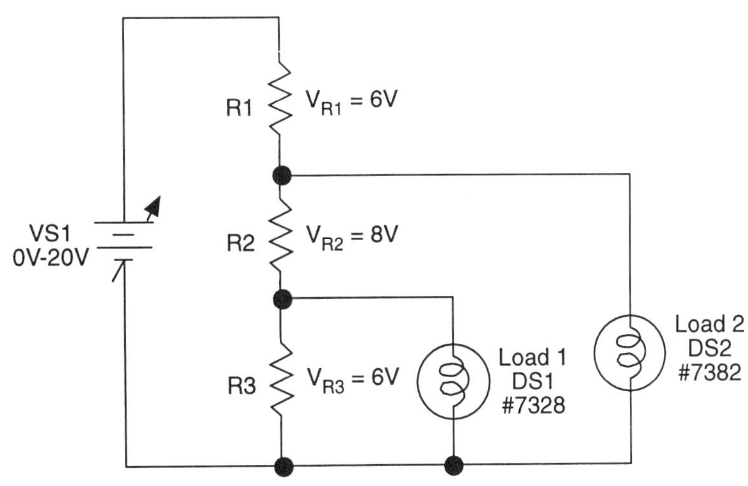

Review

After analyzing the job to be performed by the circuit you proceeded to calculate the values of the three resistors in the circuit. Your calculations took into account the load requirements, and circuit voltage and current distribution. Once you had calculated the circuit values you selected and positioned the resistors into their proper places in the circuit. Then you applied power to the circuit to find out whether your design efforts were on target. Although the voltage divider circuit is relatively simple, an understanding of how it works, and also how to design it to meet certain requirements, will be a great value when you begin your study of transistor amplifiers.

5. Return all materials to their proper places. Another job well done!

ACTIVITY SHEET EXPERIMENT 20

NAME ______________________________

DATE ______________________________

Step 1 **A.** Enter the voltage which must be dropped by each resistor in the circuit:

V_{R1} ____________________ V V_{R2} ____________________ V

V_{R3} ____________________ V

Step 1 **B.** Enter the desired current flow through each load to attain normal intensity. (Hint: Refer to the parts list.)

I_{DS1} ____________________ mA I_{DS2} ____________________ mA

Step 1 **C.** Enter the total current flow through each resistor:

I_{R1} ____________________ mA I_{R2} ____________________ mA

I_{R3} ____________________ mA

(Note: 1 bleeder is equal to $\frac{1}{10}$ total load current)

Step 1 **D.** Calculate necessary values of the resistors using desired voltage drops and estimated current flows:

R1 ____________________ Ω R2 ____________________ Ω

R3____________________ Ω

Step 4 **E.** Record measured voltage drops across the lamps:

V_{DS1} ____________________ V V_{DS2} ____________________ V

Are the lamps glowing brightly at full brilliance or does there seem to be an imbalance? ______________________________

____

F. In order to design a loaded voltage divider, what information must you know? ______________________________

____

____

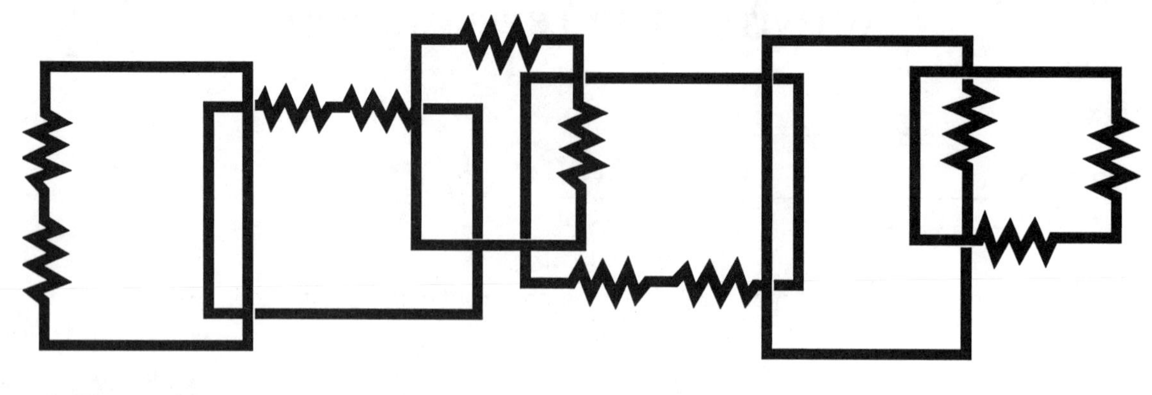

21

Bridge Circuits—Applications

Objectives

After completing this experiment, you will be able to:

1. Evaluate the operation of a bridge circuit.
2. Evaluate the thermal response of a thermistor.
3. Balance a bridge circuit by adjustment and measurement.
4. Observe the dynamics of a bridge circuit with sensory ability.

Introduction

This experiment has been designed to acquaint you with the operation and typical application of a bridge type circuit. To perform this experiment you will need:

A prototyping board
A digital voltmeter
One 1kΩ resistor, ±5% tolerance
Two 10kΩ resistors, ±5% tolerance
One 560Ω resistor, ±5% tolerance
One 820Ω resistor, ±5% tolerance
One 1kΩ potentiometer
One thermistor (10kΩ at 25°C)
A variable dc power supply
Two 8 to 12 inch lengths of #24 hookup wire
A number of $1\frac{1}{2}$ inch lengths of #24 hookup wire
Two test leads with alligator clips.

Procedure

1. Remove the activity sheet for this experiment from the back of this manual. Before constructing this experiment, measure and record the values of the fixed resistors and the thermistor.
2. Construct the circuit as shown in figure 21–1.
3. After verifying the circuit connections, turn on the digital multimeter and connect it across the bridge (points B and C). Set the DVM to measure voltage on the 10V range and adjust the power supply output to 10V.
4. The bridge is more than likely in the unbalanced condition as indicated by the voltage reading on the DVM.

FIGURE 21–1 Schematic of a bridge type circuit

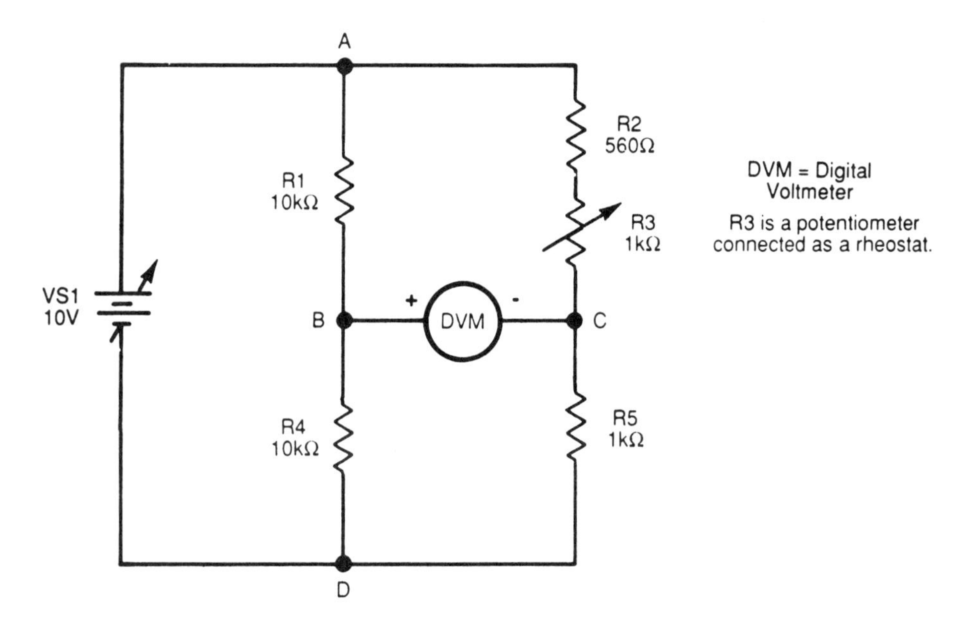

5. Adjust the variable resistor R3 to "zero" the DVM and thereby balance the bridge.
6. With the bridge balanced, vary the power supply output from 1V to 10V to verify that the bridge remains balanced despite variations in the input supply.
7. Returning the power supply output to 10V, measure and record the following voltage drops between A and B, A and C, B and D, C and D.

Review

A bridge type circuit is said to be balanced when the voltage across the output terminals, B and C, is zero volts. This was achieved in step 5 when the variable resistor R3 was adjusted to approximately 440Ω, and since it is in series with R2 (560Ω) the combined resistance of 1kΩ balanced both branches of the bridge. Once a bridge is balanced, the input voltage will be equally divided by the branches made up of R1 and R4, and R2, R3 and R5, producing the same voltage and therefore no potential difference between the output points B and C. In step 7 of the procedure, the voltage drops across all sections of the bridge were measured, and since the voltage dropped is proportional to resistance, the voltages were equally distributed resulting in a balanced condition.

8. Turn off the power supply and replace R2 and R3 with a thermistor as seen in figure 21–2. Replace R5 with the 820Ω resistor and R4 with a 1kΩ.
9. Hold the thermistor between your thumb and forefinger and then observe and periodically record the bridge output voltage and polarity as it warms up, or hold it close to your mouth and exhale warm air upon it.

FIGURE 21–2 Thermistor bridge

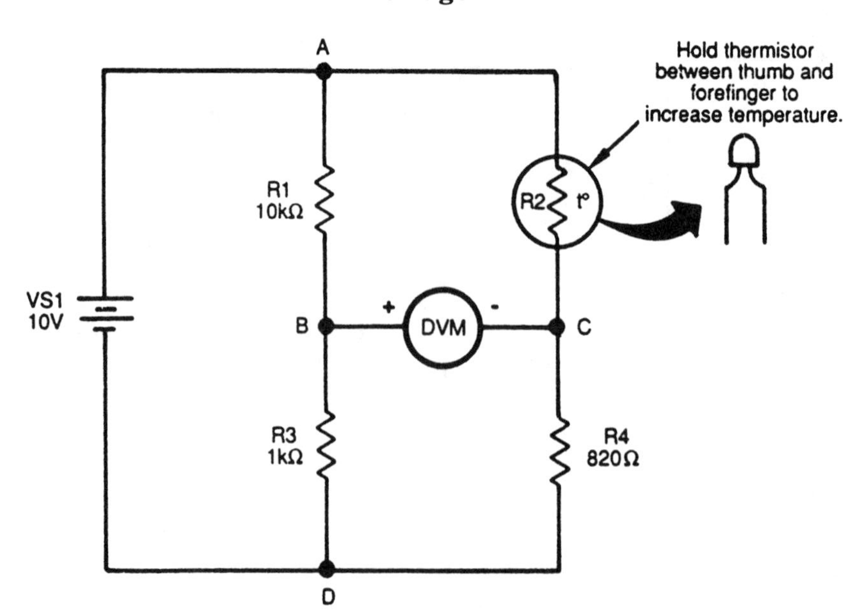

Review

A thermistor has a negative temperature coefficient of resistance, which means as temperature increases, resistance decreases. At room temperature the thermistor has a large resistance and will therefore drop a large voltage, making point C negative with respect to point B, and this will be seen on the DVM as some positive value. When the thermistor was held between your thumb and forefinger, the increase in temperature caused it to decrease its resistance. As its resistance decreased so did its voltage drop. As you recorded the voltage values on the DVM you will have noticed that it went from some positive value, to a balanced 0V, and then to some negative value as temperature increased. To produce an output voltage representing a sensed temperature is the primary application of this bridge circuit.

ACTIVITY SHEET EXPERIMENT 21

NAME ______________________________

DATE ______________________________

Step 1

A. Record measured values of the fixed resistors. Note: Do not touch or handle the thermister during this step.

10kΩ ________________ Ω 10kΩ ________________ Ω

1kΩ ________________ Ω 820Ω ________________ Ω

560Ω ________________ Ω Thermistor ________________ Ω

Step 4

B. With the DVM connected across the bridge, rotate the variable resistor fully clockwise then fully counterclockwise. Record the measured voltages and polarity of the voltages.

$V_{BRIDGE(cw)}$ ____________________ V

$V_{BRIDGE(ccw)}$ ____________________ V

Steps 5–6

C. After balancing the bridge, how does varying the source voltage affect the balanced condition? ________________

____

Step 7

D. Measure and record the following voltages:

V_{AB} ________________ V V_{AC} ________________ V

V_{BD} ________________ V V_{CD} ________________ V

Step 8

E. With the bridge circuit modified to include the thermistor and 820Ω resistor, measure and record the bridge voltage and polarity.

V_{BRIDGE} ____________________ V room temperature.

F. While holding the thermistor between you thumb and forefinger, allowing several seconds for conditions to stabilize, measure and record the bridge voltage and polarity.

V_{BRIDGE} ____________________ V body temperature.

G. Why did the bridge voltage change polarity and magnitude when you held the thermistor in your hand? ________________

H. Remove the thermistor from the circuit. Measure and record its resistance while holding it between your fingers.

$R_{THERMISTOR}$ ________________________ Ω body temperature.

Could the thermistor be used as an electrical thermometer?_____

____

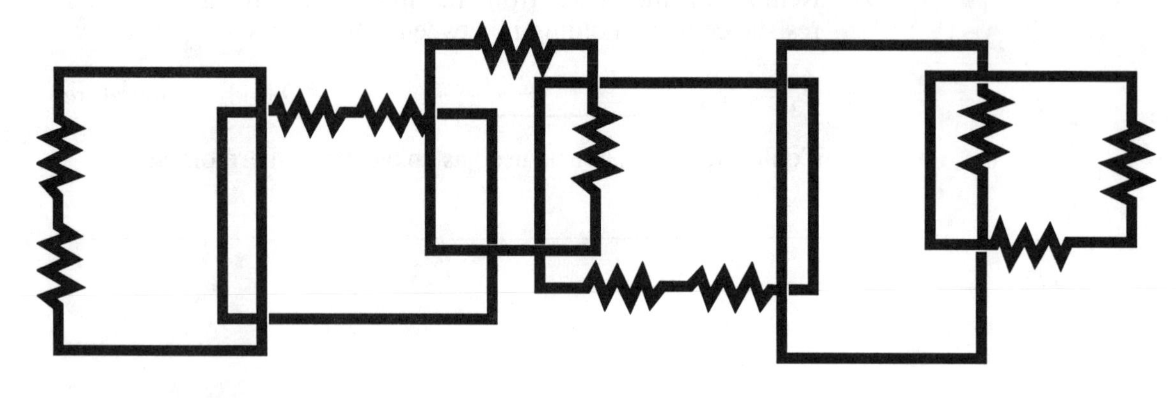

22

Measurement Limitations—The Multimeter

Objectives

After completing this experiment, you will be able to:

1. Evaluate the accuracy of the digital voltmeter as a measurement tool.
2. Prove that a voltmeter has the ability to load an electrical circuit.
3. Experience the limitations of the digital voltmeter by measurement.
4. Recognize the conditions which can affect measurement accuracy.

Introduction

This experiment will familiarize you with how difficult it is to make a measurement in an electrical circuit without affecting the circuit's operation, which may result in an inaccurate indication. To perform this experiment you will need:

A prototyping board
A variable power supply
A digital voltmeter
Two 10kΩ resistors, ±5% tolerance
Two 10MΩ resistors, ±5% tolerance
Test leads as necessary
Hookup wires as necessary

Procedure

1. Remove the activity sheet for this experiment and before constructing the experiment measure and record the values of the two 10kΩ and two 10MΩ resistors.
2. Construct the circuit shown in figure 22–1.
3. After verifying the circuit connections, turn on the power supply and set its output to 10V.
4. Measure and record the voltage drops across R1 and R2.
5. Replace the two 10kΩ resistors with two 10MΩ resistors, as seen in figure 22–2.
6. Measure and record the voltage drops across R1′ and R2′.
7. Ignoring the effect of the DVM, calculate what voltage drop should occur across R1 and R2 when both the 10kΩ and 10MΩ were in circuit and then compare the calculated values to the actual measured values.

FIGURE 22–1 Voltage divider

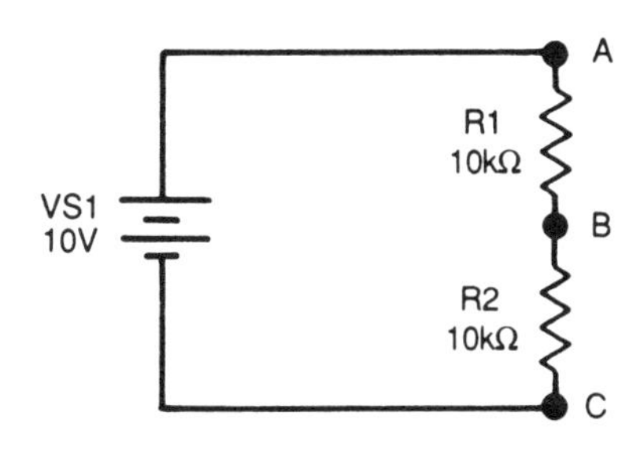

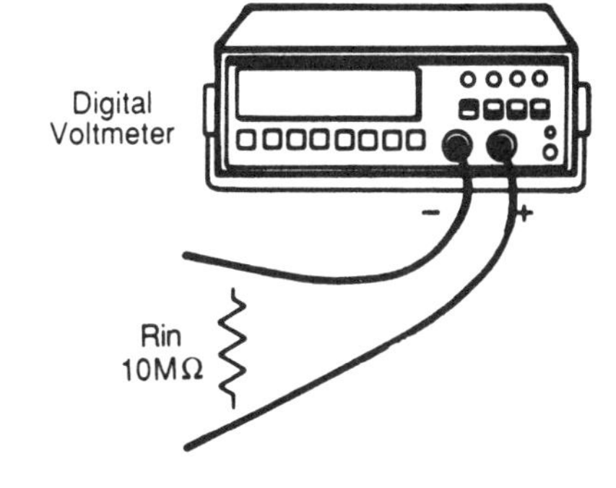

FIGURE 22–2 Voltmeter loading

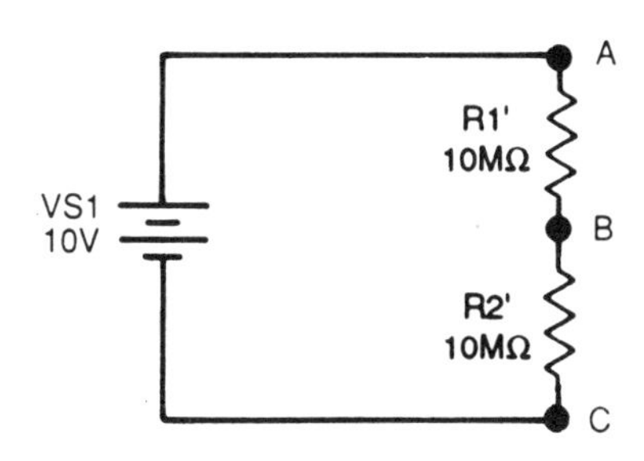

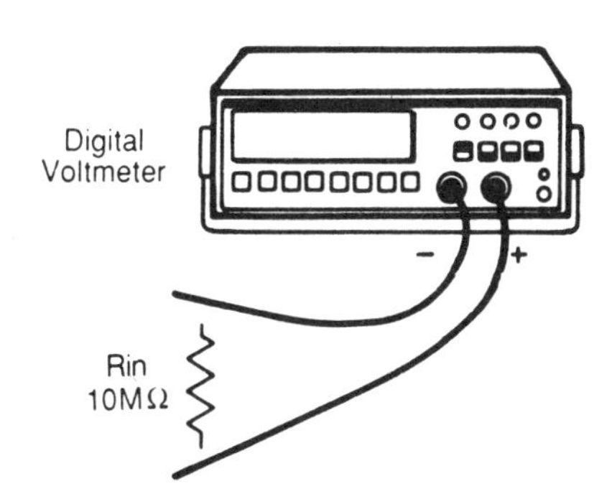

Review

When the voltmeter was connected across either R1 or R2 the circuit changed from a simple series circuit to a series-parallel circuit made up of R1, R2 and R_{in}. If the meter's internal resistance is large compared to the circuit resistance, as was the case when the two 10kΩ resistors were connected in circuit, the meter's high resistance, 10MΩ, will have little effect. However, when meter resistance is equal to or smaller than resistance in circuit, as was the case when the two 10MΩ resistors were connected in circuit, the effect will be considerable and an inaccurate reading will result. This loading effect of a voltmeter can be minimized by using a meter with a high internal resistance.

ACTIVITY SHEET EXPERIMENT 22

NAME ______________________________

DATE ______________________________

Step 1 **A.** Record the measured values of the four resistors:

10kΩ__________________ Ω 10kΩ__________________ Ω

10MΩ__________________ Ω 10MΩ__________________ Ω

Step 4 **B.** Record the measured circuit voltages:

V_S__________________ V V_{R1}__________________ V

V_{R2}__________________ V

Step 6 **C.** Record the measured voltages in the modified circuit:

V_S__________________ V $V_{R1'}$__________________ V

$V_{R2'}$__________________ V

Step 7 **D.** Using the measured resistances and the measured source voltage, calculate the voltage drops across each resistor in both circuits:

Circuit 1 V_{R1}__________________ V

V_{R2}__________________ V

Circuit 2 $V_{R1'}$__________________ V

$V_{R2'}$__________________ V

Why are the measured drops of circuit 1 very close to the calculated values?______________________________

____

____

Why are the measured voltage drops of circuit 2 different from the calculated values? ______________________________

____

____

In order to obtain the most accurate circuit voltage measurements, what conditions should be met? ____________________

____

____

Can the inaccurate loaded voltage measurement be corrected by calculation? ______________________________________

____

____

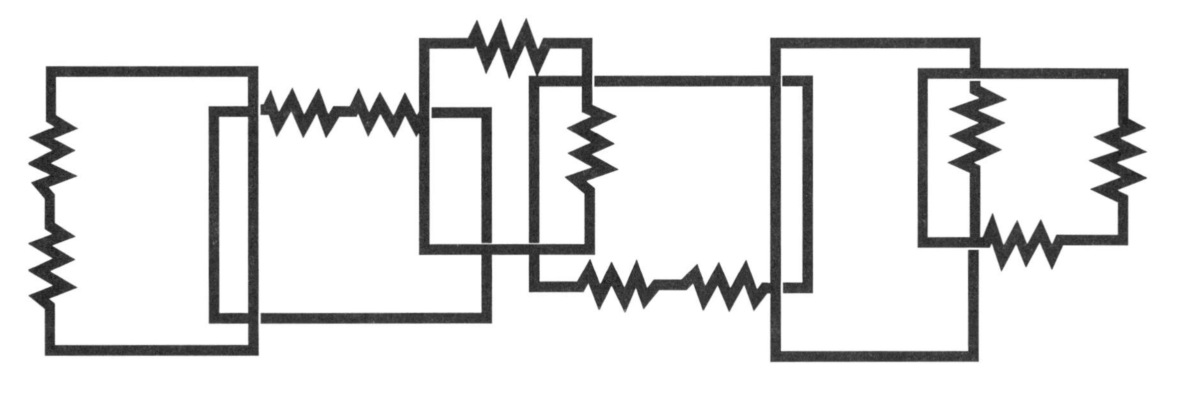

23

AC Series Circuit Measurements

Objectives

After completing this experiment, you will be able to:

1. Utilize the transformer as a low voltage ac source.
2. Measure voltage and current flow in an ac circuit.
3. Evaluate the dynamic resistance of an incandescent lamp in operation.

Introduction

This experiment will introduce you to both the transformer and the use of the digital multimeter to measure ac voltage and current. To perform this experiment you will need:

A digital multimeter
A center tapped transformer (12.6V/300mA min) with cord and plug
One 120Ω resistor
One light bulb (#7382) 14V 80mA
Hookup wires as necessary

Procedure

1. Before beginning this experiment, it may be necessary to attach a plug and line cord to the primary winding of the transformer.
2. Remove the activity sheet for this experiment and then set the transformer up as shown in figure 23–1.
3. Using the digital voltmeter (DVM), measure and record the ac voltages between the center tap and either end (between A and B, and A and C) and also across the entire secondary (between B and C).

Review

An alternating magnetic field is needed to transfer energy from the primary to the secondary of the transformer. This alternating magnetic field is provided by an alternating current of 120V/60Hz from the ac wall outlet. The step-down transformer, T1, steps the 120V input down to 12.6V which can be measured across the complete secondary. If the voltmeter is connected across only half of the transformer's output secondary, by making a connection to the center tap and either end, only half of the output voltage will be measured, 6.3V.

FIGURE 23–1 Measuring transformer voltages

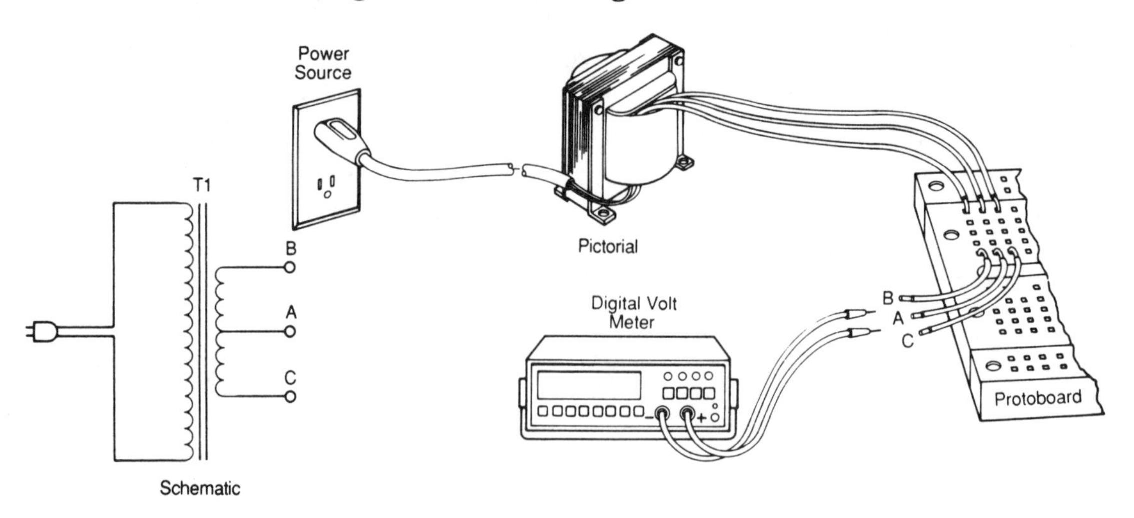

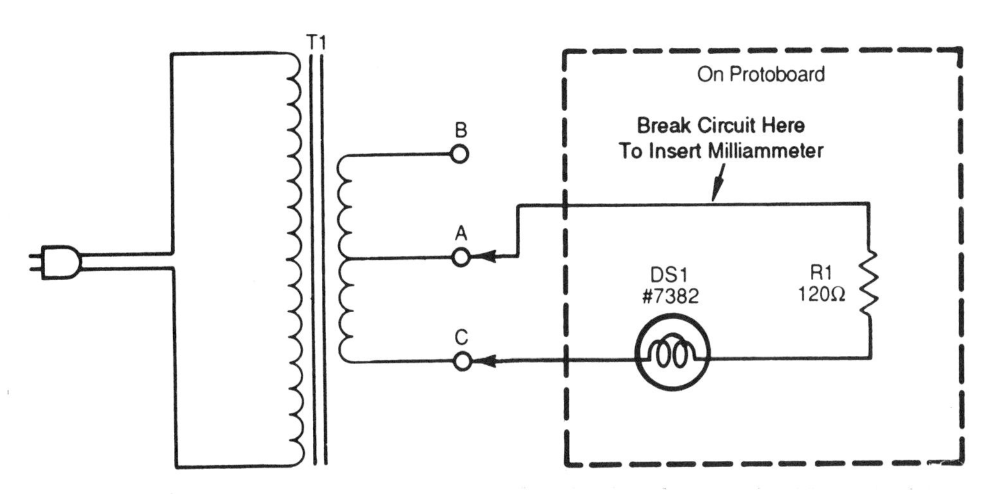

FIGURE 23–2 6.3V Transformer output

4. Construct the circuit seen in figure 23–2. Notice that the circuit on the protoboard is connected across the transformer's center tapped secondary and one end (between A and C).
5. Measure and record the voltage drops across R1 and DS1.
6. Disconnect circuit power. Insert an ammeter in the path of transformer secondary current and record the value after restoring circuit power.
7. Construct the circuit seen in figure 23–3. Notice that in this case the circuit on the protoboard is connected across the complete secondary (between B and C).
8. Measure and record the voltage drops across R1 and DS1.
9. Measure and record the total ac transformer secondary current.

Review

When the series circuit made up of R1 and DS1 was connected between the transformer's center tap and one end, the lamp glowed

FIGURE 23–3 12.6V Transformer output

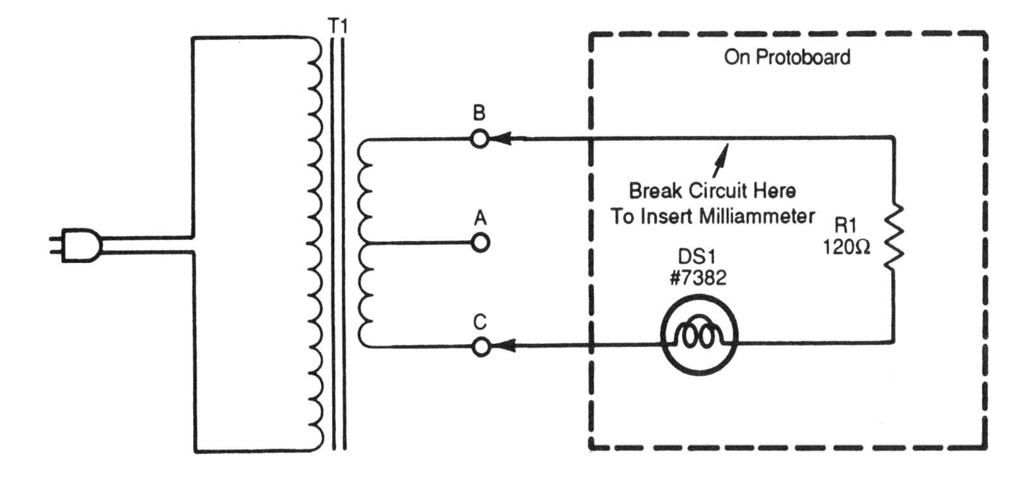

dimly since the circuit was being powered by 6.3V ac. When the circuit was connected across the full secondary voltage of 12.6V, the lamp glowed brightly due to the increased voltage causing an increased current. However, total circuit current did not double when voltage was doubled as the resistance of DS1 did not remain constant. The lamp has a positive temperature coefficient of resistance which means that as temperature increased, resistance increased.

ACTIVITY SHEET EXPERIMENT 23

NAME ______________________________

DATE ______________________________

Step 3 **A.** Record the measured transformer output voltages:

$V_{A\text{-}B}$ __________________ V $V_{A\text{-}C}$ __________________ V

$V_{B\text{-}C}$ __________________ V

Step 5 **B.** Record the measured voltage drops across R1 and DS1:

V_{R1} __________________ V V_{DS1} __________________ V

Step 6 **C.** Record the measured circuit current:

I_{TOTAL} __________________ mA

Step 8 **D.** Record the measured voltage drops across R1 and DS1:

V_{R1} __________________ V V_{DS1} __________________ V

Step 9 **E.** Record the measured circuit current:

I_{TOTAL} __________________ mA

F. Compare the measured voltage drops and circuit currents for the two circuit conditions. When the voltage applied to the circuit was doubled, why did the circuit current not increase by a factor of two also?

____

____

Did the voltage dropped across the lamp double or increase by more than double? ______________________________

Explain why this is so. ______________________________

____

____

____

____

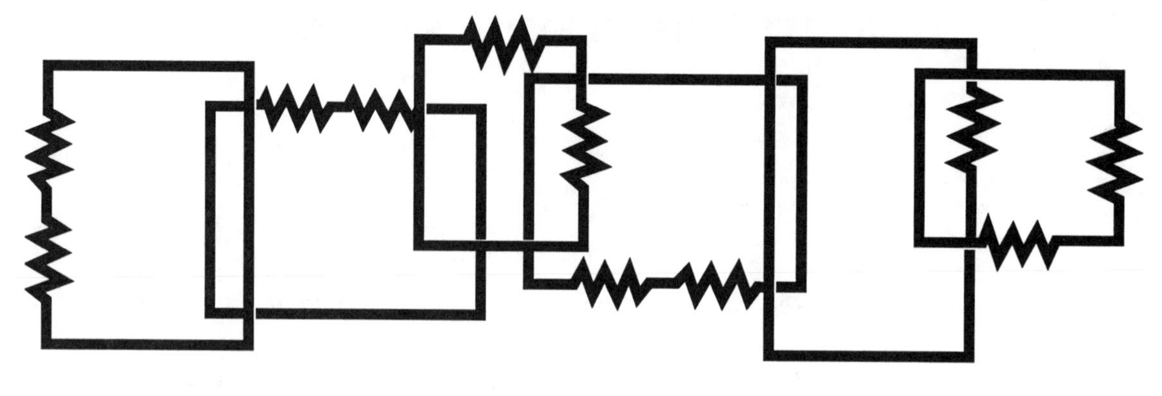

24

Waveforms—Using the Oscilloscope

Objectives

After completing this experiment, you will be able to:

1. Set up and use the oscilloscope as an accurate measurement tool.
2. Accurately measure dc voltages with the oscilloscope.
3. View ac waves and discern the differences of different waveforms.
4. Translate waveform period measurements into frequency.
5. Accurately measure voltage and time characteristics of ac waves.

Introduction

In this experiment you will demonstrate how the oscilloscope can be used to examine dc and ac wave shapes. To perform this experiment you will need:

A variable dc power supply and test leads
A function generator and test leads
An oscilloscope and probe (1 to 1)
One 10kΩ resistor, ±5% tolerance
One 1kΩ resistor, ±5% tolerance

Procedure

1. Remove the Activity Sheet for this experiment. Before constructing this experiment, measure and record the values of the resistors.
2. Construct the circuit seen in figure 24–1.
3. Turn on the oscilloscope and select the following control settings:
   dc sense
   2V/cm
   1ms/cm
4. Turn on the dc power supply and set the output to 6V.
5. Ensure that the oscilloscope ground clip is attached to point C of the circuit and then place the probe on point A.
6. Vary the power supply output and record the results.
7. Turn off the power supply and reverse the positive and negative leads so that the negative output of the power supply is now being applied to point A and the positive to point C.
8. Turn on the power supply and vary the output voltage and record the results.

FIGURE 24–1 Viewing DC voltages

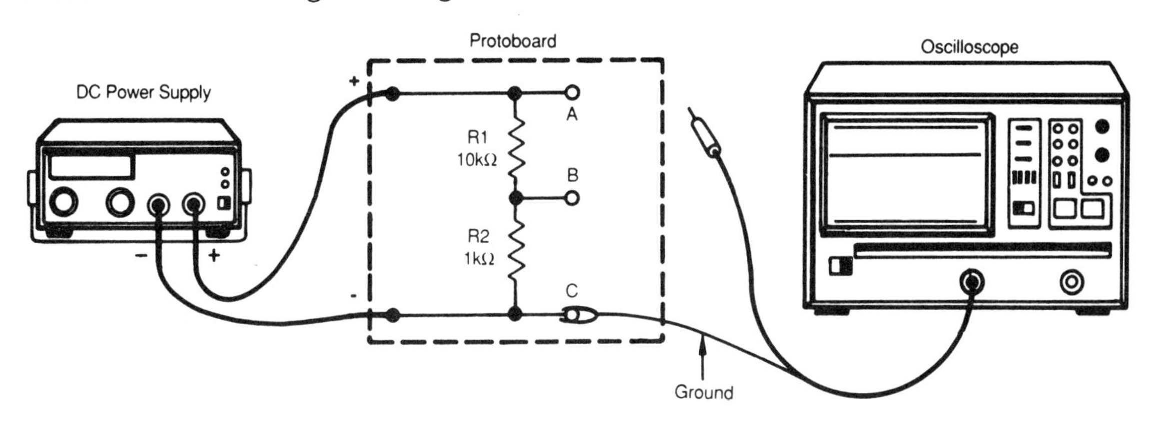

Review

A dc voltage is displayed as a horizontal line on an oscilloscope display. Its position relative to the center 0V horizontal grid line is an indication of its amplitude. A positive dc voltage appears above the center 0V line while a negative voltage appears below it.

9. Construct the circuit seen in figure 24–2.
10. Turn on the oscilloscope and select the following control settings:
    ac sense
    2V/cm
    1ms/cm
11. Turn on the function generator and set the controls for an 8V peak to peak, 500Hz sine wave.
12. Using the scope probe and making sure that an oscilloscope ground is attached to point C, touch point A and observe the waveform on the oscilloscope.
13. Change the oscilloscope's input sense from ac to dc and if the signal shifts either up or down, adjust the dc offset on the function generator until the 8V peak to peak sine wave alternates above and below the center 0V horizontal grid line.
14. Measure and record the period and peak to peak amplitude of one cycle as displayed on the scope.
15. Change the setting on the function generator from sinewave to triangular wave.
16. Measure and record the period and peak to peak amplitude of one cycle.
17. Change the function generator's output to squarewave.
18. Measure and record the period and peak to peak amplitude of one cycle.
19. Select sine wave output once again on the function generator.
20. Move the probe to point B, with the ground still connected to point C, and adjust the scope as necessary for the best display.
21. Record any period and or peak to peak amplitude changes.
22. Why are the measured voltages less?

FIGURE 24–2 Viewing AC voltages

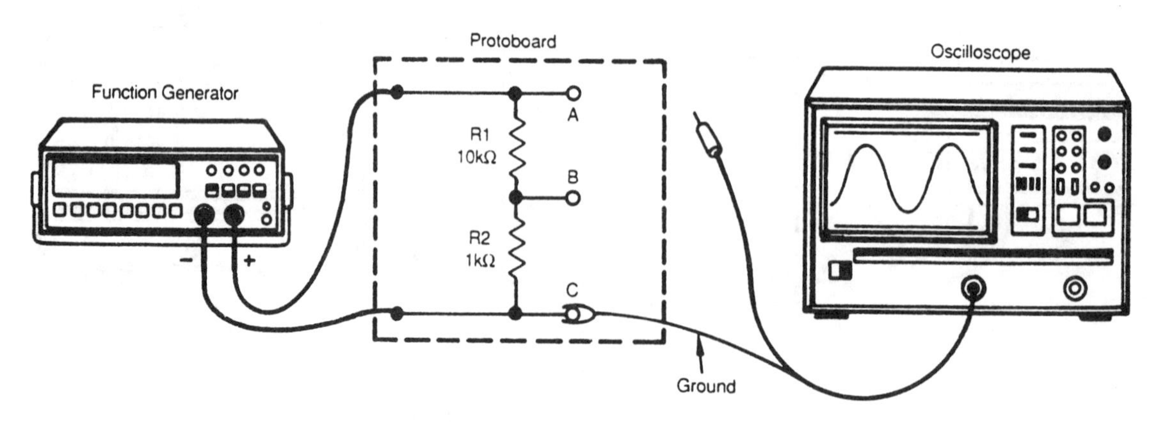

The 8V peak to peak, 500Hz sine wave from the function generator is seen as an 8V (2 divisions up, 2 divisions down at 2V/div = 8V) 500Hz (1 cycle per 2 divisions at 1ms/div = 2ms period or 500Hz frequency) sine wave on the oscilloscope display. If the dc offset is adjusted on the function generator, this sine wave will be superimposed on a dc level (dc sense must be selected to view dc offset). Waveshape, amplitude and period, and therefore frequency, can easily be calculated by observing the oscilloscope's control settings and then interpreting the display.

ACTIVITY SHEET EXPERIMENT 24

NAME ______________________________

DATE ______________________________

Step 1 **A.** Record the measured values of the resistors:

10kΩ ________________ Ω 1kΩ ________________ Ω

Step 5 **B.** With the scope probe connected to points A and C in the circuit, how many centimeters is the trace deflected from zero volts?

________________ cm Up or down? ________________

Step 5 **C.** Move the scope probe from point A to point B (leave ground wire connected to C). How many centimeters is the trace deflected now?

________________ cm Up or down? ________________

Step 6 **D.** Move the scope probe back to point A. Vary the output voltage of the power supply. How does the trace respond to the voltage variations?

____

Does the trace ever go below the zero volt reference? __________

Steps 7-8 **E.** With the polarity of the power supply connections to the circuit reversed, how does the trace respond to voltage variations?

____

Does the trace ever go above the zero volt reference? __________

____

Steps 14-18 **F.** Record measured periods and amplitudes of the waveforms:

Waveform	**Period**	**Amplitude**
Sinewave	ms	V_{pp}
Triangle wave	ms	V_{pp}
Square wave	ms	V_{pp}

Step 20 **G.** Record the voltage measured from point B to C:

________________ V_{pp}

Record the period of the waveform: ________________ ms

Step 21

H. Why is the voltage measured from point B to C less than the voltage measured from point A to C?__________________________

____

Why does the period of the waveform remain unchanged at the two measurement points? _______________________________

____

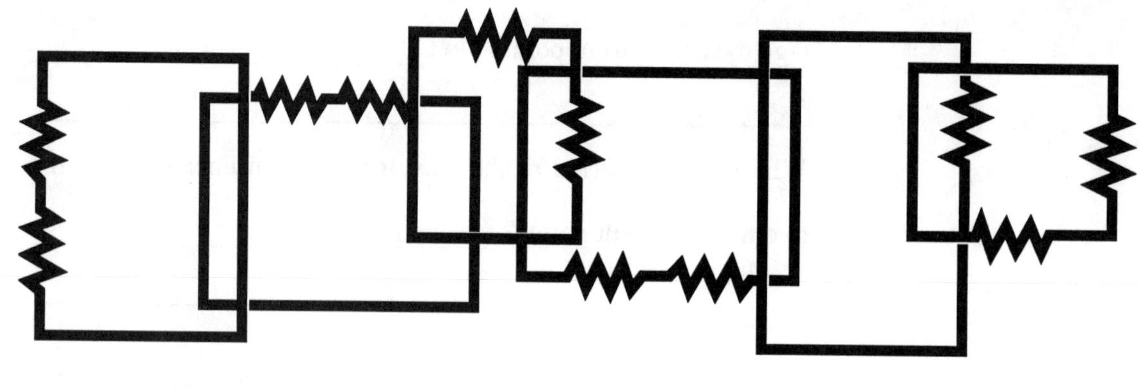

25

Capacitor Charge and Discharge

Objectives

After completing this experiment, you will be able to:

1. Measure charge and discharge slopes in an active RC circuit.
2. Calculate the time constant of an RC circuit.
3. Confirm by calculation and measurement that time constants are orderly.

Introduction

This experiment examines the dc charge and discharge characteristics of a capacitor. To perform this experiment you will need:

A function generator
An oscilloscope
One 10kΩ resistor
One 0.1 μF capacitor

Procedure

1. Remove the activity sheet for this experiment.
2. Construct the circuit seen in figure 25–1.
3. Using the oscilloscope, test the output of the function generator to verify it is producing a 100Hz 500mV peak-to-peak square wave. Adjust the dc offset so that the waveform's negative peak is at 0V and its positive peak is at +500mV (pulsating dc).
4. Connect the scope probes to circuit points A and B and then adjust the scope for the best display.
5. Measure and record the following:
   (a) Minimum voltage of waveform
   (b) Maximum voltage of waveform
   (c) Time of positive going slope
   (d) Time of negative going slope
6. Using the measured slope times, calculate the time constant of the circuit.
7. Using the values of R and C, calculate the time constant of the circuit. Is the measured time constant very close to the calculated time constant? Is the charge time equal to the discharge time?
8. Change R1 from 10kΩ to 1kΩ. Measure the charge and discharge times of the modified circuit. What affect does decreasing the resistance have?

FIGURE 25–1 RC circuit analysis

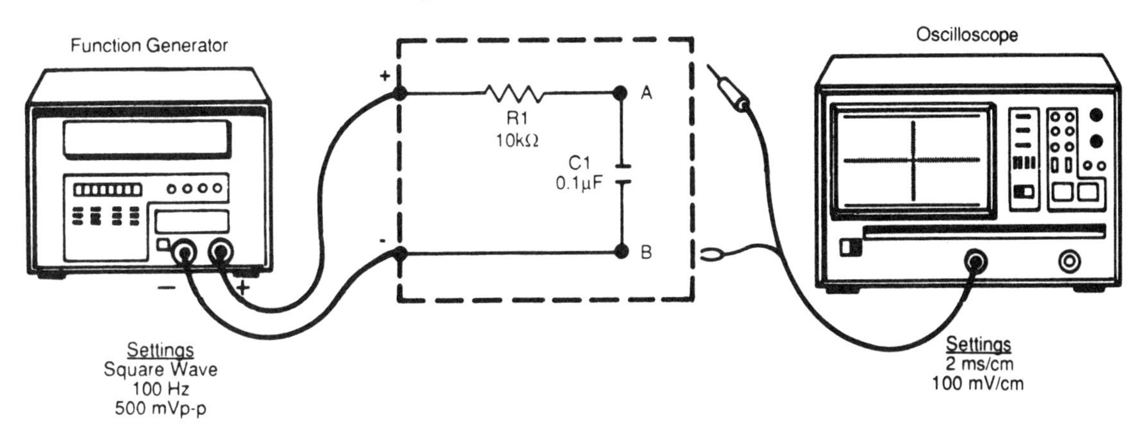

The original RC circuit, where R = 10kΩ and C = 0.1μF, had a time constant of 1ms (TC = R × C), and would therefore take five time constants (5ms) to charge to 500mV and five time constants to discharge to 0V. The square wave input had a period of 10ms ($\frac{1}{100}$ Hz) which meant that the positive alternation lasted for 5ms which was enough time for the capacitor to fully charge and the negative alternation lasted for 5ms which permitted the capacitor to fully discharge, as seen in figure 25-2.

When R1 was changed from 10kΩ to 1kΩ, the time constant also decreased from 1ms to 0.1ms (100μs) and so now the capacitor will charge and discharge at a much faster rate (5 time constants or 500μs) relative to the input.

FIGURE 25–2 RC circuit waveforms

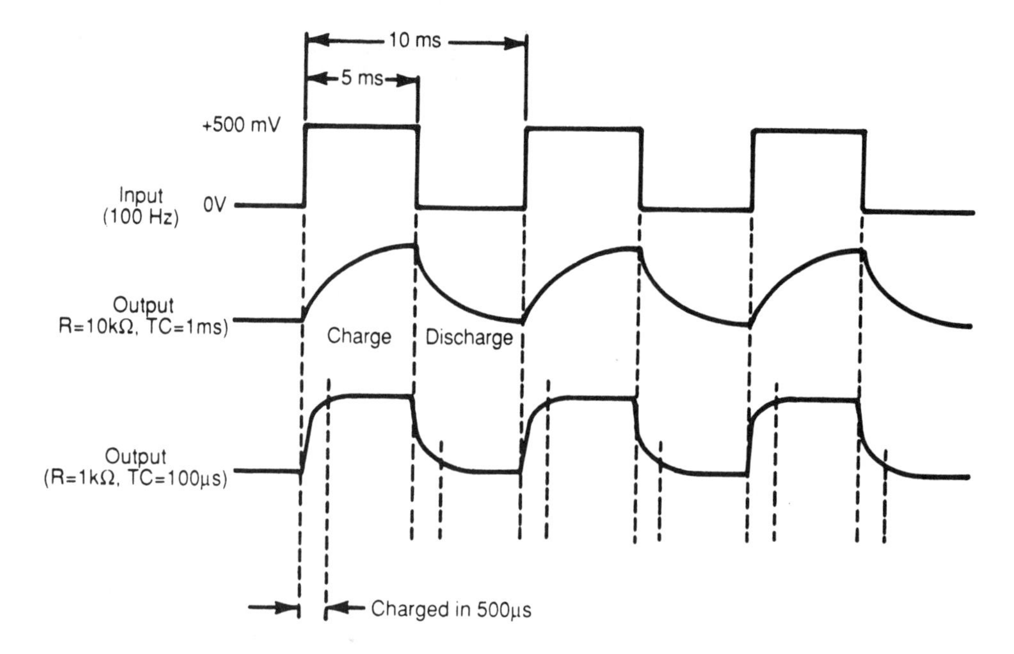

ACTIVITY SHEET EXPERIMENT 25

NAME ______________________________

DATE ______________________________

Step 3 **A.** With the scope probe connected to circuit points A and B adjust the function generator to produce a pulsating dc output which swings from zero volts to 500 mV positive. Adjustment of the output level and dc offset may interact, therefore it may be necessary to perform the adjustments two or more times.

Step 5 **B.** Measure and record the following characteristics from the scope display:

$V_{MIN\ OF\ WAVEFORM}$ ____________________ V

Time of Positive Slope ____________________ ms

$V_{MAX\ OF\ WAVEFORM}$ ____________________ V

Time of Negative Slope ____________________ ms

Step 6 **C.** Using the measured slope times calculate the time constant of the circuit:

RC = ____________________ ms

D. Using the coded values of the resistor and capacitor, calculate the

time constant of the circuit RC = ____________________ ms

Are the measured and calculated time constants equal? ________

How do you account for this? ____________________________

____

E. Are the measured charge and discharge times equal? __________

Why is this so? ____________________________________

____

Step 8 **F.** After modifying the circuit adjust the oscilloscope for the best display. Measure the charge and discharge times of the new circuit:

Time to Charge = ____________________ μs

Time to Discharge = ____________________ μs

How did reducing the series resistance affect the time constant of the circuit? ___________________________________

Using measured charge and discharge times how could you determine the capcitance of an unknown capacitor? __________

____

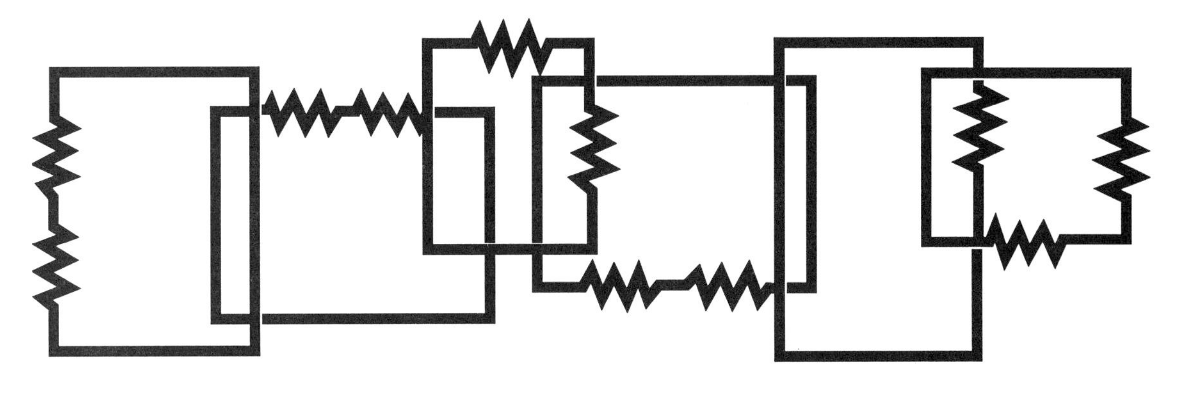

26

Series Capacitive Circuits—Voltage Distribution

Objectives

After completing this experiment, you will be able to:

1. Measure voltage distribution in a dc series capacitive circuit.
2. Measure voltage distribution in an ac series capacitive circuit.
3. Confirm that capacitive dc voltages are dependent upon charge.
4. Confirm that capacitive ac voltages are dependent upon reactance.
5. Utilize the function generator as a variable ac source.

Introduction

This experiment will highlight the dc and ac voltage distribution of a series capacitive circuit. To perform this experiment you will need:

A digital voltmeter
A dc power supply
A function generator
One 0.1μF capacitor
One 0.5μF capacitor

Procedure

1. Detach the activity sheet for this experiment.
2. Completely discharge both capacitors then construct the series circuit shown in figure 26–1. Set dc power supply output to 2.0V before applying power to the circuit.
3. With the digital voltmeter quickly measure and record the voltages across C1 and C2. Since the digital voltmeter discharges the capacitors as it measures the voltage across them, it may be necessary to (a) turn off dc supply, (b) discharge both capacitors, (c) turn on dc supply and (d) measure voltages two or more times.
4. Compare the measured voltage drops to the capacitance values of the capacitors. Why does the smaller value capacitor develop a higher voltage drop?

Review

This experiment demonstrated that although capacitors in series accumulate the same amount of dc charge, the voltages across them are quite different. $V = Q/C$

5. Construct the series ac capacitive circuit as seen in figure 26–2.
6. Set the function generator output to:
   Sinewave
   1 kHz
   2V rms

FIGURE 26–1 Series DC capacitive circuit

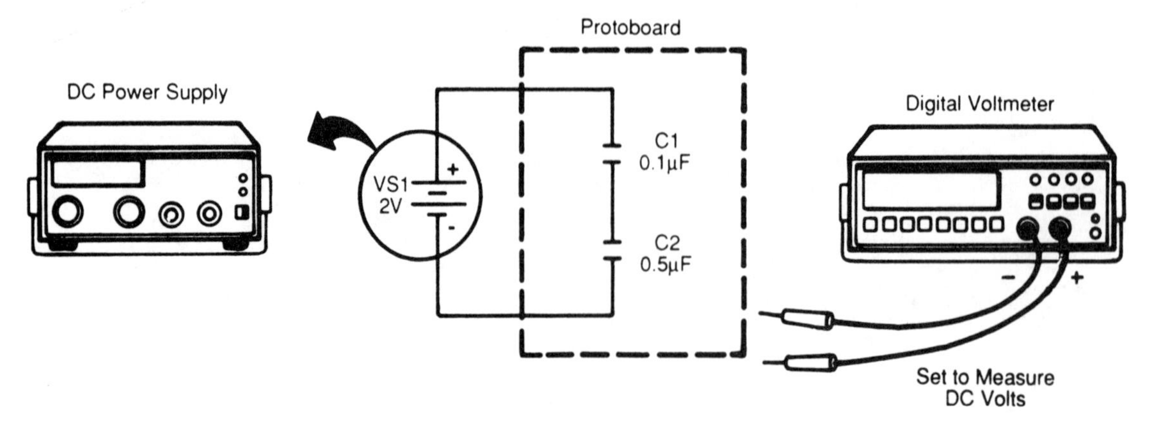

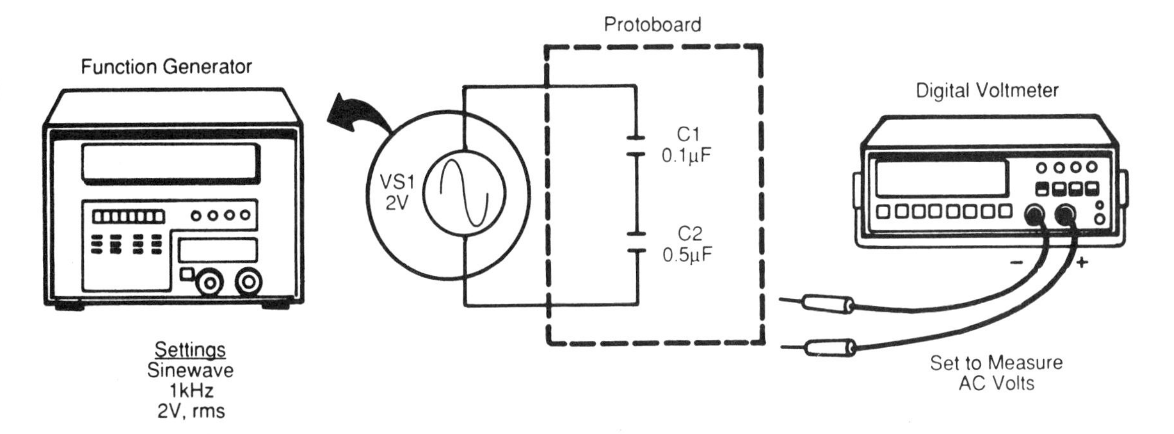

FIGURE 26–2 Series ac capacitive circuit

7. Measure and record the voltages across C1 and C2 with the DVM. Why is the voltage across C2 smaller?
8. Compare the measured ac capacitor voltage drops to the dc circuit measurements. Are the voltages measured across C1 and C2 the same for ac and dc?

Review

When the series connected capacitors are connected to an ac source, the circuit functions as a voltage divider in the same way as series connected resistors. The opposition to ac current flow in this case is the reactance of the capacitors. The voltage developed across each capacitor is directly proportional to its reactance which makes the voltage drops inversely proportional to capacitance. The voltage measurements you have just made verify these facts and further reveal that the sum of the two voltage drops is equal to the source voltage. If the frequency of the function generator was changed, would the voltage drops change?

ACTIVITY SHEET EXPERIMENT 26

NAME ______________________

DATE ______________________

Step 3

A. With the digital voltmeter connected to monitor the dc power supply output, set the power supply to 2.0 volts. Connect the supply to the series capacitors, then measure and record the voltages across the capacitors:

V_{C1} ______________ V V_{C2} ______________ V

Perform the measurements quickly because the capacitive voltages will be affected by connecting the voltmeter. After the first measurement, disconnect the dc supply and short circuit the two capacitors to discharge them. Then re-connect the supply and measure the voltages again.

Step 4

B. Analyze the measured voltages as they relate to the capacitances of the capacitors. Why does the smaller capacitance develop the larger voltage?

____

____

Do the two voltages, when added, equal the source voltage?____

Steps 5-6

C. Construct the ac circuit and while monitoring the source voltage with the digital voltmeter set it to 2.0 volts rms.

Step 7

D. Measure and record the voltages across C1 and C2:

V_{C1} ______________ V V_{C2} ______________ V

Why is the voltage measured across C2 smaller? __________

Step 8

E. Analyze your voltage measurements. Are the dc voltages across the capacitors the same as the ac voltages across the capacitors?

____

How do you account for this? ______________

____

Could the series capacitor circuit be used as a dc voltage divider?

____

Could the circuit be used as an ac voltage divider? __________

____

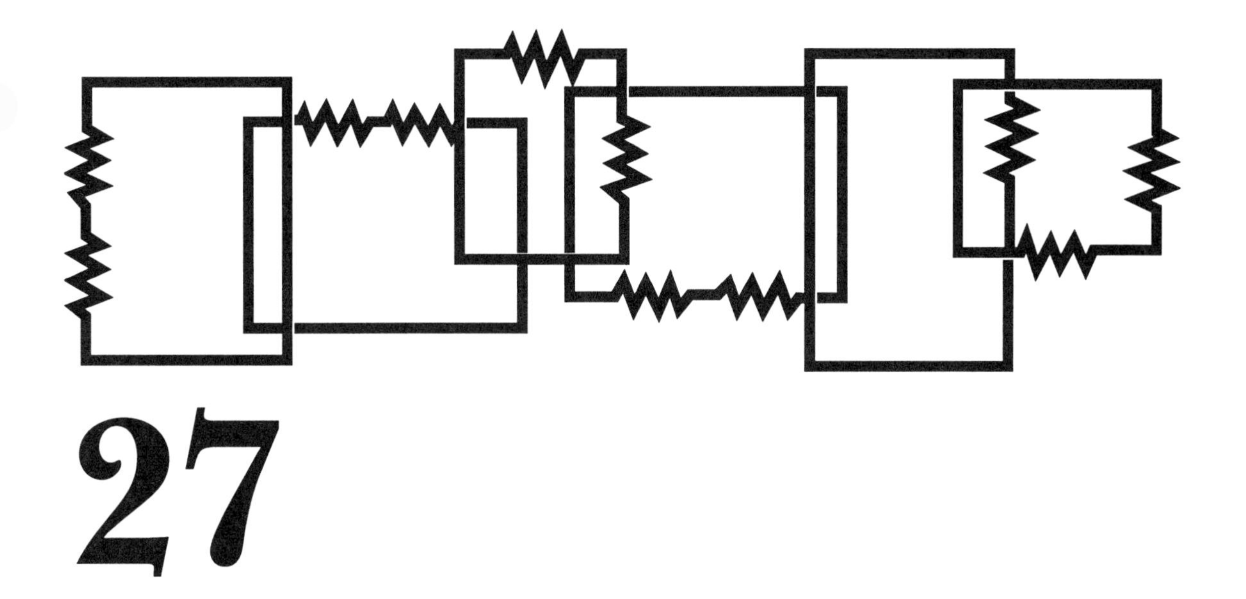

27

Parallel Capacitors

Objectives

After completing this experiment, you will be able to:

1. Verify by measurement that parallel capacitances are additive.
2. Utilize the time constant as an indicator of circuit capacitance.
3. Demonstrate that generator resistance affects measured time constants.

Introduction

This experiment demonstrates the additive capacity of parallel capacitors. To perform this experiment you will need:

An oscilloscope
A function generator
A 1kΩ resistor
Three 0.1μF capacitors

Procedure

1. Remove the activity sheet for this experiment.
2. Construct the circuit shown in figure 27–1.
3. Set the function generator output to:
   Squarewave
   1V peak-to-peak
   dc offset to make Vmin = 0V, Vmax = 1V.
   200 Hz
4. Connect the oscilloscope across C1 (points A and B), and adjust the time and volts per cm settings for the best display.
5. Measure and record the times of charge and discharge slopes from the scope display.
6. Add C2 in parallel with C1, as seen in figure 27–2, while power is still on and oscilloscope is still connected.
7. Re-adjust scope settings for best display.
8. Measure and record the charge and discharge slope times.
9. Add C3 in parallel with C1 and C2. Re-adjust display if necessary then measure the charge and discharge times from the display.

Review

You have tested the characteristics of parallel connected capacitors in three stages. The first circuit consisted of one resistor and one capacitor. You applied a pulsating dc signal to the circuit and observed on the oscilloscope the charge and discharge slopes of the capacitive voltage. When the second capacitor was added in parallel to the first, the charge/discharge times were doubled. Finally, when the third

FIGURE 27–1 Series AC capacitive-resistive circuit

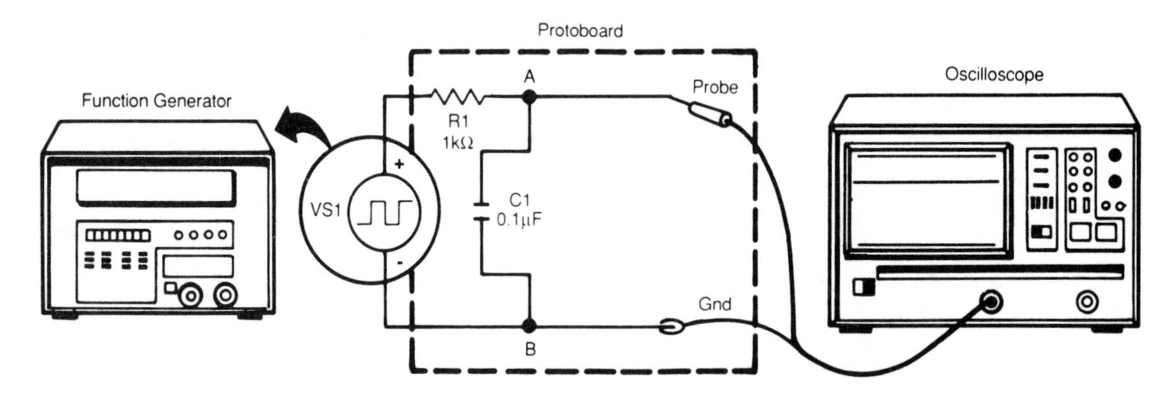

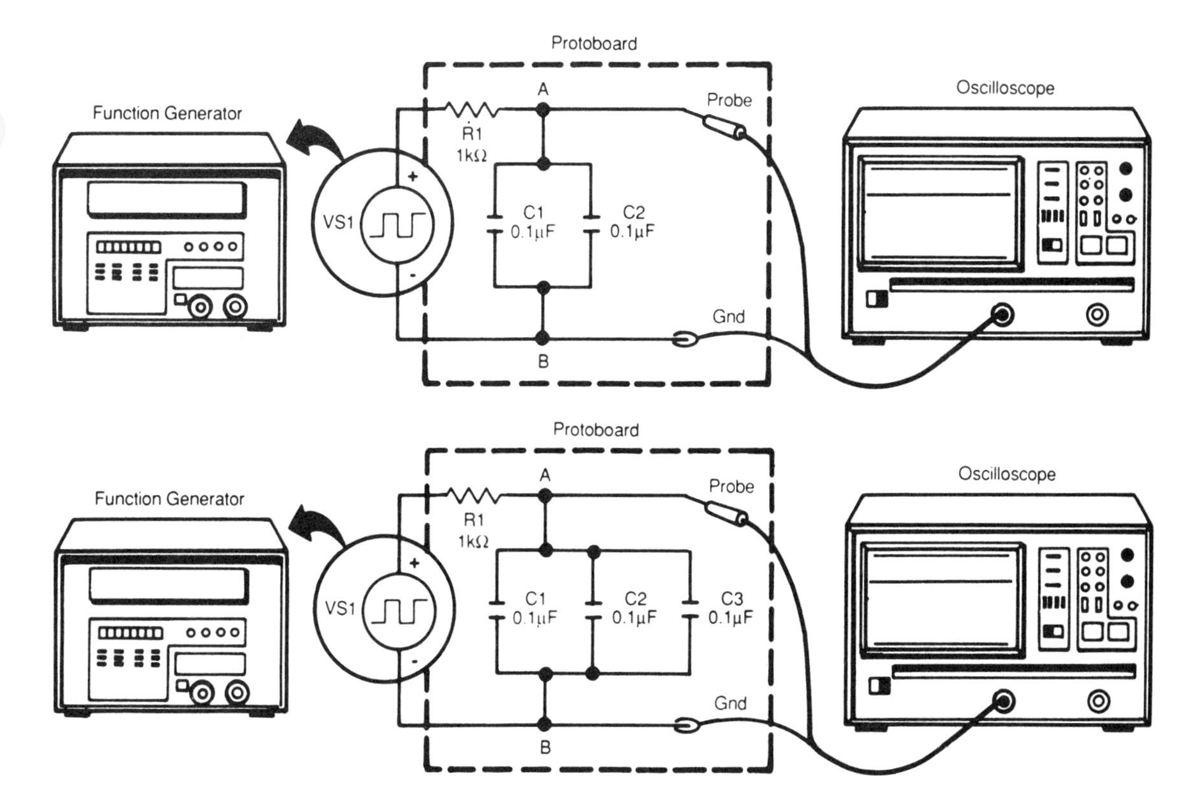

FIGURE 27–2 Two branch AC capacitive-resistance circuit

capacitor was added the charge/discharge times increased even more. Since T = RC, by transposition $C = \frac{T}{R}$. This relationship suggests that when R is held constant, an increase in the charge/discharge characteristics of a circuit is due to an increase in capacitance. Have your measurements proven that parallel connected capacitors are additive? When you calculated the time constant of the circuit using coded values of R and C you discovered a discrepancy. The total resistance of the circuit is actually in two parts, the internal resistance of the generator (600Ω) plus the external circuit resistor. Recalculate the time constants with this total resistance and the discrepancy will be resolved.

10. Review the measured data from the three circuits. Calculate the time constants for each circuit from the measured data. Could you calculate the values of capacitance in each circuit from this data? When capacitors are connected in parallel how do they electrically affect one another?

ACTIVITY SHEET EXPERIMENT 27

NAME ______________________________

DATE ______________________________

Step 3 — **A.** With the function generator connected to the series RC circuit, attach the scope probe to the points where the generator connects to the circuit. Adjust the generator's output and dc offset controls until the scope display shows pulsating dc swinging between zero volts and +1 volt.

Steps 4-5 — **B.** Place the scope leads across C1. Measure and record the charge time and discharge time of the waveform in the table below.

Steps 6-8 — **C.** Add the second capacitor, C2, in parallel with C1. Measure and record the charge and discharge times in the table below.

Step 9 — **D.** Add the third capacitor, C3, in parallel with the first two. Measure the new charge and discharge times. Record this data in the table.

Circuit	Charge Time	Discharge Time
C1	μs	μs
C1 ‖ C2	μs	μs
C1 ‖ C2 ‖ C3	μs	μs

Step 10 — **E.** Analyze the measured data that you've collected. How has the addition of each capacitor changed the charge/discharge times?

____

Does the result seem to be predictable? ____________________

From the measured data calculate the time constant of each circuit:

Time Constants of Circuits		
Circuit 1 μs	Circuit 2 μs	Circuit 3 μs

Using the known values of resistance and capacitance, calculate the theoretical time constant of each circuit:

Time Constants of Circuits		
Circuit 1 μs	Circuit 2 μs	Circuit 3 μs

Compare both sets of time constants. Does there appear to be a discrepancy?__________________________ Could the internal resistance of the generator be a factor? _____________________

Explain. ____

____

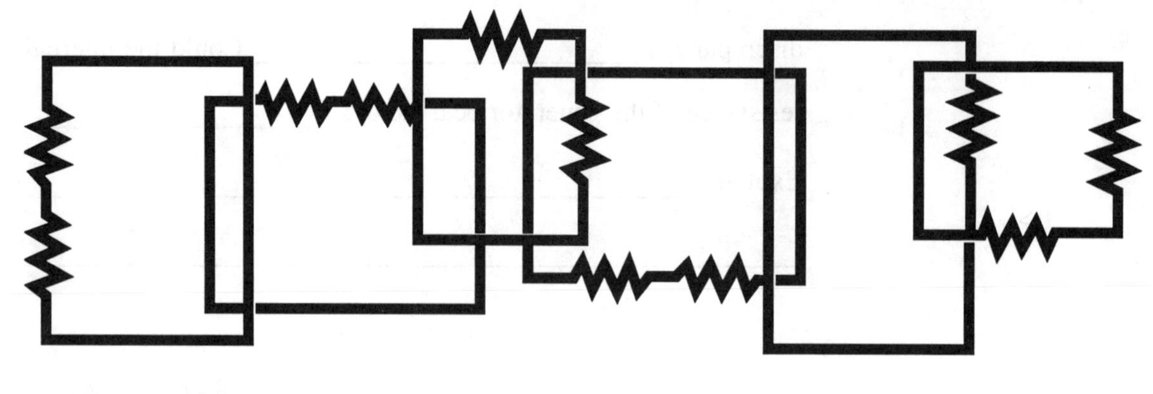

28

Integrators and Differentiators

Objectives

After completing this experiment, you will be able to:

1. Evaluate the integrator circuit response to varied waveforms.
2. Evaluate the differentiator circuit response to varied waveforms.
3. Evaluate the frequency response characteristics of differentiators.
4. Evaluate the frequency response characteristics of integrators.
5. View modified waveforms produced by the frequency selective circuits.

Introduction

In this experiment the characteristics and uses of integrators and differentiators are studied. In their simplest form these waveshaping circuits are made up of a resistor and a capacitor. To perform this experiment you will need:

- A function generator
- Test leads
- A 1kΩ resistor
- A 10kΩ resistor
- A 0.1μF capacitor
- A 0.5μF capacitor
- An oscilloscope

Procedure

1. Construct the integrator circuit shown in figure 28–1 and remove the activity sheet for this experiment.
2. Set the function generator output to:
   1 KHz
   4V p-p
   Set scope to single trace mode and adjust vertical sensitivity and sweep as necessary.
3. Connect the scope probe to monitor the waveform across the capacitor and determine, by looking at the scope's display, if the sinusoidal input waveshape from the function generator has changed.
4. Change the output waveshape of the generator to squarewave and connect the scope probe to the output of the function generator to observe its shape. Now connect the scope probe across the capacitor once again. Has the waveshape changed in any way?
5. Select a triangular waveshape. Observe the wave at the generator's output and across the capacitor. Has the waveshape changed in any way?

FIGURE 28–1 The RC integrator

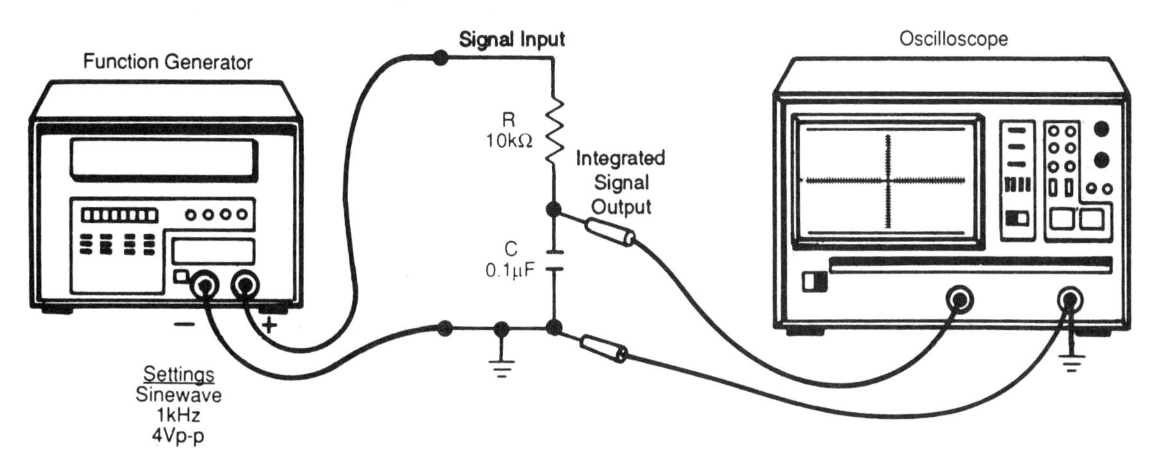

Review Break

The waveform produced by the integrator circuit is the voltage developed across the capacitor. This voltage is a function of the rate of change of the source input. When the source input is a sinusoid, the integrator output is also a sinusoid. The amplitude is reduced because of the voltage division characteristics of the circuit, however the waveshape is unchanged. When the source waveshape is non-sinusoidal then the integrated output wave shape is changed. In the case of the square source waveform, its integrated form was triangular. The triangular source waveform was integrated into a nearly sinusoidal waveshape. The phenomenon of waveshaping by integration is dependent upon charge and discharge currents, time constant to period relationships, and whether the applied waveshape is continuous or discontinuous. The integrator circuit tends to accentuate the low frequency components of the applied signal. To function effectively as an integrator, the time constant of the circuit should be at least 10 times the period of the applied wave.

6. Construct the differentiator circuit shown in figure 28–2. Do not change the function generator's frequency or amplitude settings.
7. Select a sinusoidal output from the function generator. Observe the waveforms at the generator output and at the circuit output (across the resistor). Are there any changes in the waveshape between input and output?
8. Select a squarewave output from the generator. Observe the waveform change between input and output and note the changes.
9. Select a triangular wave output from the generator. Observe the waveform change between input and output and note the changes.

Review

The differentiator circuit is made up of the same basic parts as the integrator, but the parts are reversed in their positions. The output of

FIGURE 28–2

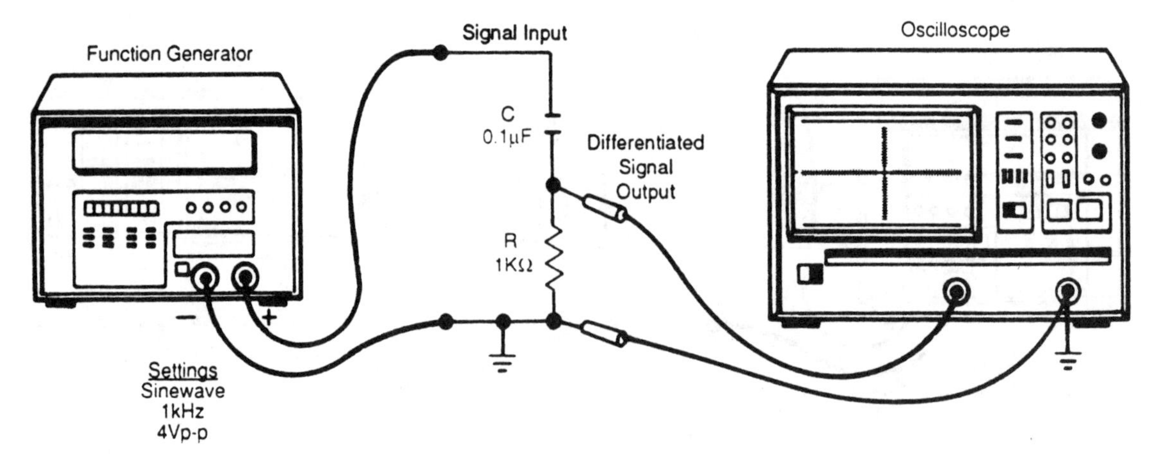

the differentiator is taken from across the resistor in the circuit. To function effectively as a differentiator the time constant of the RC combination should be no more than $\frac{1}{10}$ the period of the applied waveform. Your experimentation demonstrated that the differentiated sinusoid is a sinusoid. When a square wave is differentiated a spike waveshape is produced. A differentiated triangle wave is nearly square.

Integrators and differentiators are special circuits that are used in special applications. You will encounter them again in advanced circuit studies where you will use transistors and operational amplifiers.

The differentiator tends to accentuate the high frequency components of the input signal from the source. The time constant of the differentiator circuit should be at no more than $\frac{1}{10}$ the applied signal period. The applied signal period should be at least 10 times the time constant of the differentiator circuit.

ACTIVITY SHEET EXPERIMENT 28

NAME ______________________________

DATE ______________________________

Step 3 **A.** Observe the sinusoid waves at the generator and across the capacitor. Are there any changes in the appearance of the wave at the capacitor?

____________________ How do you account for this? _______

____

Step 4 **B.** Has the square wave been modified by the circuit as it is observed across the capacitor? ____________________

Explain. ______________________________

Step 5 **C.** How is the triangular waveform changed by the circuit?

____

Explain why this is so: ____________________

____

Step 7 **D.** How does the differentiator circuit alter the sinusoidal waveform? ______________________________

Why is this so? ______________________________

____

Step 8 **E.** How does the differentiator circuit change the square wave?

____

What does the differentiated waveform represent? __________

____

Step 9 **F.** How does the differentiator circuit change the triangle wave?

____

How do you account for this? ____________________

____

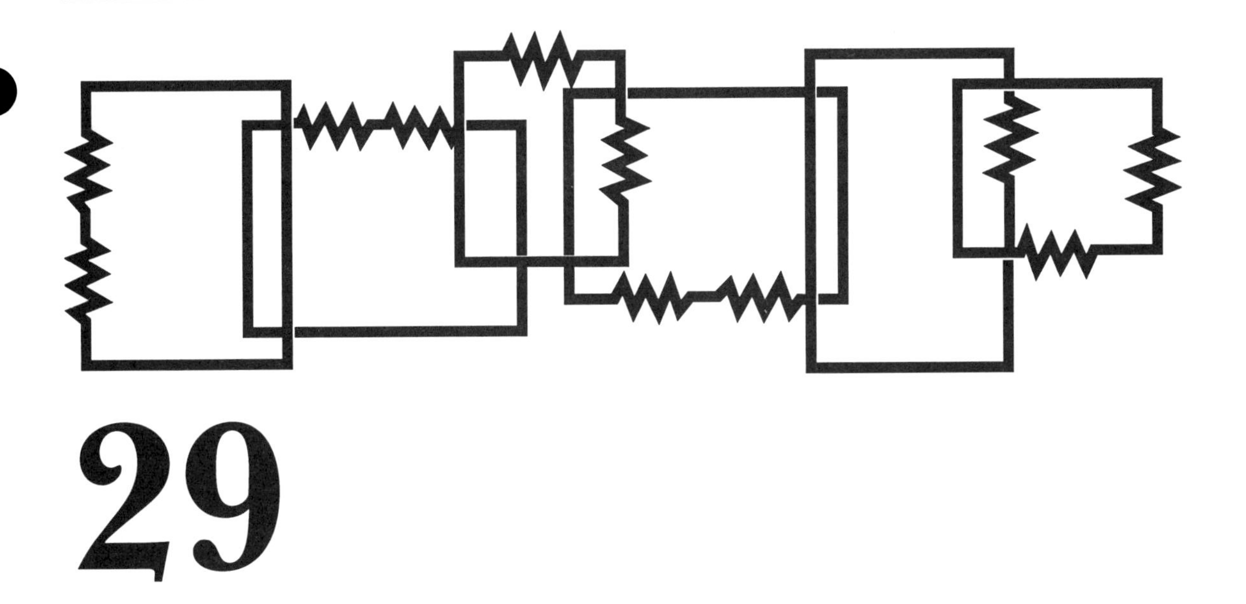

29

The Electromagnet

Objectives

After completing this experiment, you will be able to:

1. Construct an electromagnet.
2. Evaluate the magnetic characteristics of the electromagnet.
3. Verify that magnetism is produced by dc current flow.
4. Confirm that magnetic field strength is proportional to current flow.

Introduction

Without magnets, either permanent magnets or electromagnets, life would not be nearly as enjoyable as it is today. Magnets have a multitude of uses including computer memories (bubbles), electric motors, electric generators, relays for current switching, microwave sources and amplifiers, and television sets. If we were to count the number of appliances and recreational devices we have in our home which employ magnetics we would be astounded.

In this experiment you will build a small electromagnet so that you may become familiar with its characteristics. For this experiment you will need:

- A variable dc power supply
- Four test leads with alligator clips
- A sewing machine bobbin (iron or plastic)
- A roundhead metal screw about 1 inch long
- A flathead metal screw about 1 inch long
- A small flat washer to fit the flathead metal screw
- Two roundhead metal screws about $\frac{1}{2}$ inch long
- About 50 feet of #30 magnet wire
- A piece of dry pine wood 4" long, 2" wide and $\frac{3}{4}$ inch thick
- A VOM (digital, analog or both)
- A sharpened pencil
- A momentary contact pushbutton switch
- A 4 inch by $\frac{3}{4}$ inch strip of iron sheet cut from a tin can
- A small piece of sandpaper
- A pointed punch or a nail
- A pair of needle nose pliers
- A short length of transparent tape or electrical tape
- A screwdriver or two as needed

Note to Instructor: Should it be impractical to construct the magnet, a 12 volt SPDT commercially available relay may be used.

The materials list for this experiment is quite long and some of the items will have to be pre-prepared, such as the iron strip cut from a tin can. Use a pair of tin snips to cut the can or have a friend do it for you.

Procedure

1. Carefully place the sewing machine bobbin onto the end of the sharpened pencil to support it then begin winding the #30 magnet wire onto the bobbin. Leave about two inches of free wire at the beginning end for connection purposes. Wind the wire onto the bobbin until it is full then wrap the windings with tape to keep the wire from unwinding. Wrap the tape around the bobbin to cover the wire only so that no tape extends over the edges of the bobbin. Cut the end of the wire to leave two inches of wire for connection purposes. With the sandpaper gently remove the insulating enamel from the ends of the wires to expose the bare copper.
2. With the sandpaper also remove any insulating material from both sides of one end of the iron strip. This end of the strip will

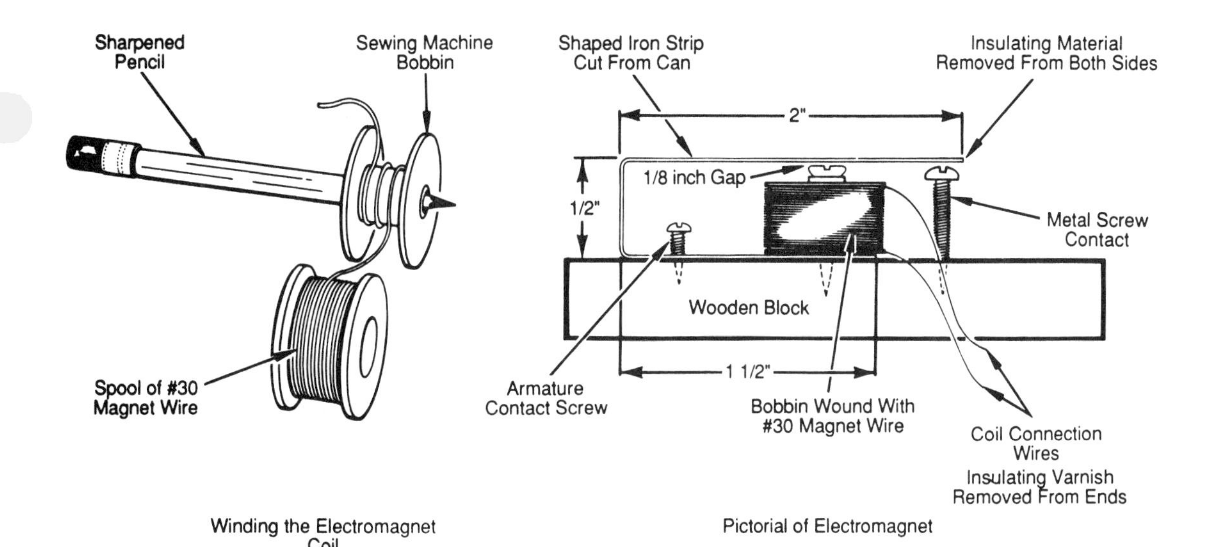

FIGURE 29–1 The electromagnet

be opposite the end to which the bobbin is attached. Mount the bobbin onto the wooden block with the iron strip and screws as shown in figure 29–1. Using the punch or a nail, punch a hole through the iron strip where the bobbin mounting screw will penetrate the strip. Do this with the iron strip in position on the wooden block. Attach the bobbin to the block over the iron strip with the flat head screw passing through the center of the bobbin. The small flat washer should be placed onto the screw before it is used to attach the bobbin to the board.

3. Punch a hole into the iron strip about $\frac{3}{8}$" behind the bobbin near the edge of the strip. Insert one of the small roundhead screws into the board through this hole and screw it in until about $\frac{1}{4}$" of the screw remains above the strip. This screw will be used for electrical connection to the iron strip. Take the 1" roundhead screw and position it about $\frac{1}{4}$ inch in front of the coil and screw it into the block so that the top of its head is level with the flathead screw running through the center of the bobbin.
4. With your needle nosed pliers shape the iron strip as shown in figure 29–1. Bend it so that the strip rests above the bobbin with a $\frac{1}{8}$" gap separating the strip and the top of the flathead screw running through the center of the coil.
5. Pull the activity sheet for this experiment then construct the first circuit as shown in figure 29–2. Set the output of the variable power supply to zero before completing circuit connections.
6. Slowly increase the output of the power supply while watching both the milliammeter and the iron strip above the magnet coil. At what current level is the strip pulled down to the magnet? Try lifting the attracted strip with your fingertip. Is the magnet fairly strong?
7. Slowly decrease the output of the power supply. At what current level does the iron strip spring free of the magnet?
8. Increase the output of the power supply until the iron strip is again pulled down to the magnet. Now turn the supply off and modify the circuit to include the switch as shown in the second

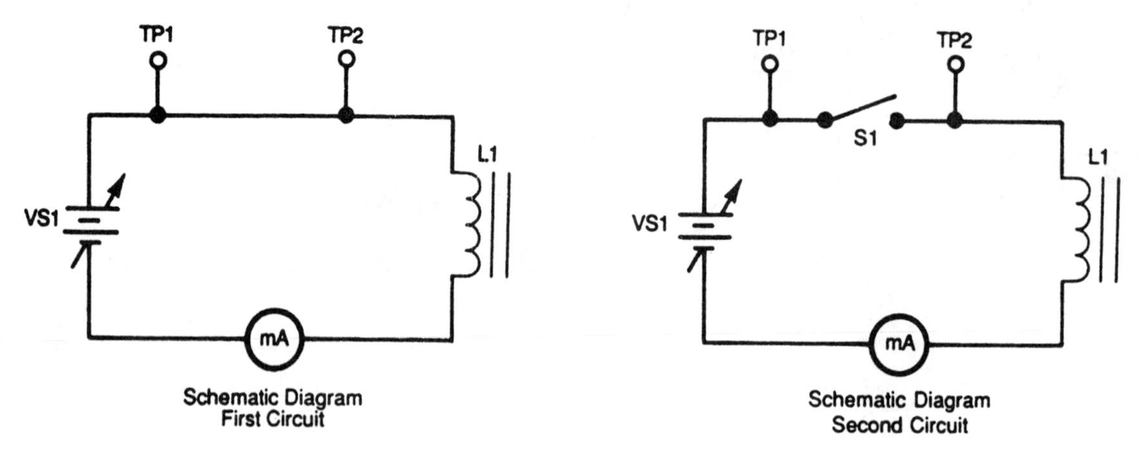

FIGURE 29–2 Schematic diagram of magnetic circuit

circuit diagram. Restore power to the circuit, then each time you push the switch the iron strip will be attracted to the magnet with a clicking sound. When you release the switch the strip will spring free. You have a device which operates on the same principle as the telegraph invented by Samuel F. B. Morse in 1844.

Review

You have made an electromagnet and with it you have proven that electricity can produce magnetism. Your electromagnet physically resembles a relay and has characteristics which are common to the commercially made relay. You discovered that it takes more current to attract the iron strip (armature) to the magnet than it does to hold it in place once it has been attracted. These characteristics are called the pulling current and the holding current. You also found that at some decreased current value the iron strip springs free. Retain your electromagnet because it will be used for more experimentation soon.

9. Return all materials to their proper places.

ACTIVITY SHEET EXPERIMENT 29

NAME ______________________________

DATE ______________________________

Record all measured data and question responses in the appropriate spaces.

Step 6 **A.** How much current flow through the magnet coil is necessary to attract the iron strip to it?____________________

Step 7 **B.** As you decrease current flow through the magnet, at what value of current is the strip released?____________________

Step 8 **C.** After installing the switch in series with the electromagnet are you able to produce controlled magnetism?____________________

____

D. As you increase current flow through the electromagnet how is its strength affected?____________________

____

E. Why is less current flow needed to hold the iron strip against the magnet than is needed to initially attract it?____________________

____

____

____

____

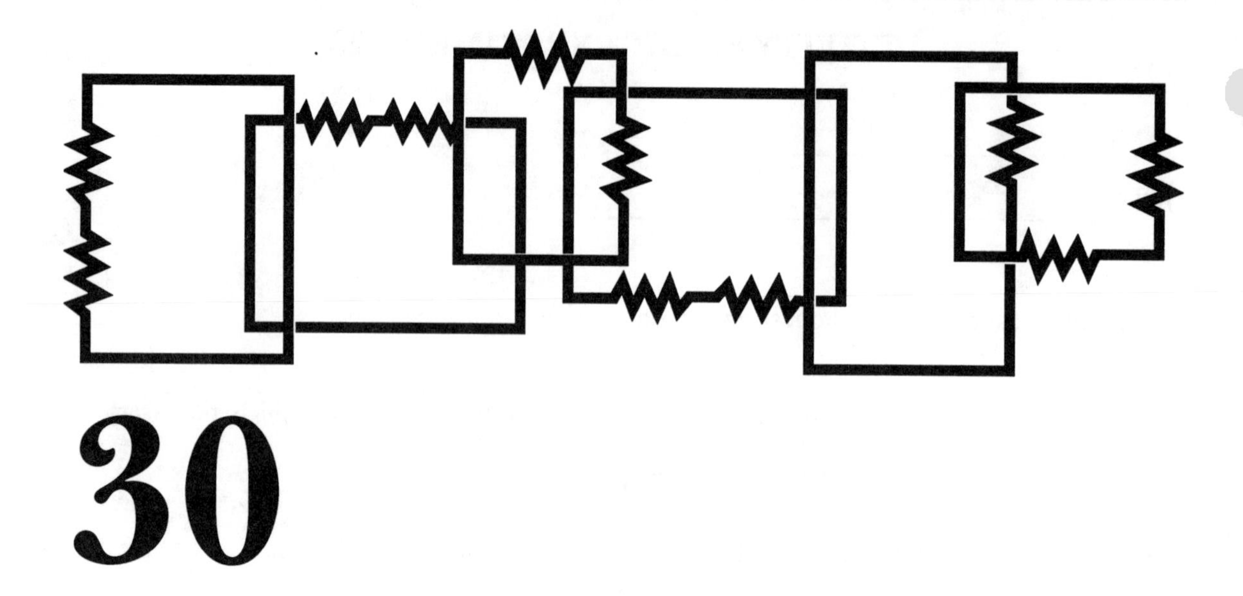

30

The Electric Generator

Objectives

After completing this experiment, you will be able to:

1. Investigate the reciprocal properties of a permanent magnet motor.
2. Evaluate the operation of the motor with varied applied voltage.
3. Confirm that increased current flow produces increased output power.
4. Evaluate the use of the motor as a generator.
5. Verify that increased rotational speed increases electrical output.

Introduction

There are certain electric devices which are capable of doing two related things. The permanent magnet dc electric motor is one of these. Although designed to convert electrical energy into rotating mechanical energy, the motor also has the ability to function as a generator. This ability is in accordance with the law of reciprocity. You will learn that there are many more devices which exhibit dual capabilities. The loudspeaker, for example, makes a rather good microphone. You will verify this in more advanced experiments in the future. For now let us examine the ability of the motor to produce electricity. To perform this experiment you will need:

A variable dc power supply
A VOM (analog is best for this experiment)
Four test leads with alligator clips
A toy dc permanent magnet electric motor

Procedure

1. Pull the activity sheet for this experiment then construct the circuit as shown in figure 30–1. Set the output of the variable dc supply to zero. You will first test the motor.
2. Hold the motor in your hand and slowly increase the output of the power supply. At what voltage does the motor just begin to spin? Grab the motor shaft with your other hand. Is the motor producing much power?
3. Continue to slowly increase the power supply output up to the rated voltage of the motor. What do you see? Is the motor producing more power? Note the direction of motor rotation.

FIGURE 30–1 Motor and schematic of motor test circuit.

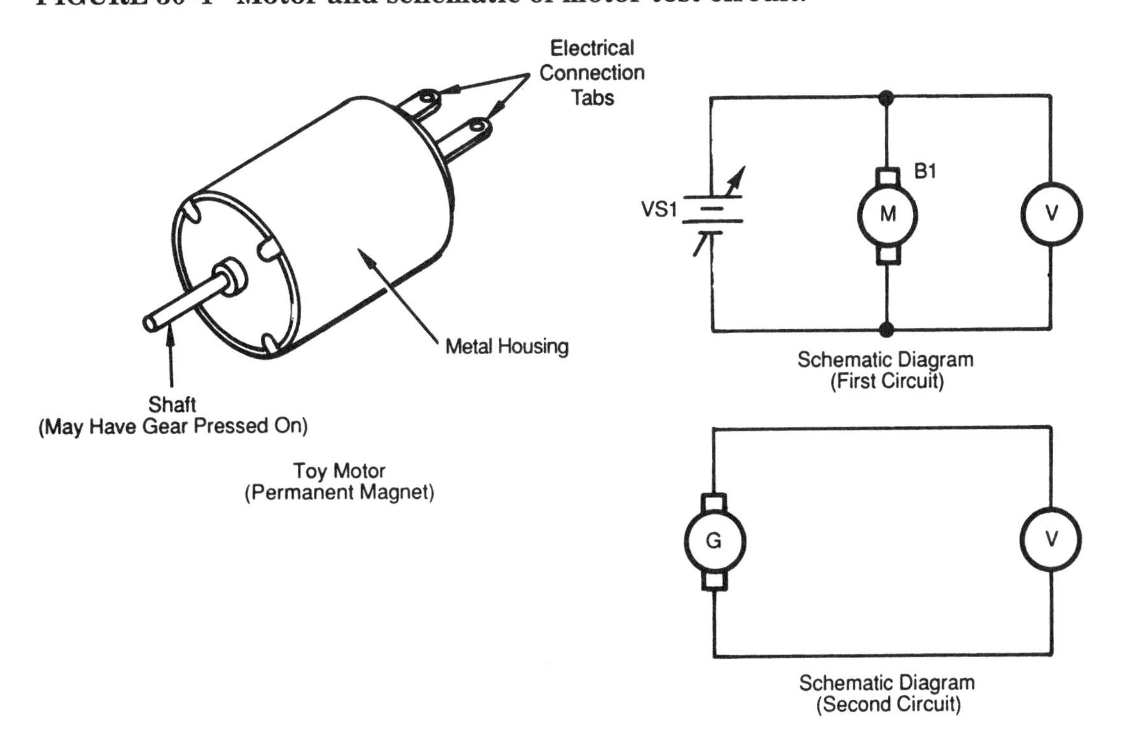

4. Reverse the wires connecting to the motor so that current will flow through it in the opposite direction. What do you observe?
5. Disconnect the power supply from the circuit but leave the voltmeter connected across the motor. While holding the motor in one hand spin its shaft between your fingers of the other hand. What does the voltmeter show you? Spin the motor in the opposite direction. What difference do you observe? Connect milliammeter across the motor terminals and spin the shaft again in both directions. What did you observe?

Review

You verified that your motor does convert electrical energy to mechanical energy. You found that when the applied voltage is small the motor rotates slowly and doesn't produce much power. As the applied voltage is increased the motor rotates faster and the power produced is increased. When you reversed the direction of current flow through the motor you found that the rotation of the motor also reversed. Varying the voltage applied to the motor is one way to vary the speed of the motor, but as you observed, it is not very efficient. In later studies you will learn how to control motor speed electronically so that even while it is turning slowly it will have a high torque characteristic.

Then you tested the ability of the motor to act as a generator. You found that when the motor is spun manually it generates a voltage. The faster you spin it the higher the voltage. When you spin the motor in the opposite direction the polarity of the output voltage is reversed. The dc permanent magnet motor will function as a dc generator!

If you have never seen the internal parts of the dc motor you can carefully disassemble yours (with your instructor's approval). There are some parts of the motor which you must be cautious not to damage. Once you have satisfied your curiosity put your motor back together and test it to see that it still works. Then put all materials back to their proper places.

ACTIVITY SHEET EXPERIMENT 30

NAME ______________________________

DATE ______________________________

Record measured data and question responses in the appropriate spaces.

Step 2 **A.** What is the voltage measured across the motor as it just begins to spin? ____________________

Step 3 **B.** After increasing the power supply voltage to equal the rating of the motor is the motor spinning rapidly? __________________

Does it produce much more power than previously? __________

____

Step 4 **C.** When the connections to the motor are reversed what happens?

____

Is there any loss of speed or power? ______________________

____

Step 5 **D.** How much voltage are you able to generate when you spin the motor shaft in one direction? ____________________

What happens when you spin the motor shaft in the opposite direction? ____

____

How much current do you measure while spinning the shaft?

____

____

E. Why does the polarity of the generator's output change as the direction of rotation is changed? __________________________

____

____

F. Why does the output increase as the generator is turned faster?

____

____

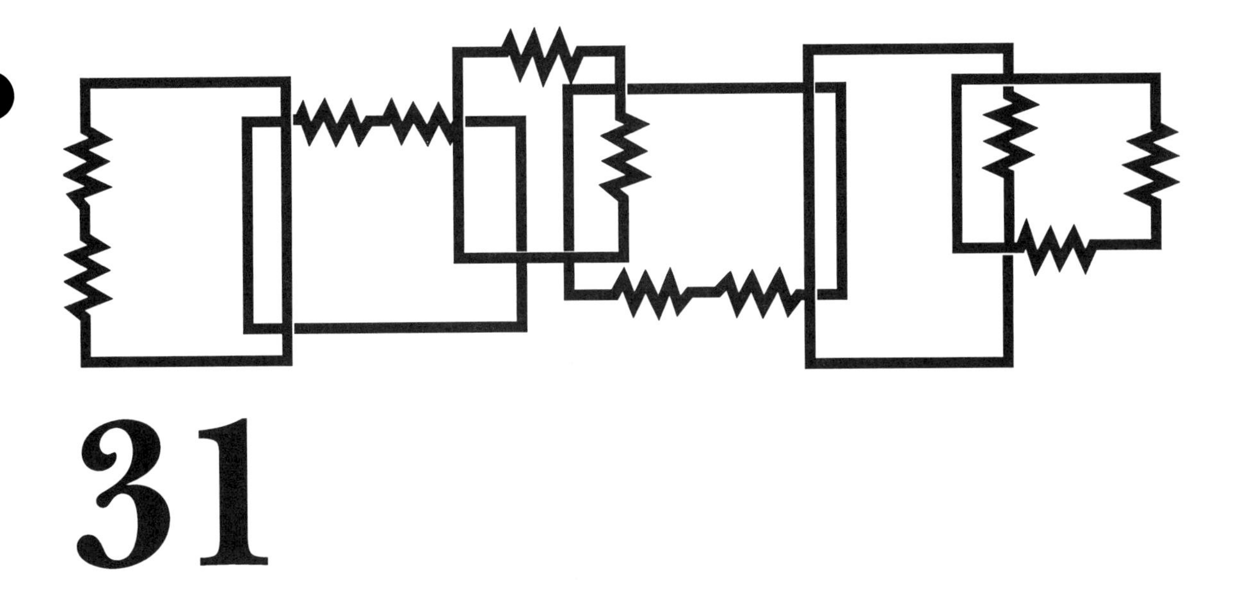

31

The Electric Relay

Objectives

After completing this experiment, you will be able to:

1. Evaluate the operation of the electrical relay.
2. Evaluate the operation of the contacts of the relay.
3. Operate the relay at varied input current levels.
4. Operate the relay as an electrical switch.

Introduction

It is now time to modify the electromagnet you made previously to make it into a simple relay with two sets of contacts. You will need:

A dual output variable dc power supply
A momentary contact pushbutton switch
Six test leads with alligator clips
A #7382 bi-pin lamp (14V 80 mA)
A paper clip
A $\frac{1}{2}$ inch metal screw
A pair of needle nose pliers

Note to Instructor: If construction of the relay is impractical, a commercially available 12 volt SPDT relay may be used.

Procedure

1. Modify your electromagnet by adding the shaped paperclip as shown in figure 31–1. Put a loop in the paperclip for the mounting screw. Once the paperclip is securely mounted to the board, shape it and position it with the needle nosed pliers to push the iron strip down about $\frac{1}{16}$ inch from the top of the magnet. The paperclip will serve as an electrical contact as will the iron strip (armature). These contacts are normally closed since they are closed while the relay is de-energized.
2. With your finger, push the armature down to the top of the magnet. The armature should make contact with the top of the screw just in front of the magnet coil. This set of contacts is normally open since they are not in contact with one another until the armature is pulled down to the magnet. Note that when this occurs, the normally closed contacts will open.
3. Pull the activity sheet for this experiment then connect the circuit as shown in figure 31–2. Increase the output of source 1 from zero until the armature is pulled to the magnet each time the switch is pushed. Set the output of source 2 to provide normal illumination intensity of the lamp.

FIGURE 31–1 Electromagnet modification and schematic symbol of result

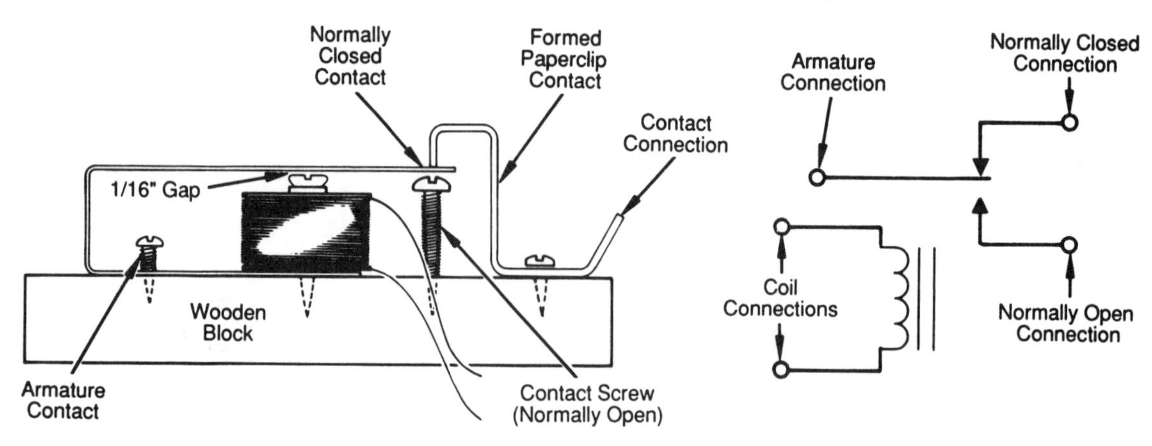

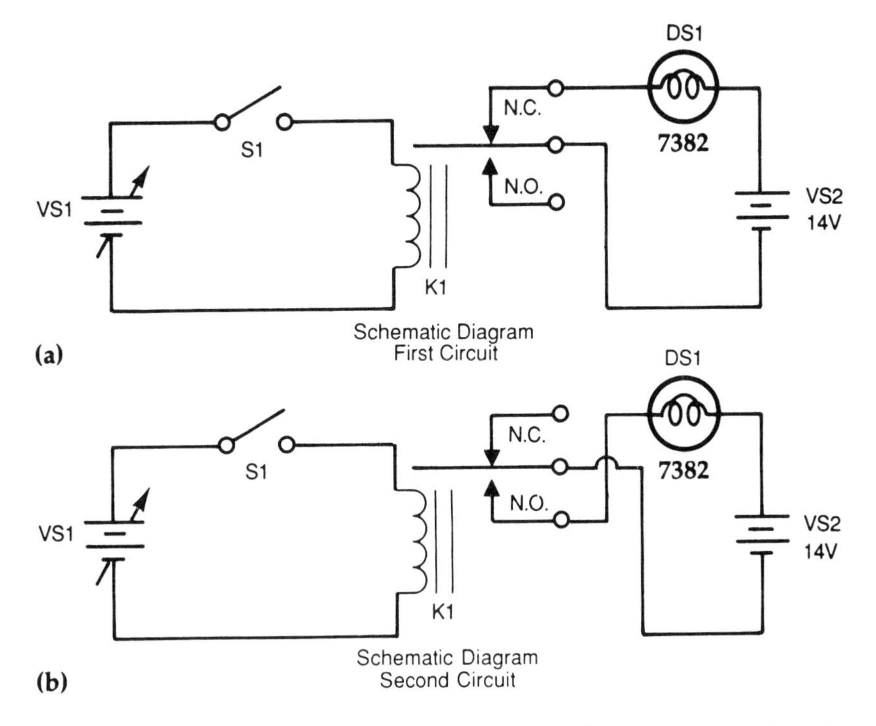

FIGURE 31–2 Schematic diagram of relay circuit using (a) normally closed contacts and (b) normally open contacts

4. Each time you push the switch does the relay respond? When is the lamp illuminated? Why?
5. Move the clip lead connected to the normally closed contact to the normally open contact as seen in figure 31–2b. Push the switch several times. When does the lamp light now? Why?
6. Remove the switch from the circuit by disconnecting the test leads from the switch and connecting them together.
7. Slowly increase the power supply output until the relay just activates then slowly decrease the power supply output until the relay just deactivates. Does this suggest any uses for the relay?

Review

The relay you have made duplicates electrically the operation of commercially made relays. The primary use for relays is switching of electrical current. It allows one electrical circuit to control another electrical circuit with isolation. Both circuits are separate electrically because there is no common connection between them. This may not seem an important characteristic now, but in the future you will study applications where it is essential.

You found that the relay can also be used as a current sensor because it takes a certain minimum value of current to activate it. It is this characteristic which makes the voltage regulator possible. The automotive voltage regulator controls the output of the alternator at varying engine speeds to assure that it is a constant 14 to 15V. This is necessary to avoid damage to the car's battery and electrical system. Although the relay you have constructed is comparatively crude, it works. Retain it. We will use it again.

8. Return all materials to their proper places.

ACTIVITY SHEET EXPERIMENT 31

NAME ______________________________

DATE ______________________________

Write the answers to the questions in the spaces provided.

Step 4

A. When you push the switch what happens to the relay?

____

Does the lamp light when the relay is activated or de-energized?

____

____

Why is this so? ______________________________

____

Step 5

B. After modifying the connection to the relay and testing circuit operation, what is now different? ______________________

____

Why is this so? ______________________________

____

C. Would you be able to modify the circuit again so that it has two lamps which turn on and off alternately as you work the relay?

____

Draw a diagram of the circuit below.

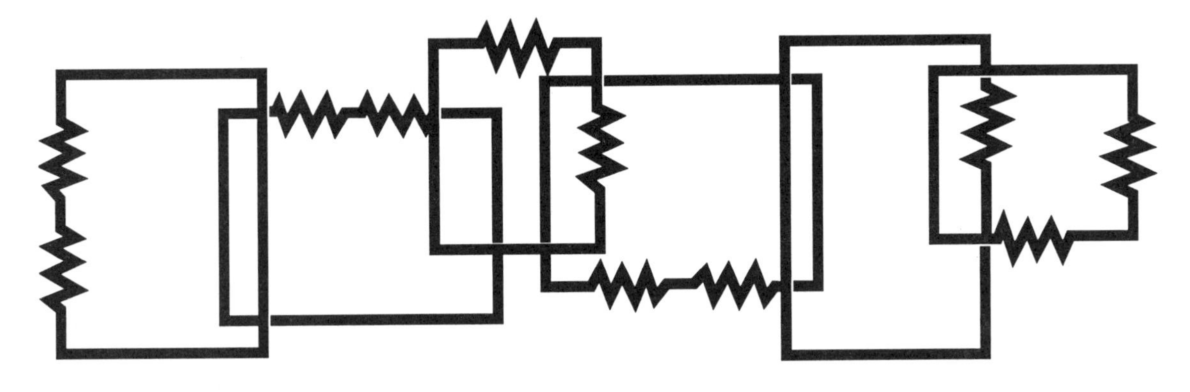

32

Application of an Electromagnet

Objectives

After completing this experiment, you will be able to:

1. Operate the electrical relay as a buzzer.
2. Determine the effect of varied current flow on the buzzer.
3. Experience the sound of the buzzer.

Introduction

The purpose of this experiment is to make noise. You will find that the relay which you made in an earlier experiment can be wired to work as a buzzer. In order to do this you will need:

A variable dc power supply
Your relay
Four test leads with alligator clips

Procedure

1. Set the output of the power supply to zero and connect the relay as shown in figure 32–1. Pull the activity sheet for this experiment.
2. Slowly increase the output from the power supply until the relay begins to buzz. Can you see any sparking at the contacts of the relay?
3. Increase the output from the power supply slightly. Does this cause the sound from the buzzer to change?
4. With your finger, push the contact made from the paper clip downward slightly towards the coil. Does this cause the sound to change? Turn the output of the power supply back to zero.

Review

What causes the relay to operate as a buzzer when it is connected this way? Were you able to see the armature moving rapidly while the buzzer was energized? The armature is the source of the sound made by the buzzer. While the armature is contacting the paper clip (the normally closed set of contacts) there is current flow through the magnet coil. The magnetism attracts the armature down towards the coil away from the paper clip. When the armature is pulled away from the paper clip, the circuit is opened temporarily and there is no current flow through the coil. Since there is no current flow through the coil, there is no magnetism, so the armature is no longer attracted to the coil. The armature then springs back to its original position making contact with the paper clip, and the process starts all over

FIGURE 32–1 Schematic of buzzer connected relay

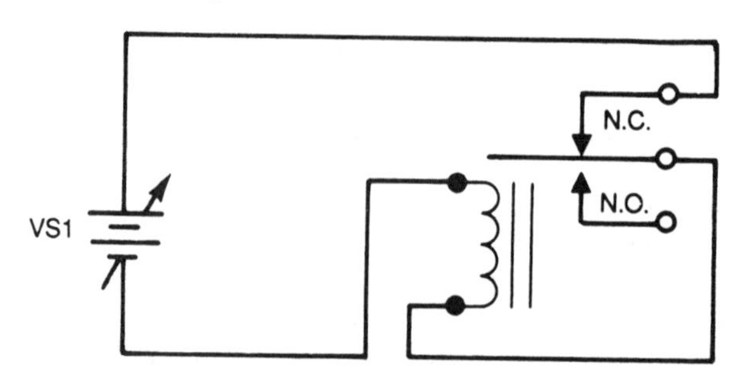

again. The armature is vibrating rapidly because the magnet is pulsating, energized for an instant, then de-energized for an instant. These pulsations continue as long as power is applied to the buzzer.

5. Return all materials to their proper places. You have demonstrated that direct current can be used to make noise. It's just a matter of knowing how to make magnetism work for you.

ACTIVITY SHEET EXPERIMENT 32

NAME ______________________________

DATE ______________________________

Step 2

A. While the buzzer is buzzing do you see sparks at the contacts?

____

What is causing the sparks? ______________________________

____

Step 3

B. Does increased current flow change the sound of the buzzer?

____

Why? ______________________________

Step 4

C. Does pushing the upper contact slightly downward change the tone of the buzzer? ____________________

Why? ______________________________

D. While the buzzer is operating, is the current flow through the electromagnet continuous or pulsating? ____________________

Why? ______________________________

____

E. What possible uses are there for your buzzer? ______________

____

____

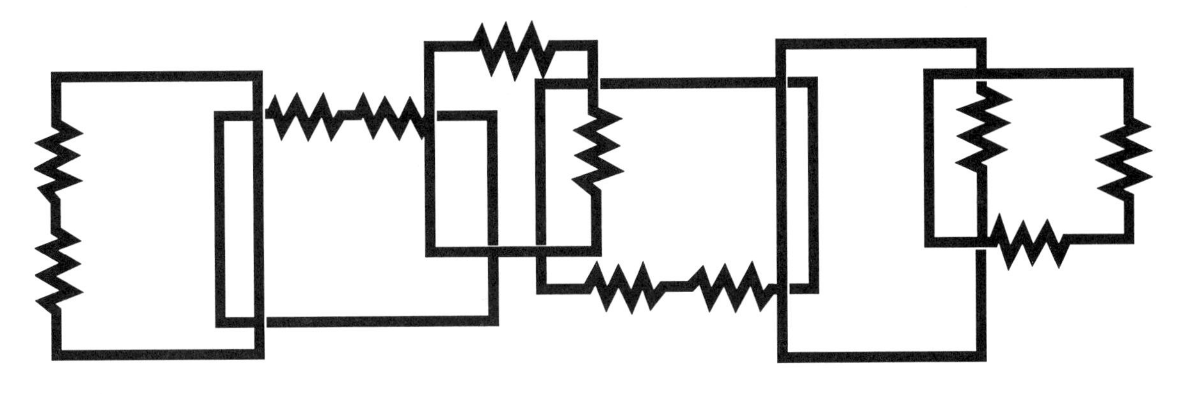

33

Listening to Alternating Current

Objectives

After completing this experiment, you will be able to:

1. Evaluate the operation of the loudspeaker.
2. Hear the sound of different ac waveforms and frequencies.

Introduction

In this experiment we will put together a small device which contains an electroacoustical transducer, more commonly called a loudspeaker. To perform this experiment you will need:

A function generator
A $2\frac{1}{4}$ inch loudspeaker (8Ω to 40Ω)
A plastic margarine dish and top
Some test leads with alligator clips
Some tissue paper.

Procedure

1. Analyze the construction of the loudspeaker in figure 33–1.
2. Obtain a small margarine dish with a plastic top. Cut a slot pattern in the top similar to the one seen in figure 33–1.
3. Attach alligator clip test leads to the speaker terminals.
4. Pack the dish with tissue (not too firmly), and then place the speaker on top of the tissue facing up.
5. Place the lid onto the dish to hold the speaker in place and then connect the alligator clips to the output of the function generator, as seen in figure 33–2.
6. Remove the activity sheet for this experiment.
7. Select different waveform shapes at varying volume levels and apply the signals directly to the speaker terminals. Record your observations.

Review

This experiment demonstrated how an alternating electrical output from the function generator could be transformed into an audible sound wave output from the loudspeaker. You discovered that the different waveshapes produce different sounds from the speaker

FIGURE 33–1 Construction of an audible transducer

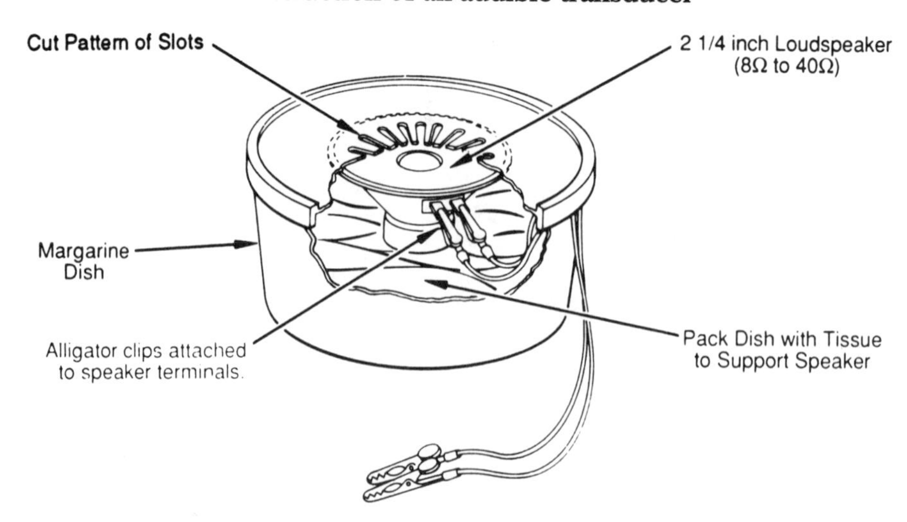

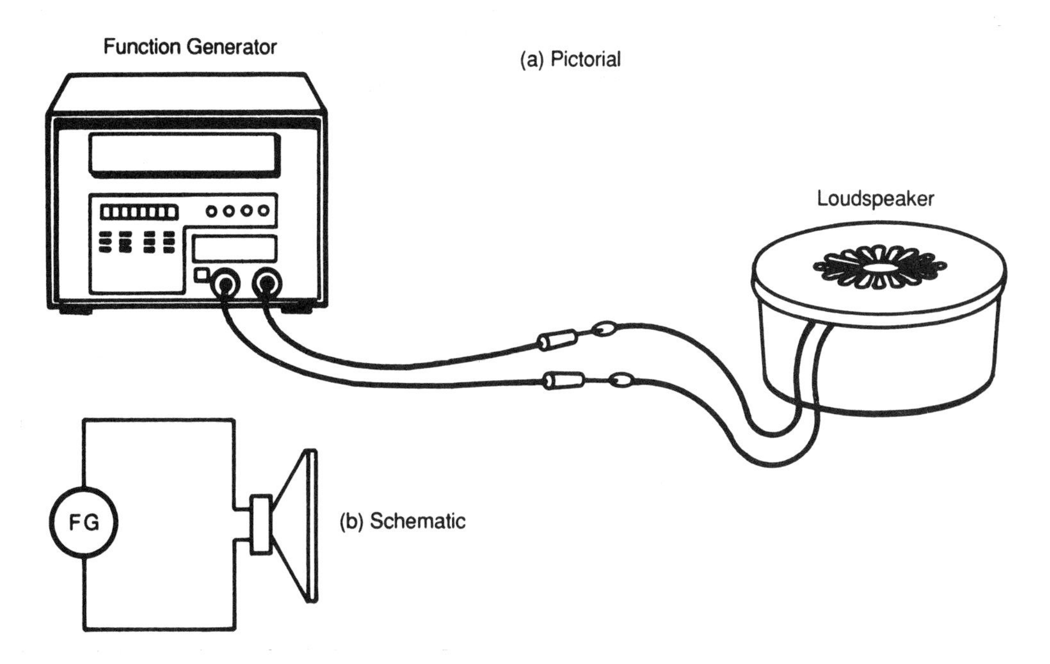

FIGURE 33–2 Hooking-up the loudspeaker

although the frequency of the signal remains the same. The power content and harmonic frequency content of the different types of waves produce different sound and volume characteristics even though the signal is applied to the speaker at the same amplitude. Which of the three waveshapes produces the most interesting sound?

ACTIVITY SHEET EXPERIMENT 33

NAME ______________________

DATE ______________________

Steps 1-5 **A.** Mount the loudspeaker with test leads as outlined in lab manual.

Step 7 **B.** Connect speaker to output terminals of the function generator. Select each of the output waveforms in succession and vary the output level and frequency while listening to each. Can you hear a difference in the sound characteristics of the different waveforms?

____

____

At a constant output level (frequency approximately 500 Hz)

which waveform seems loudest?______________________

Why do you suppose this is so?______________________

____

Vary the output frequency and record the lowest and highest audible frequencies for each waveform:

Waveform	Lowest Audible Freq.	Highest Audible Freq.
Sine wave	Hz	KHz
Square wave	Hz	KHz
Triangle wave	Hz	KHz

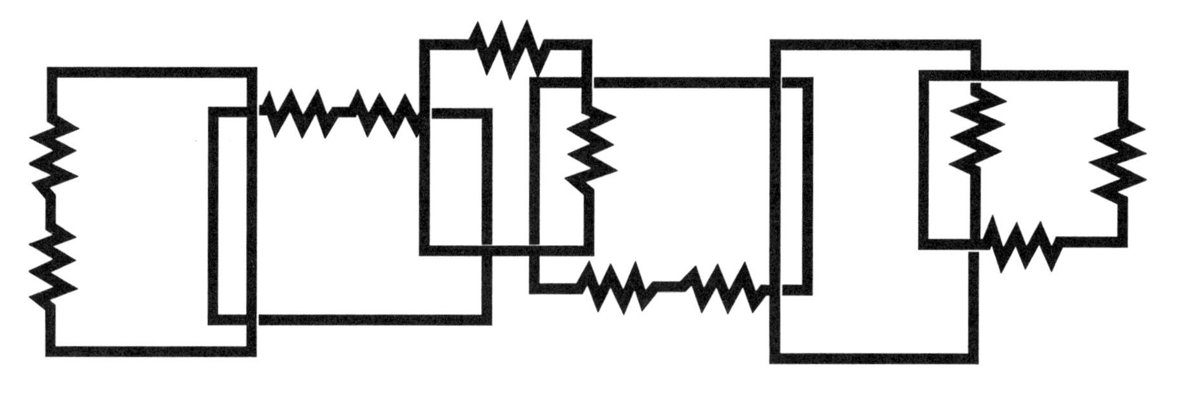

34

The Electromagnet Revisited

Objectives

After completing this experiment, you will be able to:

1. Operate the electromagnet with alternating current.
2. Operate the buzzer with alternating current.
3. Distinguish the differences in operation with ac as compared to dc.
4. Experience the sound and feel of an ac operated electromagnet/buzzer.

Introduction

This experiment will explore how the previously constructed electromagnet will react when an alternating current is applied. To perform this experiment you will need:

A center tapped transformer (12.6V/300mA min.)
The relay/buzzer
Some test leads with alligator leads

Note to Instructor: A 12 volt SPDT relay may be used in lieu of the hand built.

Procedure

1. Remove the activity sheet for this experiment.
2. Construct the circuit seen in figure 34–1.
3. Connect the 6.3V ac output of the transformer to the electromagnet of the relay.
4. Record your observations.
5. Gently place your finger under the armature of the relay and describe the characteristic of magnetic attraction.
6. Connect the relay as a buzzer as seen in figure 34–2 and then describe the sound produced.

Review

When the ac voltage is connected to the relay the alternating current passing through the electromagnet generates a magnetic field of one polarity, reduces to zero, increases to a maximum in the opposite

FIGURE 34–1 AC electromagnet

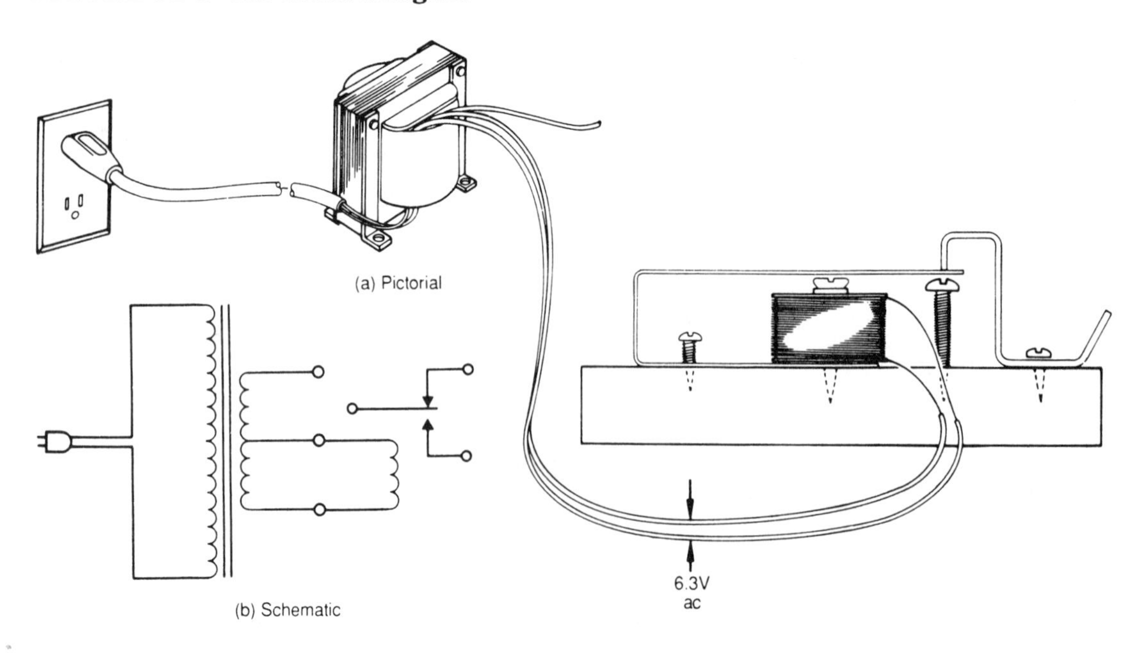

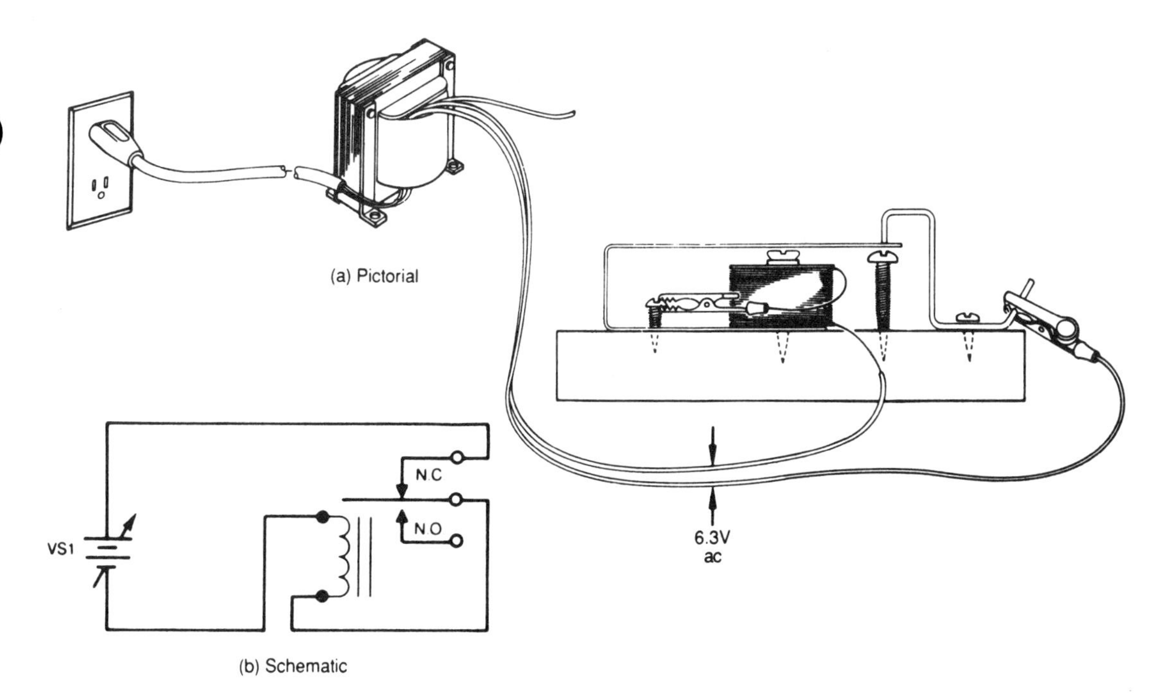

FIGURE 34–2 AC buzzer

polarity, decreases again to zero, and then repeats. The armature is attracted whenever a magnetic field is present and is released when it is absent, hence the buzzing sound of the armature.

When connected as a buzzer, almost the same operation occurs except the pulling of the armature breaks the circuit and causes the magnetic field to collapse momentarily at a nonuniform rate.

ACTIVITY SHEET EXPERIMENT 34

NAME ______________________________

DATE ______________________________

Step 4 **A.** With the electromagnet (relay coil) connected to the ac source, what is the armature doing? ______________________________

____

Is the relay making a sound? ______________________________

Describe what you hear? ______________________________

____

Step 5 Place your forefinger on the underside of the armature and gently push up. What do you feel? ______________________________

____

Why is the magnetic attraction pulsating? ____________________

____

What is the frequency of the sound? (Careful!)

____________________ Hz

Step 6 **B.** With the relay connected as a buzzer and connected to the ac source, has the sound it makes changed? ____________________

How does it differ from the previous sound? ____________________

Do you see sparking at the relay contacts? ____________________

Why? ______________________________

____

In your opinion, which is better for powering electromagnets, ac or dc?

____

____

For what reasons?____

____

____

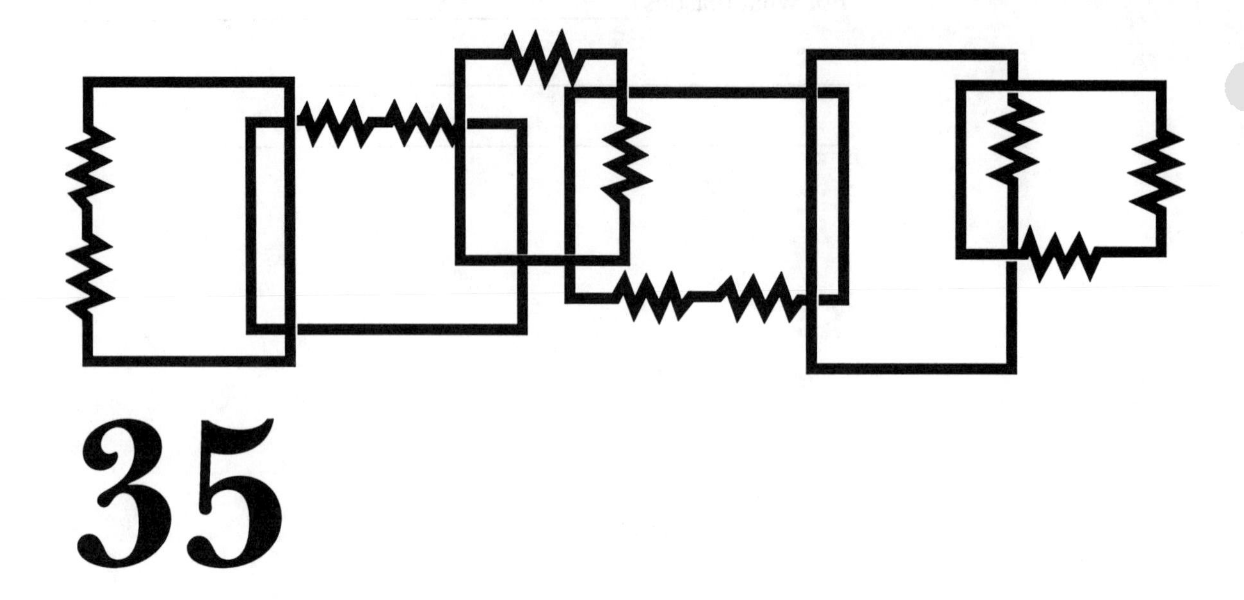

35

Inductive Kick

Objectives

After completing this experiment, you will be able to:

1. Demonstrate how the inductor can be used to create high voltage.
2. Use the neon lamp as an indicator of high voltage.
3. Display the inductive kick waveform on the oscilloscope.

Introduction

One of the most important abilities of an inductor is its energy storage capacity. In this experiment you will charge an inductor and then allow it to discharge, with an illuminating result. To perform this experiment you will need:

A 120V:12.6V step down transformer
One NE-2 or NE-2H neon lamp
Alligator test leads
Hookup wire
A momentary contact pushbutton switch
A variable dc power supply

Procedure

1. Remove the activity sheet for this experiment.
2. Construct the circuit shown in figure 35–1. Since a transformer is simply two inductors in close proximity to one another we can use the 120V primary of T1 as our inductor. DS1 is a neon lamp that requires approximately 70V or more to fire it.
3. Set the power supply output to about 2V dc.
4. Depress the switch S1 for a couple of seconds to allow L1 to charge. Then release the switch and observe the neon lamp, DS1. What did you see?

SAFETY FIRST

Once the inductor has been charged, be careful not to connect yourself across its two leads as the discharge through your body will be painful.

FIGURE 35–1 Energy storage of an inductor

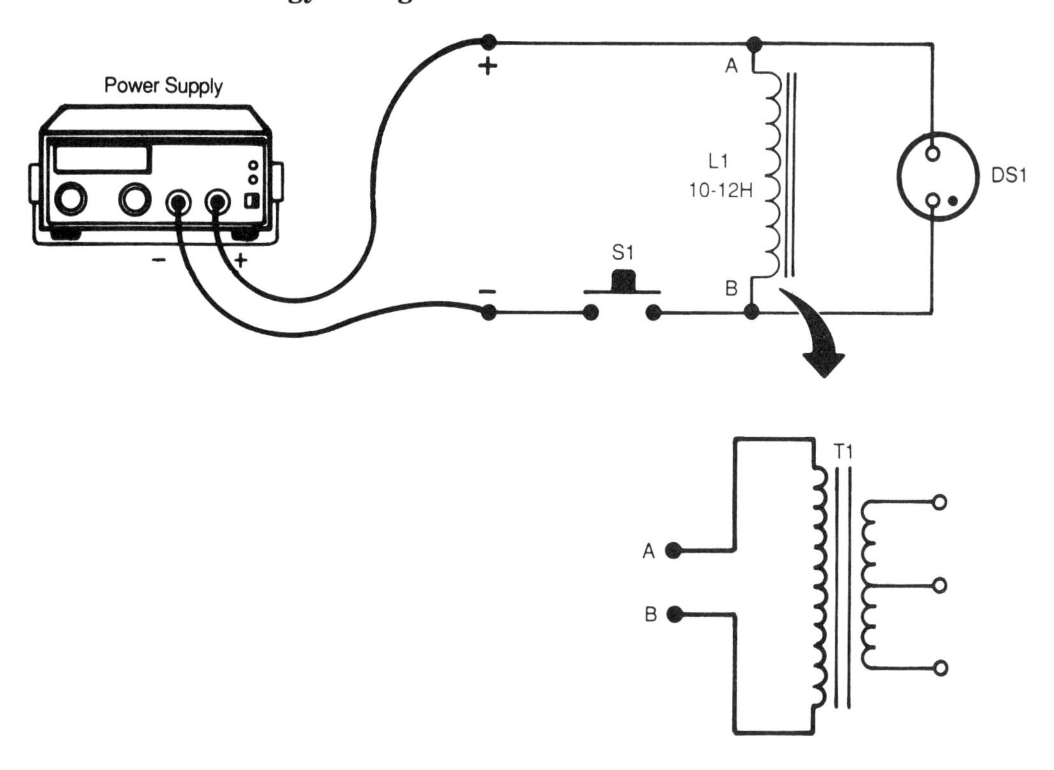

5. Increase the output of the supply to about 5V dc. Depress the switch once again for a couple of seconds and then release the switch. What did you see? Was there a difference when a different supply voltage was used?
6. Turn off the power supply. Disconnect DS1 from the circuit and connect an oscilloscope in its place.
7. Turn on the power supply. Press the switch for a couple of seconds and then release it. Observe the discharge impulse from L1 on the scope display. Can you measure the amplitude of the pulse? Adjust scope sensitivity and sweep as necessary for best presentation.

Review

Inductors, which are specialized electromagnets, have the ability to store energy magnetically. Current flow through the inductor "charges" it by producing a magnetic field. When current flow is interrupted suddenly the magnetic field collapses very quickly creating a surprisingly high voltage pulse. Because the discharge impulse can cause a painful shock it has come to be known as a "kick." You constructed a simple circuit to generate inductive kick. You found that a small dc voltage can be converted to a surprisingly large voltage impulse which easily illuminates a neon lamp. As you might have expected, increasing inductor charge current ultimately increases the inductive kick. When you measured the magnitude of the impulse with the scope I am sure you were amazed.

Inductive kick is a very useful phenomenon which you will encounter in future studies of power supplies, pulsed microwave systems and industrial control circuits. It also presents hazards to switching circuits and you will learn how to incorporate protection into such circuits to avoid component burnout.

ACTIVITY SHEET EXPERIMENT 35

NAME ______________________________

DATE ______________________________

Step 4 **A.** While manipulating S1 does the lamp flash when the switch is closed or when it is opened? ____________________

Why does the lamp flash? ____________________

Step 5 **B.** Does the lamp flash more brightly when the dc supply voltage is increased?____________________ Why is this so? ________

____

Step 7 **C.** Adjust the scope's triggering level to most reliably reveal the voltage impulse. What is the magnitude of the pulse?

____

Why is the impulse predominantly dc? ____________________

____

How is it possible that an inductor can step up a dc voltage?

____

____

____

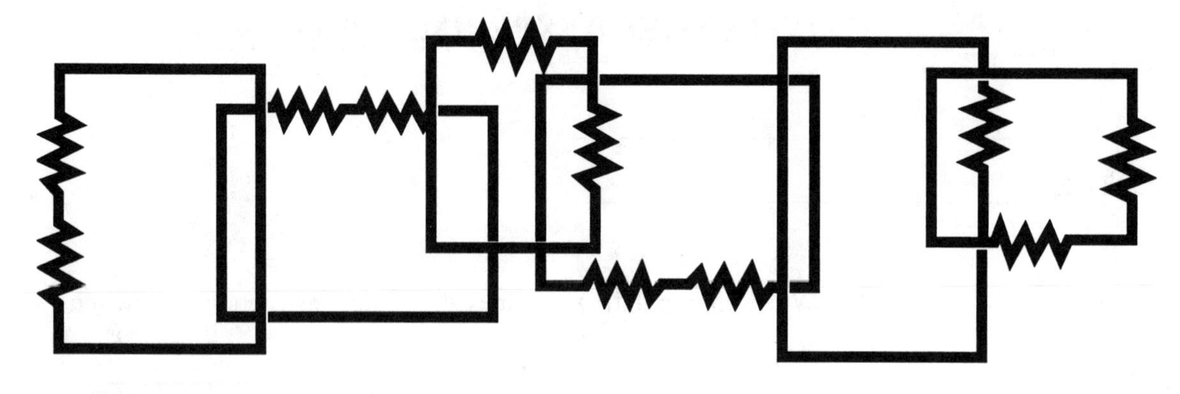

36

RL Circuit Measurements

Objectives

After completing this experiment, you will be able to:

1. Evaluate the characteristics of the resistor inductor series circuit.
2. Determine the inductance of an inductor by electrical measurements.
3. Determine the impedance of an inductor resistor series circuit.
4. Calculate the phase angle of an inductor resistor series circuit.

Introduction

This experiment will demonstrate the similarities and differences between RL and RC circuits. To perform this experiment you will need:

A function generator
A 120V:12.6V transformer
One 330Ω resistor
An oscilloscope with test probes
One 1kΩ potentiometer

Procedure

1. Remove the activity sheet for this experiment.
2. Construct the circuit shown in figure 36–1. Set scope to A-B mode.
3. Monitoring the voltage across the inductor with the oscilloscope, adjust the frequency of the function generator until 3.0 Vpp appears across L1.
4. Determine the frequency of the applied signal.
5. Measure the voltage across R1.
6. Calculate the inductance of L1.
7. Using the measured voltages and known resistance, calculate the circuit's impedance and phase angle.
8. Modify the existing circuit so that it now includes a series connected 1kΩ rheostat as seen in figure 36-2.
9. Alternate the scope leads from across R2 to across the RL circuit, back and forth, while adjusting R2 until the two voltages are equal.
10. Remove R2 from the circuit and measure its resistance, which will closely approximate the impedance of the RL circuit. Does the measured impedance compare favorably with the calculated impedance?

FIGURE 36–1 RL circuit voltages

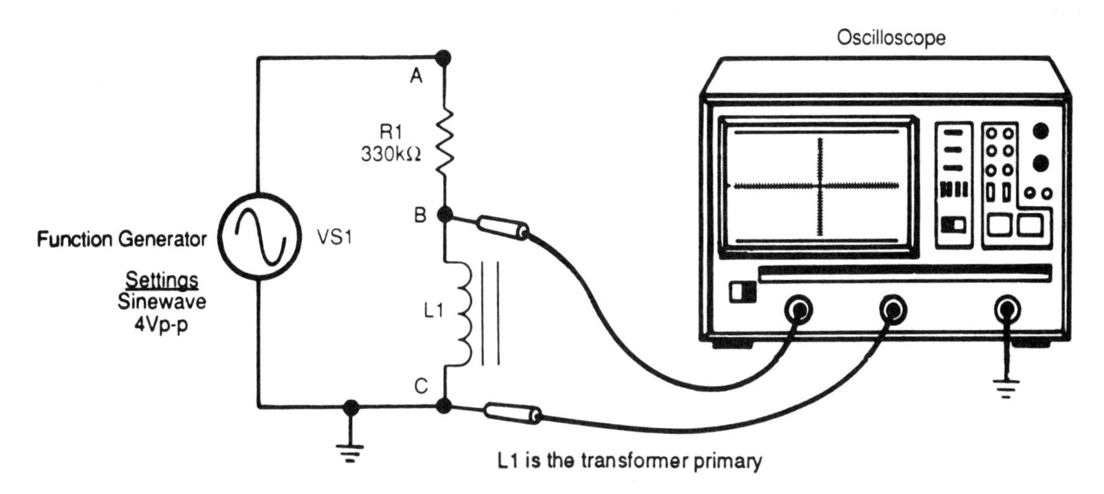

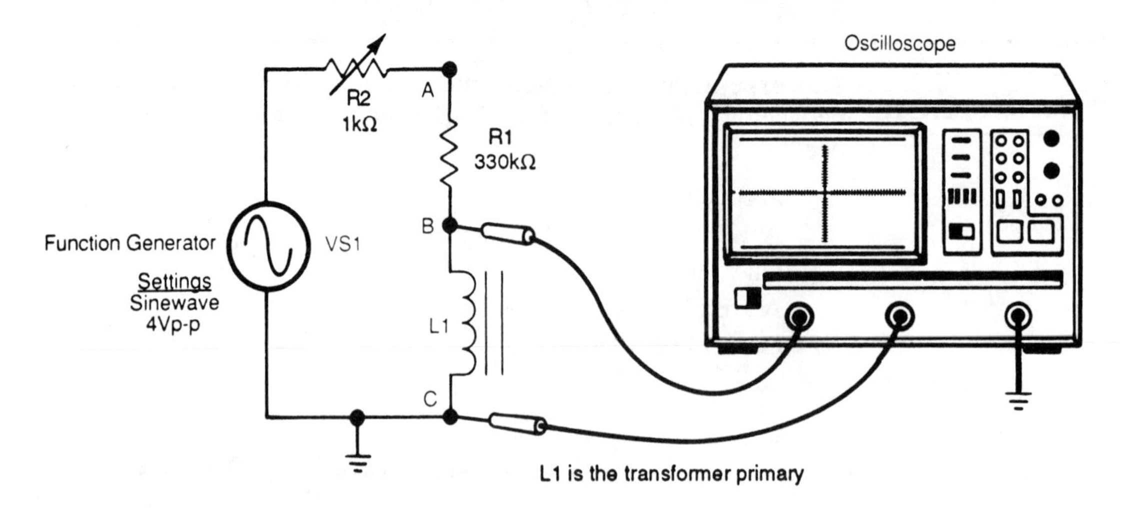

FIGURE 36–2 RL impedance measurement

Review

As you explored the characteristics of the series RL circuit, you found in some respects it was similar to the series RC circuit, but it also had differences.

You constructed the circuit and applied to it a signal from the function generator whose output was set to 4Vpp. You adjusted the frequency of the generator until the measured voltage across L1 was 3.0 Vpp. You then measured the voltage drop across R1, and discovered that the direct sum of the voltages exceeded the applied source voltage. This indicates that the circuit is reactive and that the two component voltages are not in phase. Using the measured voltages, OHM's law and the transposed formula for XL you were able to determine the inductance of L1. You were also able to calculate the impedance and phase angle of the circuit.

You then modified the circuit to include a variable resistor in order to use it as an impedance verification device. You adjusted the variable resistor until its resistance was equal to the circuit impedance, as indicated by the respective voltages being equal, then measured the resistance. Your measured results should have been reasonably close to your calculation.

Using the variable series resistance technique you can measure the impedance of much more complex circuits than the circuit used in this experiment.

ACTIVITY SHEET EXPERIMENT 36

NAME ______________________

DATE ______________________

Step 3 **A.** Vary the output frequency until the voltage measured across the inductor is 3.0 Vpp.

Step 4 **B.** Measure from the oscilloscope the period of the displayed waveform as accurately as possible and from the period determine the frequency.

f = ______________ Hz

Step 5 **C.** Measure the voltage across R1 with the scope as accurately as possible.

V_{R1} ______________ V

$I_S = \frac{V_R}{R} =$ ______________ mA

Using the measured voltage calculate circuit current, then determine the reactance of L1: $X_L =$ ______________ Ω

Step 6 **D.** Transpose the inductive reactance formula to solve for L. Plug in the inductive reactance from C above and calculate the inductance of L1.

$X_L = 2\pi fl$

L = ______________

Step 7 **E.** Use the measured circuit voltages and the known resistance value, calculate the circuit impedance and phase angle.

$\angle\theta = \text{ARCTAN}\ \frac{V_L}{V_R} =$ ______________ °

$Z = \frac{R}{\cos\theta} =$ ______________ Ω

Step 8 **F.** Modify the circuit by adding the variable resistor R2.

Steps 9–10 **G.** Adjust R2 until the voltages indicated across R2 and the series RL circuit are equal. Remove R2 from the circuit and measure its resistance with the digital ohmmeter.

R2 measured ______________ Ω

Compare the measured the calculated impedances. Are the two figures reasonably close? ______________

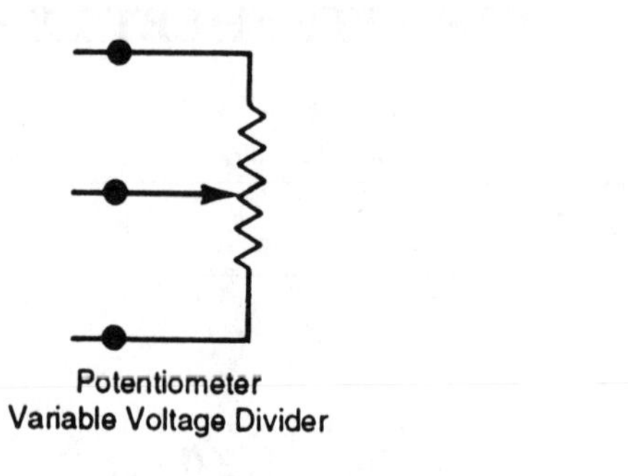
Potentiometer
Variable Voltage Divider

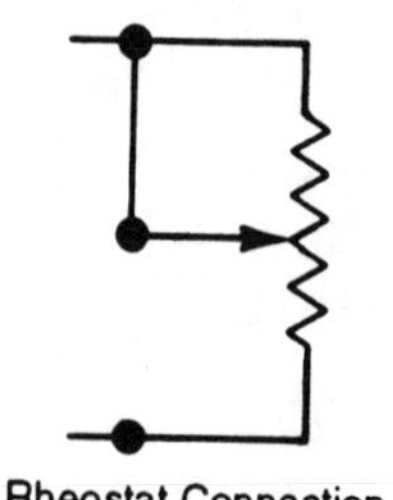
Rheostat Connection
Variable Resistor

FIGURE 36–3

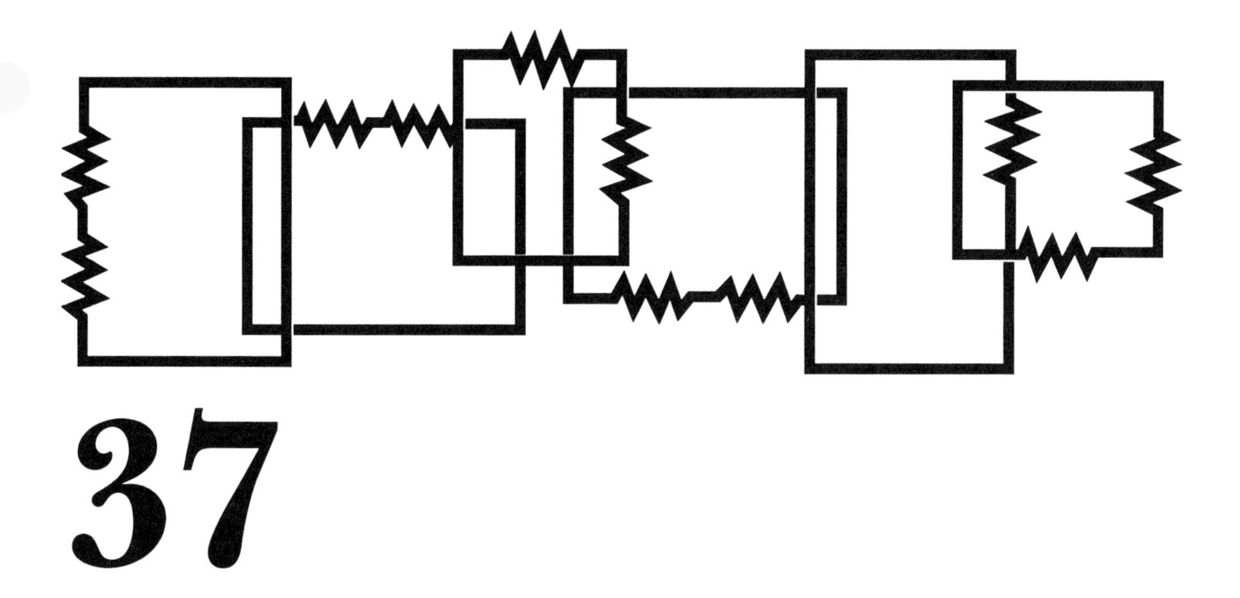

37

Transformers—Stepping Up

Objectives

After completing this experiment, you will be able to:

1. Demonstrate how a transformer can be used to step-up voltage.
2. Describe how the step-up transformer will respond to different input voltages and frequencies.
3. Explain how the step-up transformer can be used to drive a buzzer and neon lamp.

Introduction

In a previous experiment the transformer was used to step down a dangerously high 120VAC to a safe low voltage 12.6VAC. The reverse is also possible: to step up a low voltage to a much higher voltage. This experiment will demonstrate how voltages can be stepped-up.

SAFETY FIRST

There is a potential shock hazard in this experiment so proceed carefully.

To perform this experiment you will need:

A function generator.
A 120V:12.6V transformer.
An oscilloscope.
The previously constructed buzzer.
A dc power supply.
A 0.5μF capacitor.
A NE-2 or NE-2H neon lamp.

Note to Instructor: A 12 volt SPDT commercially available relay may be used in lieu of the handbuilt.

Procedure

1. Remove the activity sheet for this experiment and then construct the circuit shown in figure 37–1. Note that the function generator is connected to what would normally be the low voltage secondary winding of the transformer.

FIGURE 37–1 Voltage step-up

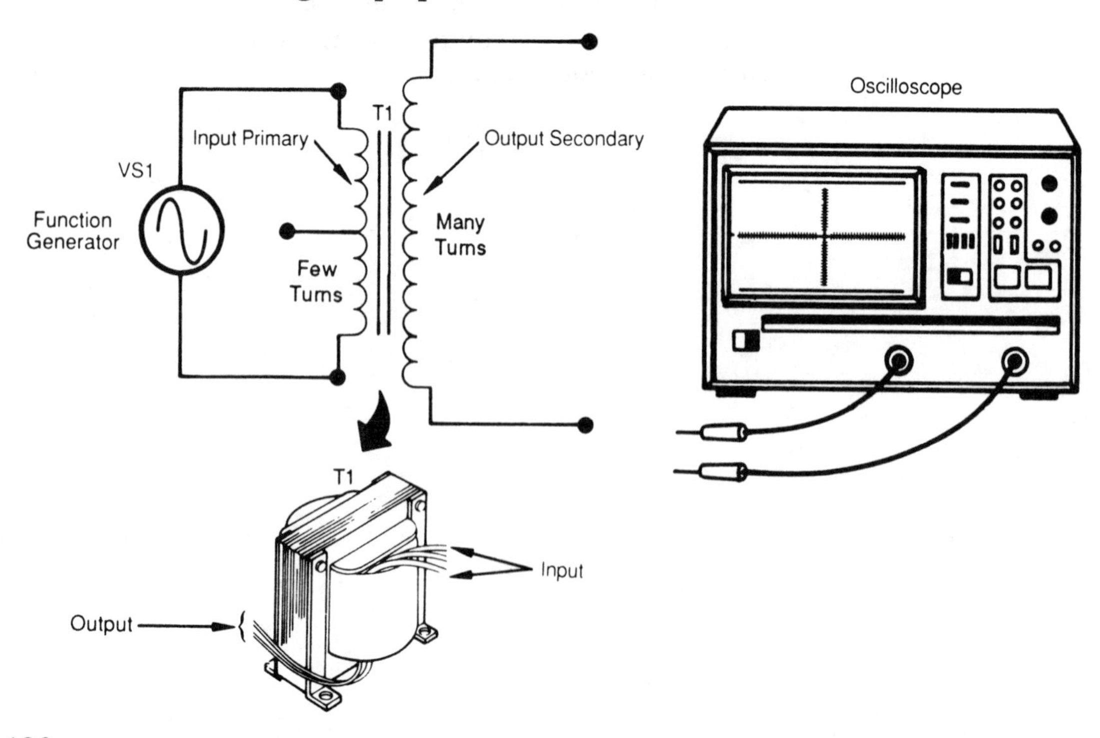

2. Set the function generator output to sinewave and using the oscilloscope to monitor the voltage across the primary, adjust the function generator's output level for a maximum output with minimum distortion.
3. Connect the oscilloscope to the transformer secondary winding, and adjust its settings for the best display while adjusting the frequency of the function generator to obtain a maximum output. Is the output waveform the same as the input?
4. Select, in succession, the triangular and square wave outputs from the generator. While viewing the transformer's output waveform on the scope, vary the generator's output frequency in each case to obtain the maximum peak to peak output. Is the output waveform the same as the input waveform in both cases? Is the maximum output voltage the same frequency as the input in both cases?
5. Turn off the function generator and then connect the neon lamp across the circuit transformer's secondary winding. Turn on the function generator. Does the lamp light up, and if it does, why do you think this happens?

CAUTION

High voltages will be present.

6. Construct the circuit shown in figure 37–2. **Be very careful as you proceed because electrical shock may result.**
7. Increase the output of the dc supply from 0v until the buzzer starts buzzing steadily. **Turn off dc supply.**
8. Carefully connect the oscilloscope to the secondary of the circuit transformer. Turn dc supply on and adjust the scope settings for the best display and then measure the amplitude of the output.
9. **Turn off the dc supply.** Connect the neon lamp across the secondary winding while leaving the scope lead attached.
10. Turn on the power supply and then study the scope waveform. Why has it changed? Is the neon lamp illuminated?

FIGURE 37–2 Stepping-up oscillations

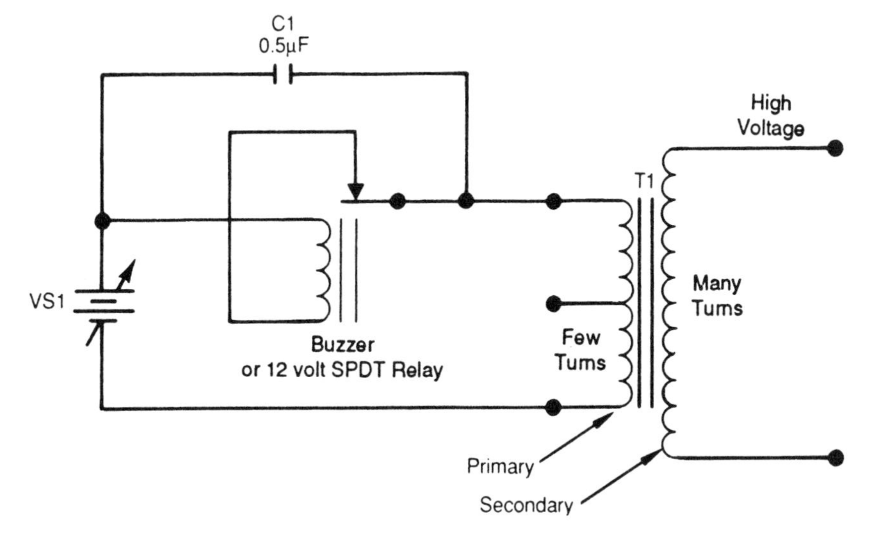

Review

You used the transformer, which is usually used to step 120V down to 12.6V, to prove that reverse operation is also possible. By using what is normally the secondary winding of the transformer as the primary in the test circuit it is possible to step voltage up.

You first connected the function generator to the transformer while using the scope to monitor primary and secondary waveforms. You varied the frequency of the generator and found that the transformer does havc definite frequency response characteristics. The frequency response of transformers will be measured with purpose in more advanced experiments.

You applied three waveforms to the transformer and discovered that the transformer responded differently to the three. You observed different peak to peak voltages in the secondary, at different frequencies and with some distortion of the waveform when the input was non-sinusoidal. But the main point to be made was that the secondary voltage was considerably greater in magnitude than the primary voltage.

When you connected the neon lamp across the secondary you may have seen a faint glow, but whether or not you did was dependent upon the characteristics of your function generator and the transformer.

In the second phase of the experiment you used the transformer with another kind of oscillator, the buzzer. You used the buzzer to produce pulsating dc, at the buzzer frequency, and discovered that pulsating dc transforms very well. When the neon lamp was connected across the secondary it illuminated quite brightly. While monitoring the secondary voltage with the scope you should have measured pulses of at least 100 volts, perhaps even as much as 300 volts. With the neon lamp connected the pulses were clipped to about 70 volts.

Option: Connect a small fluorescent tube across the transformer output—up to a 12" lamp.

ACTIVITY SHEET EXPERIMENT 37

NAME ______________________________

DATE ______________________________

Step 3 **A.** What is the maximum attainable output?

____________________ $V_{p\text{-}p}$

What is the frequency which produces the maximum output?

____________________ Hz

Is the output waveshape identical to the input?______________

Step 4 **B.** With the triangle wave what is the maximum output?

____________________ $V_{p\text{-}p}$

At what frequency is the maximum produced?

____________________ Hz

Is the output waveform identical to the input? ______________

With the square wave what is the maximum output?

____________________ $V_{p\text{-}p}$

At what frequency is the maximum developed?

____________________ Hz

Is the output waveform identical to the input? ______________

Step 5 **C.** Connect the neon lamp across the circuit transformer secondary winding. Use the three generator waveshapes at varying frequencies to see if you can make the lamp light up. Indicate below whether the lamp glowed:

Sinewave ____________________

Trianglewave ____________________

Squarewave ____________________

Step 8 **D.** What is the amplitude of the transformer output?

____________________ $V_{p\text{-}p}$

Step 10

E. With the neon lamp connected to the transformer secondary is it glowing brightly?____________________ Does one electrode in the lamp glow more brightly than the other?______________

____

Why has the shape of the waveform, as viewed on the oscilloscope, changed?____________________________________

Why is the amplitude reduced?___________________________

____

____

____

____

____

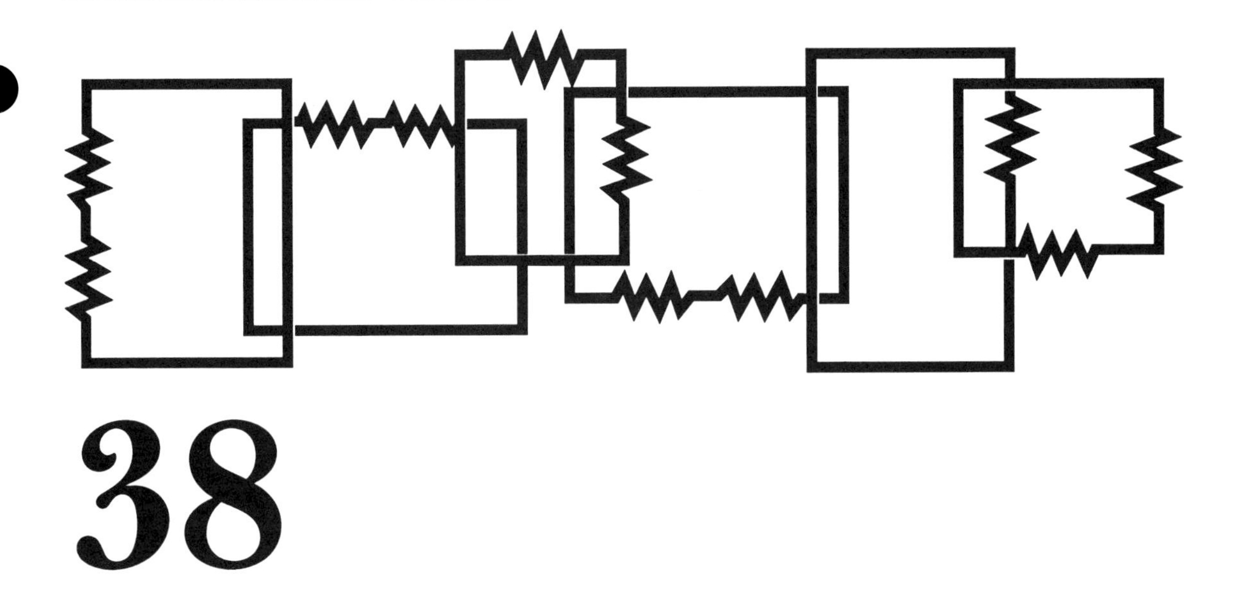

38

Impedance Matching

Objectives

After completing this experiment, you will be able to:

1. Demonstrate how a transformer can be used as an impedance matching device.
2. Show how a step-down transformer can be used to match the high impedance output from the function generator to the low impedance of the speaker.
3. Describe how impedance matching the source to the load is a more efficient power transfer.

Introduction

When it is important to transfer maximum power from a source to a load, it is necessary that the impedances of the two be closely matched. In this experiment you will observe how the transformer can serve as an impedance matching device. To perform this experiment you will need:

A function generator
A 120V:12.6V transformer
A loudspeaker

Procedure

1. Remove the activity sheet for this experiment. Connect the speaker directly to the function generator's output as shown in figure 38–1. Vary the generator's output amplitude, frequency and waveshape, and discern the audible characteristics.
2. Modify the circuit to include the transformer as shown in figure 38–2. The transformer is being used to step down the source voltage.
3. Vary the function generator's output amplitude, frequency and waveshape. How does the volume of the speaker compare to that produced by the previous circuit connection?

Review

The function generator has a relatively high output impedance (600Ω), so that when the speaker is connected directly to its output the resultant sound level is rather feeble. The impedance of the speaker is much lower (8Ω to 40Ω) which results in a very inefficient power transfer.

When the transformer is used as a matching device the volume at the speaker is significantly greater. The context of the generator power is changed from high voltage/low current in the primary circuit to low voltage/high current in the secondary, better matching the source to the load.

FIGURE 38–1 Impedance mismatch

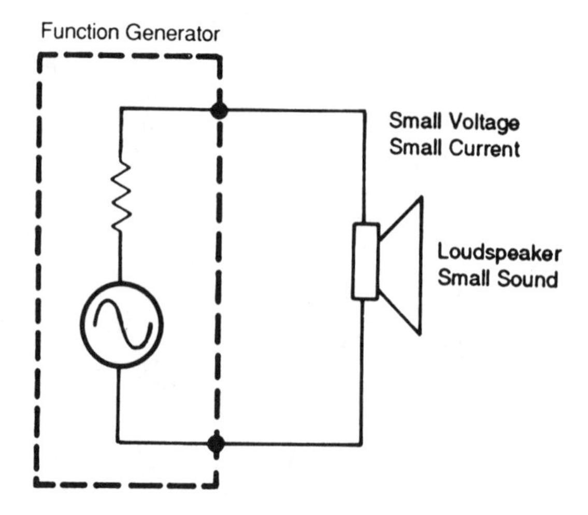

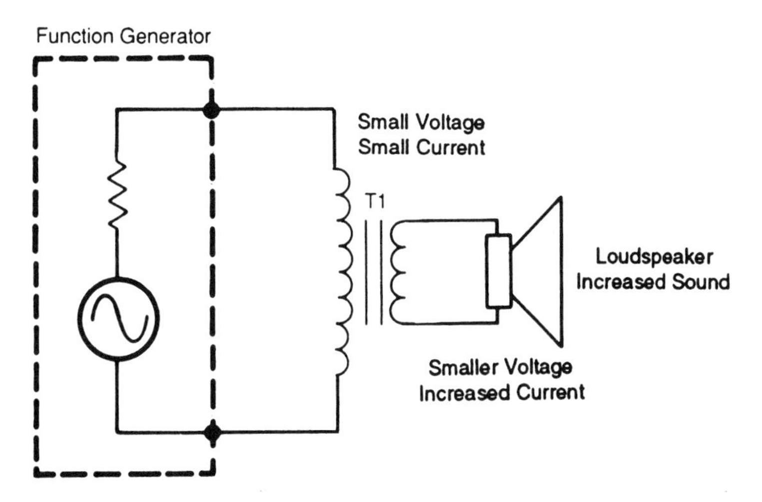

FIGURE 38–2 Close impedance match

ACTIVITY SHEET EXPERIMENT 38

NAME ______________________________

DATE ______________________________

Step 1

A. With the speaker connected directly to the function generator, is the audio signal loud? ____________________ Does the volume of sound produced vary with the output amplitude of the generator? ____________________ Why do you suppose the volume of the speaker is not too loud? ____________________________

____

____

Step 3

B. When the transformer is used as a matching device is the volume of the speaker louder? ____________________

Why is this so? ____

____

____

Would it be more efficient to use the transformer to step up the generator voltage? ____________________________

Why? ____

____

____

____

____

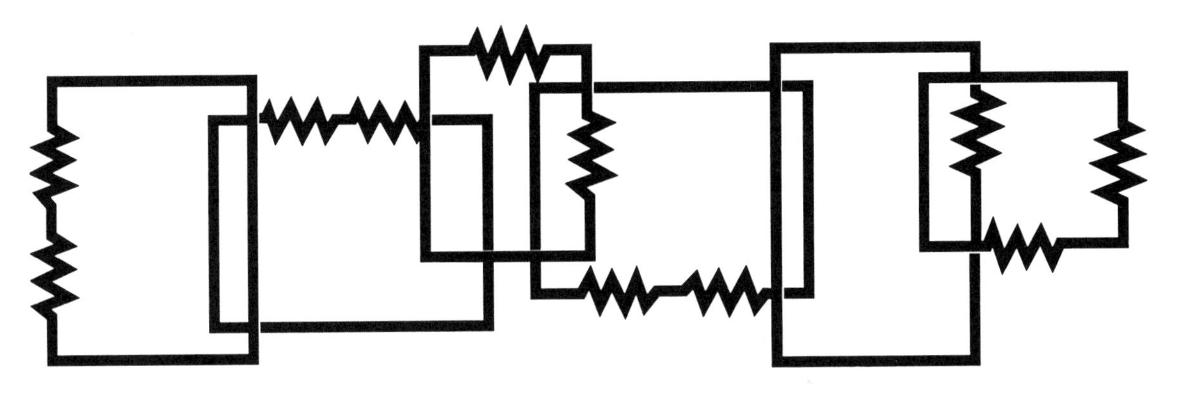

39

The RC Circuit— AC Characteristics

Objectives

After completing this experiment, you will be able to:

1. Demonstrate the ac characteristics of an RC circuit.
2. Show how the ac characteristics of an RC circuit can be measured with the oscilloscope.

Introduction

This experiment demonstrates the ac characteristics of an RC circuit and how these characteristics can be measured with the oscilloscope. To perform this experiment you will need:

A function generator
A dual trace oscilloscope
One 1kΩ resistor
One 0.1μF capacitor

Procedure

1. Remove the activity sheet for this experiment.
2. Construct the circuit seen in figure 39–1.
3. Set the function generator output to:
   Sinewave
   1Vpeak to peak
   500Hz
4. Set the oscilloscope controls as follows:
   A – B Mode
   AC sense
   A and B channels set to 200mV/cm
   0.5ms/cm
   (NOTE—When the oscilloscope is operated in the A minus B mode of operation (A - B) the waveform from channel B is subtracted from the channel A waveform and the difference is displayed as a single trace.)
5. When measuring circuit voltages with the oscilloscope do not relocate the ground connections. Use both of the probes as illustrated in figures 39–2, 39–3 and 39–4.
6. Measure and record the following voltages:
   V_{AC}–The source voltage across the circuit (figure 39–2)

FIGURE 39–1 The RC circuit with an ac input

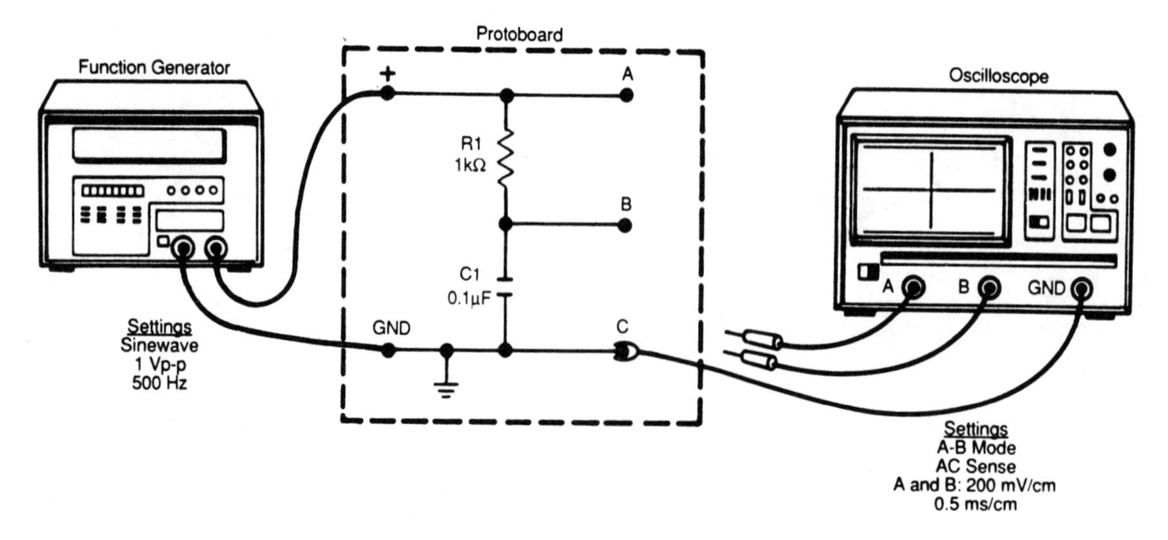

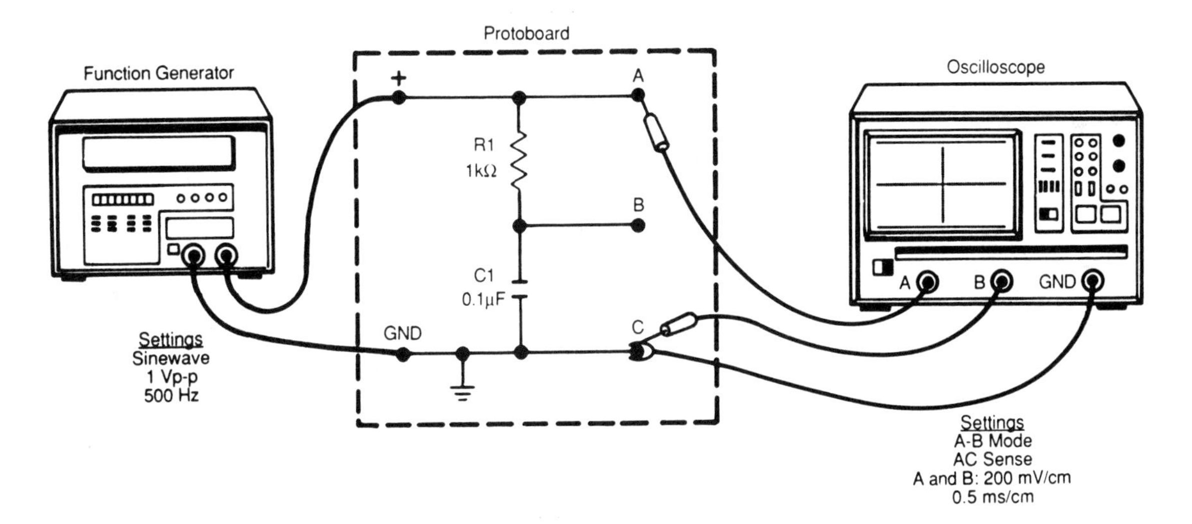

FIGURE 39–2

V_{AB}–The voltage across R1 (Figure 39–3)
V_{BC}–The voltage across C1 (Figure 39–4)
Does the sum of the voltage drops across R1 and C1 equal the source voltage?

7. Increase the function generator's frequency to 1KHz. Measure and record the voltages V_{AC}, V_{AB} and V_{BC}. Why has the voltage distribution changed?
8. Using the measured voltage drops across the resistor R1, calculate the circuit current at the two previously selected frequencies. Why does current flow increase with frequency?
9. Increase the function generator's frequency to 2KHz. Measure and record the voltages V_{AC}, V_{AB} and V_{BC}.
10. Using the measured voltage drop across the resistor, calculate the circuit current at 2KHz.
11. Compare circuit current with the different input frequencies. What have you noticed?

FIGURE 39–3

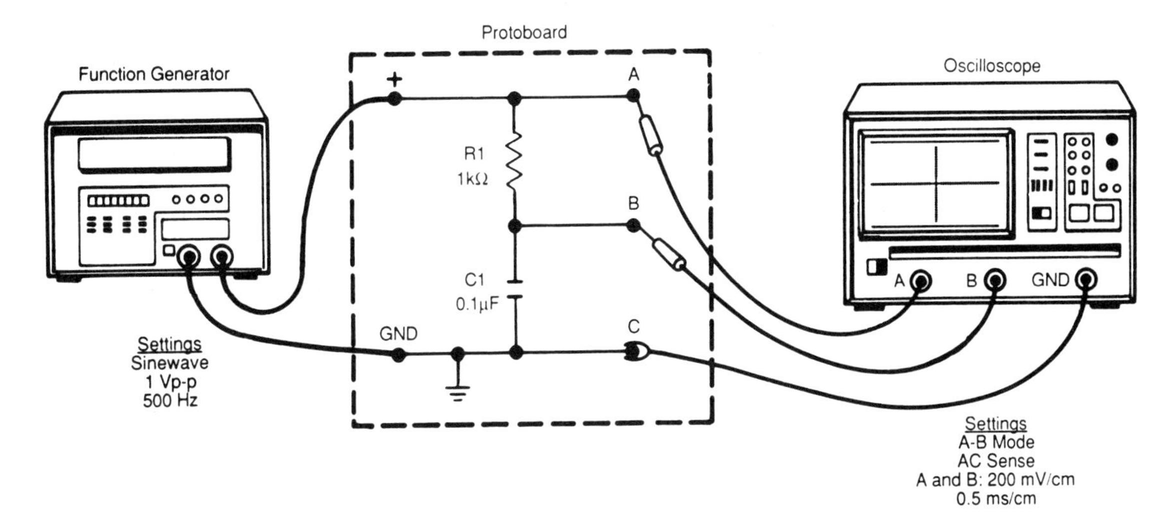

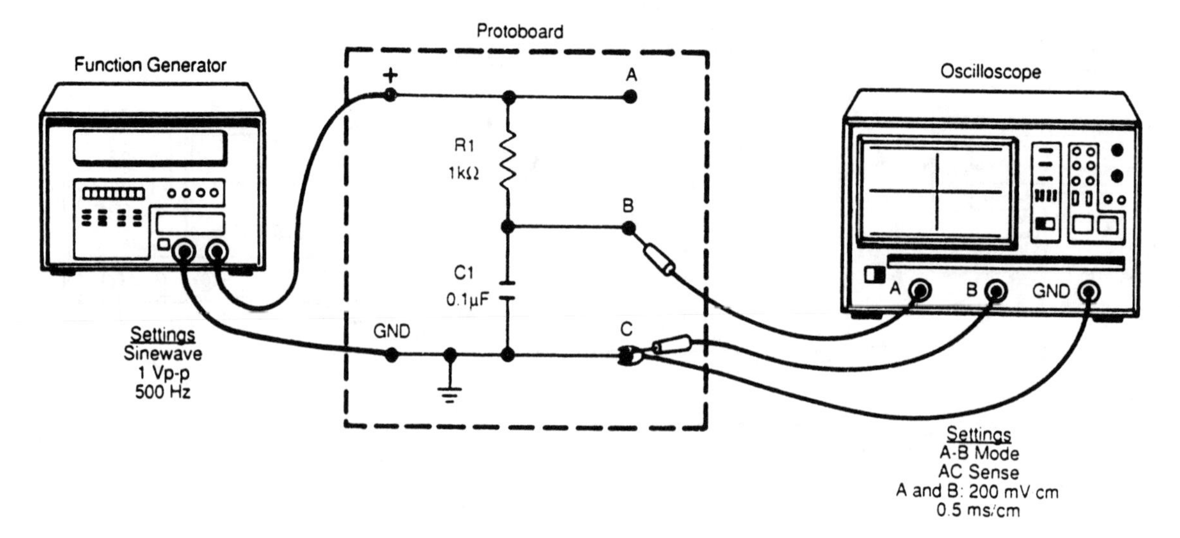

FIGURE 39–4

Review

You have constructed an RC series circuit and you have learned a new way to use the oscilloscope to accurately measure voltages across any component in the circuit. The differential mode method is especially valuable when measuring voltages in circuits which are referenced to ground. Using this method allows you to safely measure the voltages without any harm to the circuit since the scope ground connection is always connected to the circuit ground.

After measuring the circuit voltages you discovered that the sum of the resistive and capacitive voltages was greater than the source voltage. Because the two voltages are out of phase this is a normal characteristic. To correctly add the voltages you must incorporate the phase angle into the addition. Then the sum will be equal to the source and Kirchhoff's voltage law will be found to apply to ac circuits as well as dc circuits. Tricky but true.

You increased the frequency of the generator and again measured the circuit voltages. You noted that the voltage distribution in the circuit had changed. This indicates that the circuit is frequency sensitive—at different frequencies the voltages across the series components will change.

You calculated circuit current at the two frequencies by use of Ohm's Law. You found that at the higher frequency the circuit current was increased.

You increased the generator frequency once again and duplicated the previous measurements, observing another change in voltage distribution and a further increase in circuit current. It all has to do with capacitive reactance. Can you figure it out?

ACTIVITY SHEET EXPERIMENT 39

NAME ______________________________

DATE ______________________________

Step 2 **A.** Measure and record the actual resistance of the 1000Ω resistor:

R = ____________________ Ω

Step 3 Adjust the output of the function generator to 1 volt peak to peak while using the oscilloscope to display the signal. The generator should be connected to the circuit with its frequency set to approximately 500 Hz.

Step 6 **B.** With the scope set up to operate in the differential mode, measure and record all of the circuit voltages:

V_{R1} ____________________ V_{p-p} V_{C1} ____________________ V_{p-p}

V_{source} ____________________ V_{p-p}

Is the sum of the voltage drops equal to the source voltage?

____________________ Why is this so? ____________________

____

Step 7 **C.** At the new generator frequency, measure and record all circuit voltages:

V_{R1} ____________________ V_{p-p} V_{C1} ____________________ V_{p-p}

V_{source} ____________________ V_{p-p}

Why has the voltage distribution changed? ____________________

____

Step 8 **D.** Using the measured resistive voltage drops, and the actual value of R1, calculate circuit current at the two frequencies.

I_{500Hz} ____________________ mA_{p-p}

I_{1000Hz} ____________________ mA_{p-p}

Step 9 **E.** Measure and record all circuit voltages at 2KHz

V_{source} ____________________ V_{p-p}

V_{R1} ____________________ V_{p-p}

V_{C1} ____________________ V_{p-p}

Step 10

F. Calculate circuit current: I_{2000Hz} ______________________ $mA_{p\text{-}p}$

G. How does circuit circuit current vary with frequency? ________

Why is this so? ____

____

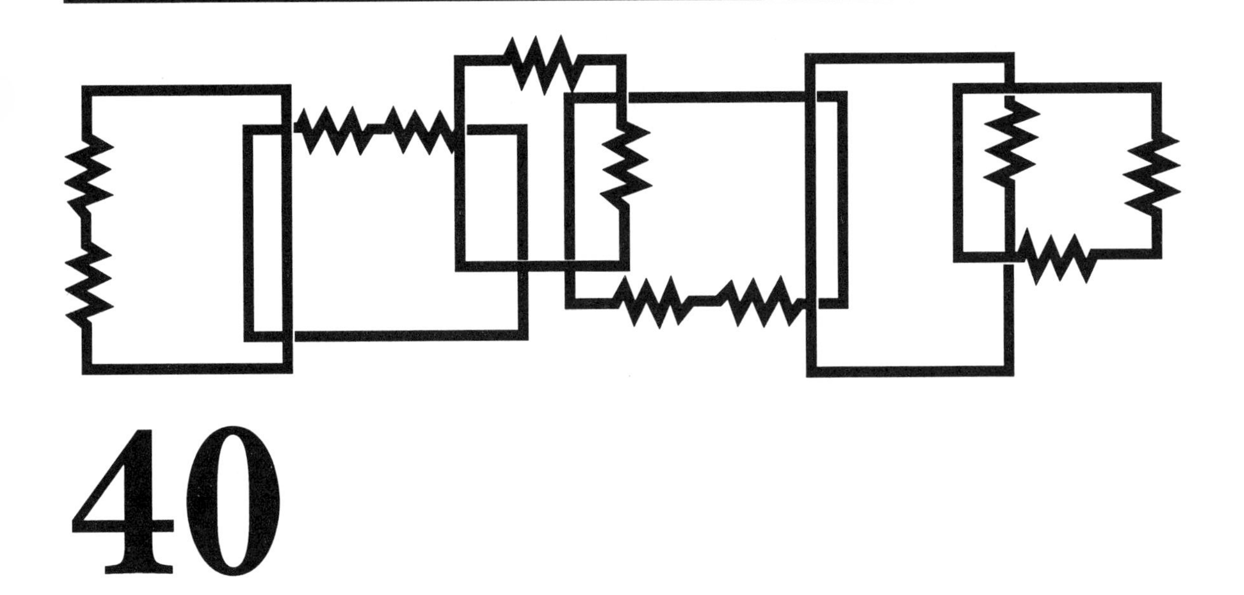

40

Impedance Measurement

Objectives

After completing this experiment, you will be able to:

1. Demonstrate by measurement the impedance of an RC circuit.
2. Prove that Ohm's law applies to both resistive and reactive circuits.
3. Describe how a reactive circuit is frequency sensitive.

Introduction

To see something with your own eyes is in most cases the best way to understand. This experiment is designed to verify by measurement the impedance of an RC circuit. To perform this experiment you will need:

A function generator
An oscilloscope with two probes
One 1kΩ potentiometer
One 560Ω resistor
One 0.1μF capacitor
Hookup wire as necessary

Procedure

1. Remove the activity sheet for this experiment.
2. Construct the circuit seen in figure 40–1.
3. Connect the function generator to the circuit and select a 2 V_{p-p} (not critical) sinewave output at a frequency of 2KHz.
4. Set oscilloscope up to function as differential voltmeter (A-B). Connect the scope probes between points A and C and then adjust the settings for the best possible display.
5. Adjust the frequency of the function generator until one cycle of the displayed waveform is exactly equal to the calculated period. (This will accurately set the output of the function generator to 2KHz.)
6. Switch the two scope leads between A and B, and B and C, while adjusting R1. The objective is to produce two equal amplitude waveforms between points A to B and from B to C, by adjusting R1. Once the voltage across A-B and B-C are equal, the resistance of R1 will equal the impedance of the RC test circuit.

FIGURE 40–1 RC circuit impedance

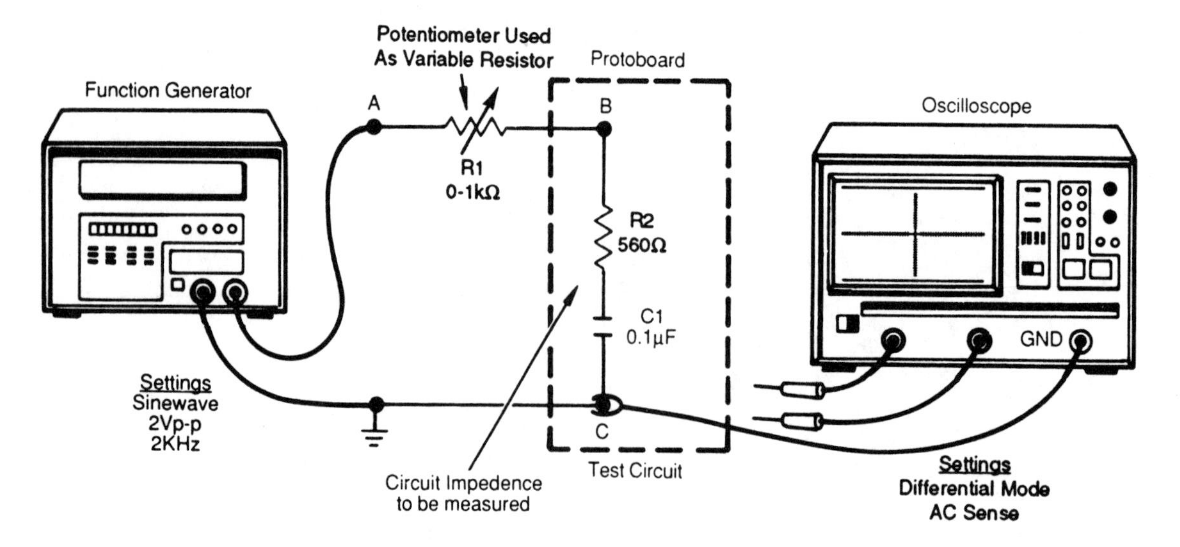

7. Now that the measured voltages are equal, carefully disconnect R1 from the circuit and measure its resistance with the digital ohmmeter. Record the measured value.
8. Using the coded values of R2 and C_1, calculate the impedance of the circuit. Compare the calculated and measured values.
9. Connect up the original circuit with R1 and then set the generator output to 3KHz accurately by using the oscilloscope. Repeat steps 6, 7 and 8.
10. Set the generator output to exactly 4KHz and repeat the process once again.

Review

In this experiment you utilized a simple technique to find the impedance of a series RC test circuit. The success of this method proves that ohm's law applies to both resistive and reactive circuits (V = IR, V = IZ). In a series connected circuit a common current flows throughout and when the voltages measured across two different parts of the circuit are equal, the ohmic opposition to current flow in both parts of the circuit is also equal. This measurement technique will work with any circuit, series or parallel, simple or complex, capacitive or inductive. In this experiment you found that the measured impedance and the calculated impedance were very close. Another important fact that you observed was that a circuit containing reactance is frequency sensitive. This means that the impedance of the circuit varies with the frequency of the applied source power. Remember this impedance measurement technique as it will serve you well in more advanced circuit studies.

FIGURE 40–2

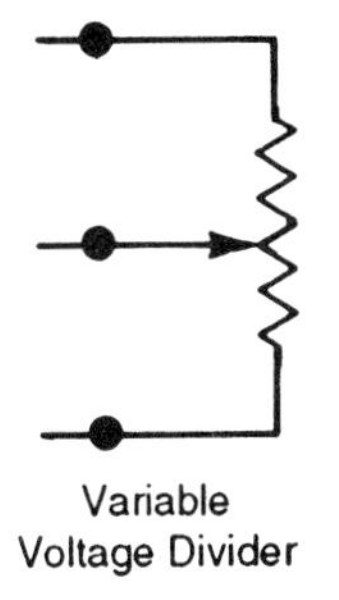

Variable
Voltage Divider

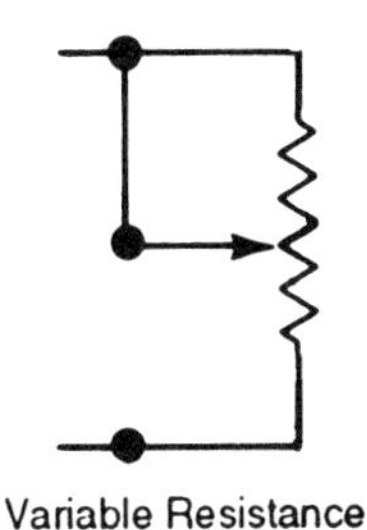

Variable Resistance

ACTIVITY SHEET EXPERIMENT 40

NAME ______________________________

DATE ______________________________

Step 3 **A.** Set the output of the generator to about $2V_{p\text{-}p}$ while monitoring the output of the generator with the oscilloscope. The amplitude of the signal applied to the circuit is not critical but the frequency is.

Step 5 **B.** Calculate the period of one cycle at 200 Hz. Adjust the frequency of the generator while monitoring the waveform on the scope until the time of one cycle is exactly equal to the period of 2 KHz.

Step 6 **C.** Using the scope in the differential mode alternately measure the voltage across R1 and the series RC circuit. Adjust R1 until the two voltages are exactly equal. R1 resistance is then equal to the impedance of the RC series circuit.

Step 7 **D.** Carefully remove R1 from the circuit and measure its resistance with the digital ohmmeter. Record in the table below.

Step 8 **E.** Using the coded values of R2 and C1, and the generator frequency, calculate the impedance of the series RC circuit. Record the calculated value in the table below.

Steps 9-10 **F.** Repeat the above steps at exact frequencies of 3 KHz and 4 KHz. Record data in table below.

Frequency	Measured Impedance	Calculated Impedance
2 KHz	Ω	Ω
3 KHz	Ω	Ω
4 KHz	Ω	Ω

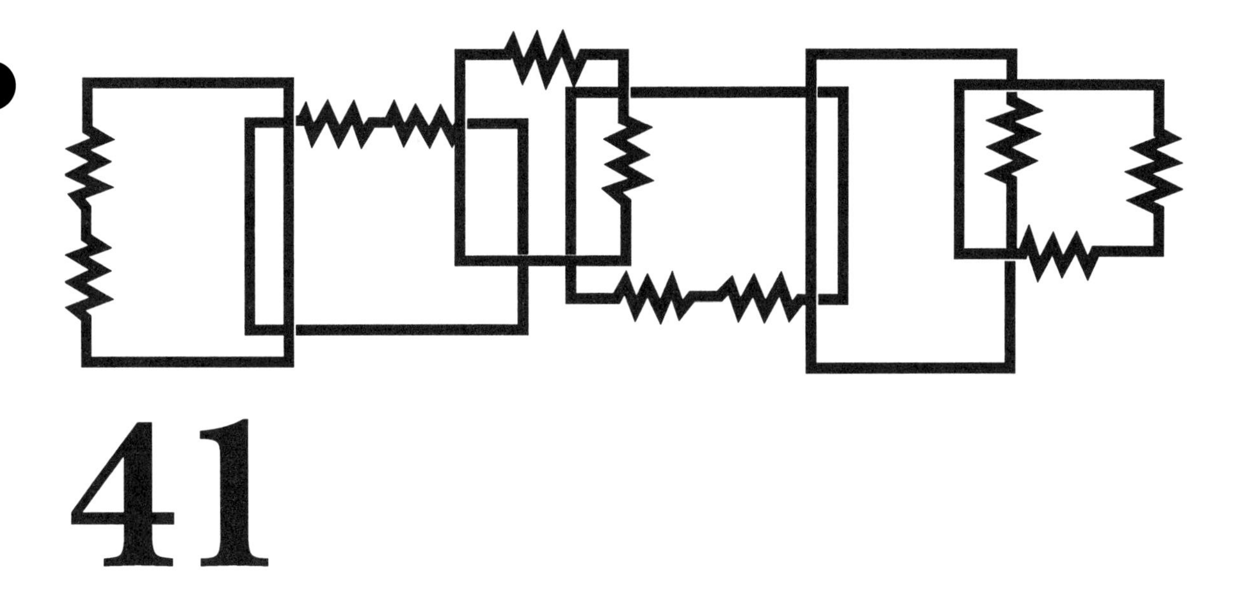

41

Phase Angle

Objectives

After completing this experiment, you will be able to:

1. Demonstrate how to determine the phase angle of an RC circuit through the use of:
   a. A dual trace oscilloscope
   b. Trigonometric calculation based on accurate voltage measurements.
2. Prove that a series RC circuit with an applied ac signal has a characteristic which is common to all series circuits.

Introduction

There are two ways to determine the phase angle of an RC test circuit. The first is by direct measurement using a dual trace oscilloscope and the second is by trigonometric calculation based on accurate voltage measurements. This experiment will explore both methods. To perform this experiment you will need:

A dual trace oscilloscope
A function generator
One 560Ω resistor
One 0.1μF capacitor

Procedure

1. Remove the activity sheet for this experiment and construct the circuit seen in figure 41–1.
2. Set the function generator output to:
   Sinewave
   3KHz
   2Vp-p
   Use the oscilloscope to verify these settings and once they are all accurate, do not change them for the remainder of this experiment.
3. Set the oscilloscope to:
   Dual trace mode (Alt or chopped)
   0.5V/cm
   100μS/cm
4. Connect the A oscilloscope probe to circuit point A and the B oscilloscope probe to circuit point B. Ground is connected to point C.
5. Set the oscilloscope's internal trigger select switch to input A (oscilloscope sweep is triggered by the channel A signal input).
6. Adjust the oscilloscope's triggering level adjustment for a stable display and use the vertical positioning controls for both chan-

FIGURE 41–1 RC test circuit

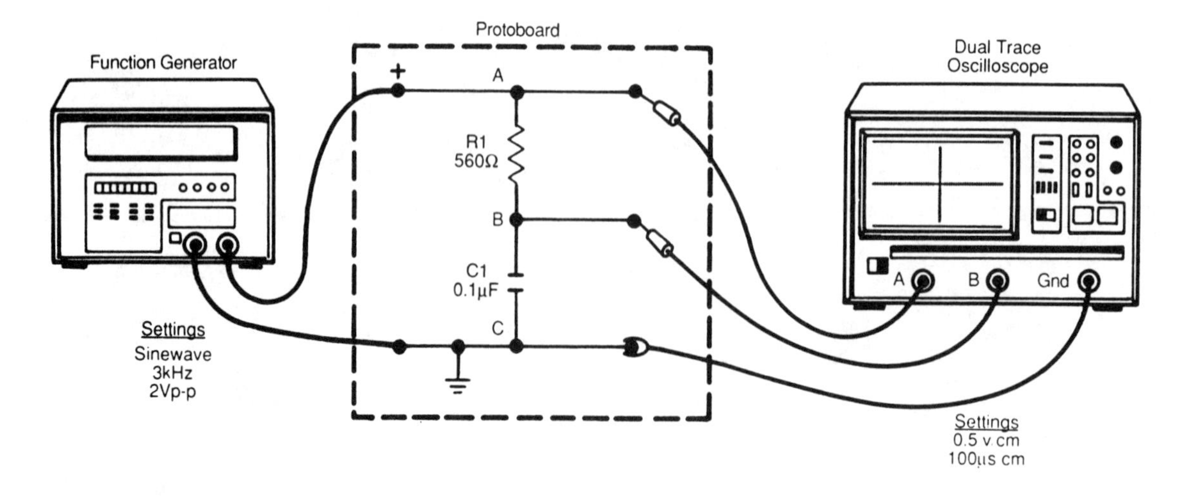

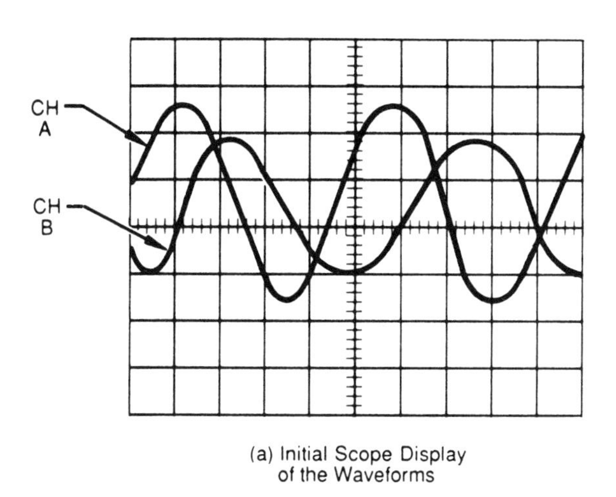

(a) Initial Scope Display of the Waveforms

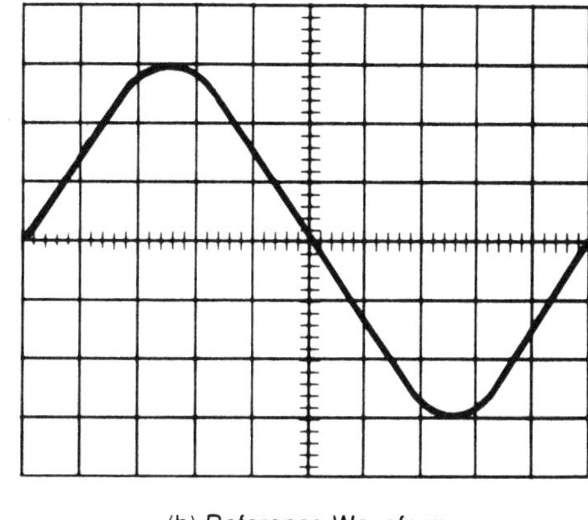

(b) Reference Waveform Channel A

FIGURE 41–2 Scope displays

nels to align the waveforms A and B on the scope display as seen in figure 41–2(a). Both waveforms should be clearly visible and it should be possible to see a delay between the waveforms indicating that they are not in phase with one another.

7. Select channel A only on the oscilloscope to make a reference adjustment in preparation for measuring the phase difference between the two signals. Adjust the vertical sensitivity so that the displayed source voltage is about 6 cm peak to peak on the scope graticule, as seen in figure 41–2(b).

8. With channel A only still selected, adjust the time base to cause one cycle of the displayed waveform to just fill the 10 cm width of the screen graticule, as seen in figure 41–2(b).

9. Switch back to the dual trace mode and adjust the vertical sensitivity of channel B until the displayed peak to peak value is exactly the same as that of channel A. Looking at the scope display you should be able to see that the channel B (capacitive voltage) waveform is lagging the channel A (source voltage) waveform, as seen in figure 41–3.

FIGURE 41–3 Capacitive voltage/source voltage lag

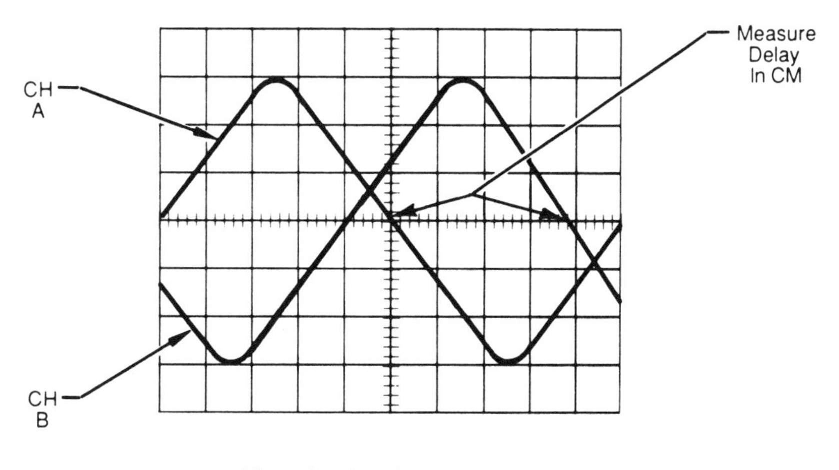

Measuring delay between waves from scope display.

10. Using the graduated baseline, which runs horizontally through the center of the scope, measure the displacement between waveform A and B, as seen in figure 41–3.
11. Follow the instructions on this experiment's activity sheet to determine the phase angle of the circuit using your measurements.

Review Break

In this section of the experiment you used the oscilloscope to view two waveforms from an ac circuit so that the phase angle of the circuit could be measured. The oscilloscope was used in the dual trace mode so that the two waveforms could be viewed simultaneously. Before measuring the delay between the waveforms, a reference had to be established. The source voltage waveform A was used as a reference and the display of this waveform on the scope was adjusted so that one cycle of the wave filled the 10cm graticule of the screen horizontally. The vertical size of the waveform on the display was also adjusted for increased accuracy of the subsequent measurement. Both waveforms were then displayed, and channel B was adjusted to make its vertical size on the scope display equal to the size of waveform A on the display to get the most accurate delay measurement. The displacement of wave B from wave A along the horizontal axis of the scope graticule was measured and by following instructions on the activity sheet your measurement was converted to the circuit phase angle.

In the last part of this experiment you will use another method to determine the phase angle and the result can then be used to check your first result.

12. Change the mode setting of the oscilloscope so that it acts as a differential voltmeter. Be sure to put the vertical sensitivity adjustments back to their calibrated positions. Also put the time base back into its calibrated position.
13. Using the A and B probes, accurately measure all circuit voltages.
14. Follow the instructions on the activity sheet to determine the phase angle of the circuit trigonometrically from the measured voltages.
15. Compare the results of both methods. Are the phase angles the same? Which method best shows the phase relationship? Which method is easiest to use?

Review

The series RC circuit with an applied ac signal has a characteristic which is common to ALL series circuits: circuit current through each component in the series string is exactly the same. Therefore in the experimental circuit the current flow is exactly the same in all respects as it flows from the source through the resistor on through the capacitor. Since the capacitor is a reactive component, the voltage developed across the capacitor lags current flow by 90°. A plot of circuit values in phasor form is shown in Figure 41–4.

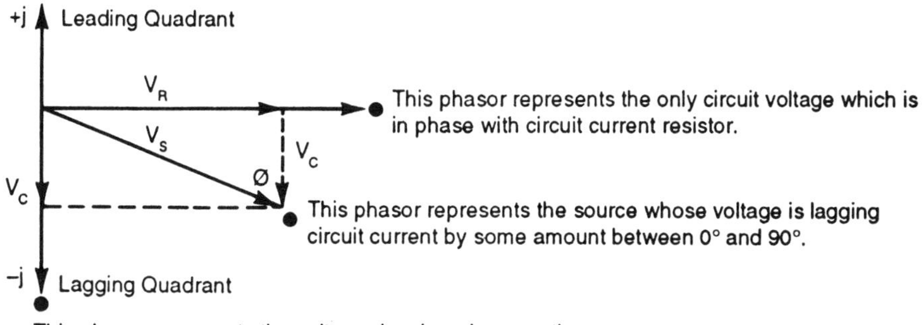

FIGURE 41–4

The circuit phase angle θ is the amount of angular displacement of source voltage from source current. Source voltage magnitude, in accordance with Kirchhoff's law, is the phasor sum of the capacitive and resistive voltages.

ACTIVITY SHEET EXPERIMENT 41

NAME ______________________________

DATE ______________________________

Step 10 **A.** The measured displacement in centimeters from waveform A to waveform B: ____________________ cm

Step 11 **B.** The reference wave is exactly 10 cm long on the scope display therefore 10 cm = 360°. To convert the measured delay to an angle plug your measurement into the formula

$$\frac{10\text{ cm}}{360°} = \frac{\text{delay cm}}{\angle\phi} \quad \text{By transposition } \angle\phi = \frac{\text{delay cm}(360°)}{10\text{ cm}}$$

Calculate angle ϕ: ____________________ °

The circuit phase angle, $\angle\theta = 90° - \angle\theta$.
This is because angle ϕ is the complement of angle θ.

Calculate angle ϕ: ____________________ °

Step 13 **C.** Measure and record all circuit voltages:

V_S ____________________ V_{pp}

V_R ____________________ V_{pp}

V_C ____________________ V_{pp}

Step 14 **D.** By trigonometry $\angle\theta = \arctan \frac{V_C}{V_R} = \arcsin \frac{V_C}{V_S} = \arccos \frac{V_R}{V_S}$

With your calculator determine the phase angle, $\angle\theta$, by all three methods.

$\angle\theta = \arctan \frac{V_C}{V_R} =$ ____________________ °

$\angle\theta = \arcsin \frac{V_C}{V_S} =$ ____________________ °

$\angle\theta = \arccos \frac{V_R}{V_S} =$ ____________________ °

E. Are the results of both methods equivalent? ____________________

Why is this so? ____

Which method is easier to use?___________________________

Which method shows the phase relationships best?___________

____

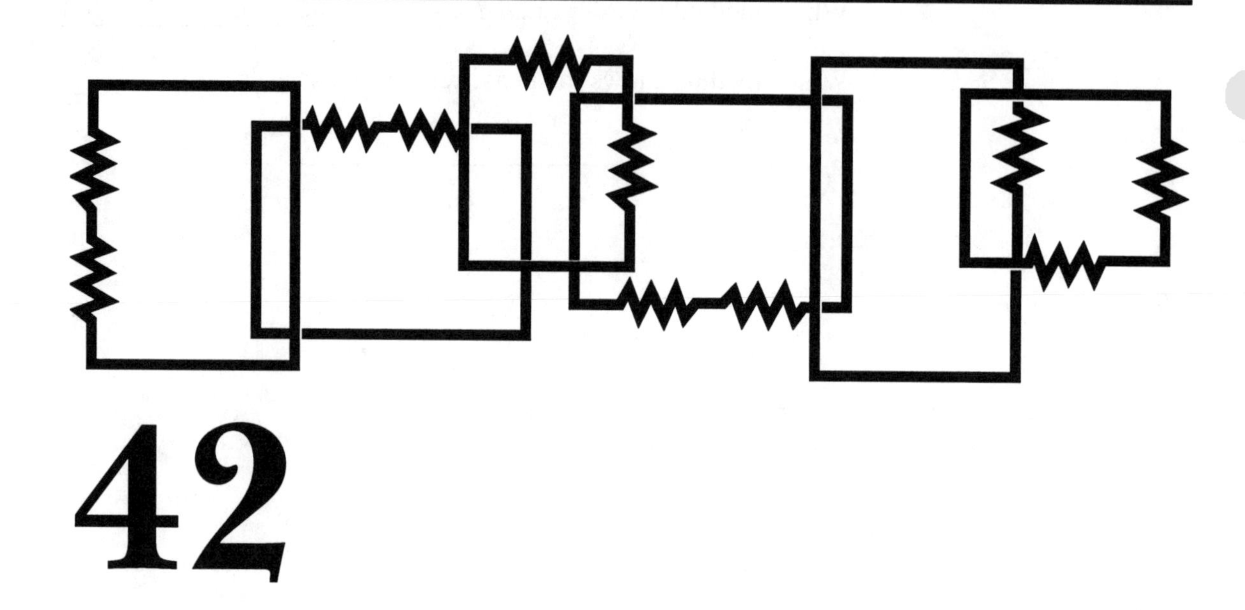

42

The RLC Circuit

Objectives

After completing this experiment, you will be able to:

1. Demonstrate the properties of a non-resonant RLC circuit.
2. Compare a series RLC circuit to a parallel RLC circuit.

Introduction

In this experiment you will examine the properties of non-resonant RLC circuits, both series and parallel. To perform this experiment you will need:

A function generator
An oscilloscope
A 120V:12.6V transformer
A digital multimeter
A 330Ω resistor
A 0.5μF capacitor
A 5kΩ potentiometer

Procedure

1. Pull the activity sheet for this experiment and then construct the circuit shown in figure 42–1.
2. Connect the scope leads to the function generator terminals. Set the function generator's frequency to about 200 Hz and then adjust the output level to produce a $4V_{pp}$ sinewave.
3. Measure and record all the circuit voltages on the activity sheet.
4. Connect the scope leads to the function generator once again and then adjust the frequency and amplitude to 600Hz, $4V_{pp}$.
5. Measure and record all circuit voltages.
6. Connect the scope leads to the function generator once again. Adjust frequency and amplitude to produce a 2000Hz, $4V_{pp}$.
7. Measure and record all circuit voltages.
8. Follow instructions on the activity sheet.
9. Construct the circuit shown in figure 42–2.
10. Set the potentiometer R1, which is functioning as a rheostat to 0Ω.
11. Connect the scope leads to the function generator's output and then set the function generator frequency to 200Hz. Adjust the output level to produce a 4Vpp sinusoid.
12. Alternately monitor the voltages across R1 and the parallel RLC circuit as you adjust R1. Once the voltage drop across R1 is

FIGURE 42–1 Series RLC measurement

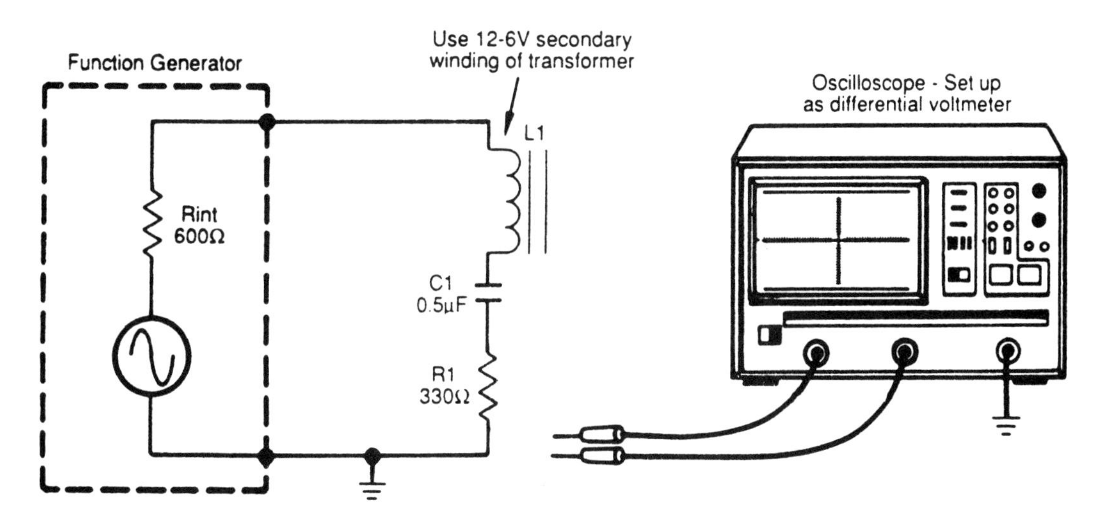

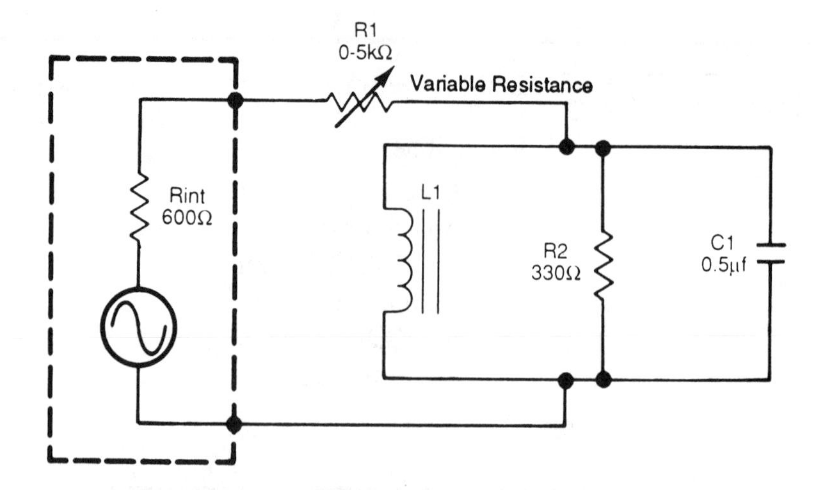

FIGURE 42–2 Parallel RLC measurement

equal to the voltage drop across the RLC circuit, carefully disconnect R1 and measure its resistance with the digital ohmmeter.

13. Record this resistance, which is equal to the RLC circuit impedance, on the activity sheet.
14. Repeat steps 10 through 13 for frequencies of 600Hz and 2000Hz. Record these measurements on the activity sheet.

Review

You have, through a series of voltage measurements, evaluated the performance of a series RCL circuit at three different frequencies. You found through your measurements that the circuit is frequency sensitive, and that the reactive voltages varied in proportion to the respective reactances. As the frequency of the signal applied to the test circuit was increased you found that the capacitive voltage decreased as the inductive voltage increased.

By use of OHM'S Law, and the voltage dropped across the series resistance, you were able to calculate circuit current at each of the different frequencies. Then you were able to calculate the circuit impedance based upon the total circuit voltage and the circuit current, also by use of OHM'S Law.

After analyzing the series connected circuit, you connected the same circuit components in parallel, with a series connected rheostat to use in determining the new circuit impedance. Then you evaluated the impedance of the parallel connected circuit at the same three frequencies. You discovered that even though the circuit was made up of the same components, just connected differently, the measured impedances were very much different.

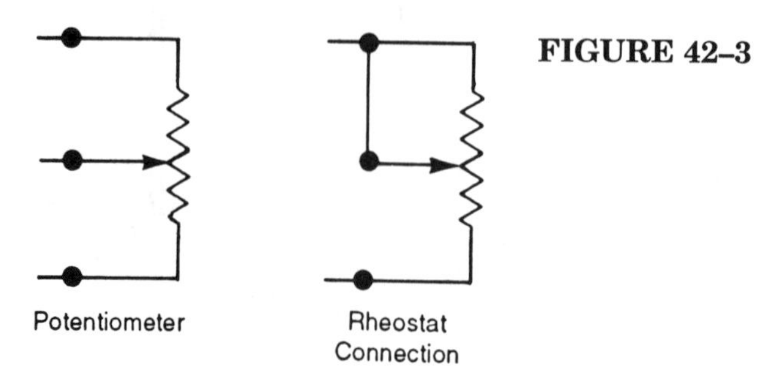

FIGURE 42–3

ACTIVITY SHEET EXPERIMENT 42

NAME ______________________

DATE ______________________

Steps 2, 5, & 7 **A.** Record measured voltages below:

Frequency	V_{GEN}	V_{L1}	V_{C1}	V_{R1}
200 Hz	4 $V_{p\text{-}p}$	$V_{p\text{-}p}$	$V_{p\text{-}p}$	$V_{p\text{-}p}$
600 Hz	4 $V_{p\text{-}p}$	$V_{p\text{-}p}$	$V_{p\text{-}p}$	$V_{p\text{-}p}$
2000 Hz	4 $V_{p\text{-}p}$	$V_{p\text{-}p}$	$V_{p\text{-}p}$	$V_{p\text{-}p}$

Step 8 **B.** Calculate circuit current for the three conditions by Ohm's Law:

$I = \frac{V}{R}$:

$I_{circuit}$: ______________ $mA_{p\text{-}p}$ ______________ $mA_{p\text{-}p}$
200 Hz 600 Hz

______________ $mA_{p\text{-}p}$
2000 Hz

Calculate circuit impedance for the three conditions by Ohm's Law: $Z = \frac{V}{I}$:

Z circuit 200 Hz ______________ Ω

600 Hz ______________ Ω

2000 Hz ______________ Ω

Steps 12-14 **C.** Record measurements below:

Z circuit 200 Hz ______________ Ω

600 Hz ______________ Ω

2000 Hz ______________ Ω

D. Compare the impedance measurements above. Both circuits are composed of the same components and are operated at the same frequencies in each case. Explain why the impedances are not

the same: ____

____

____

____

____

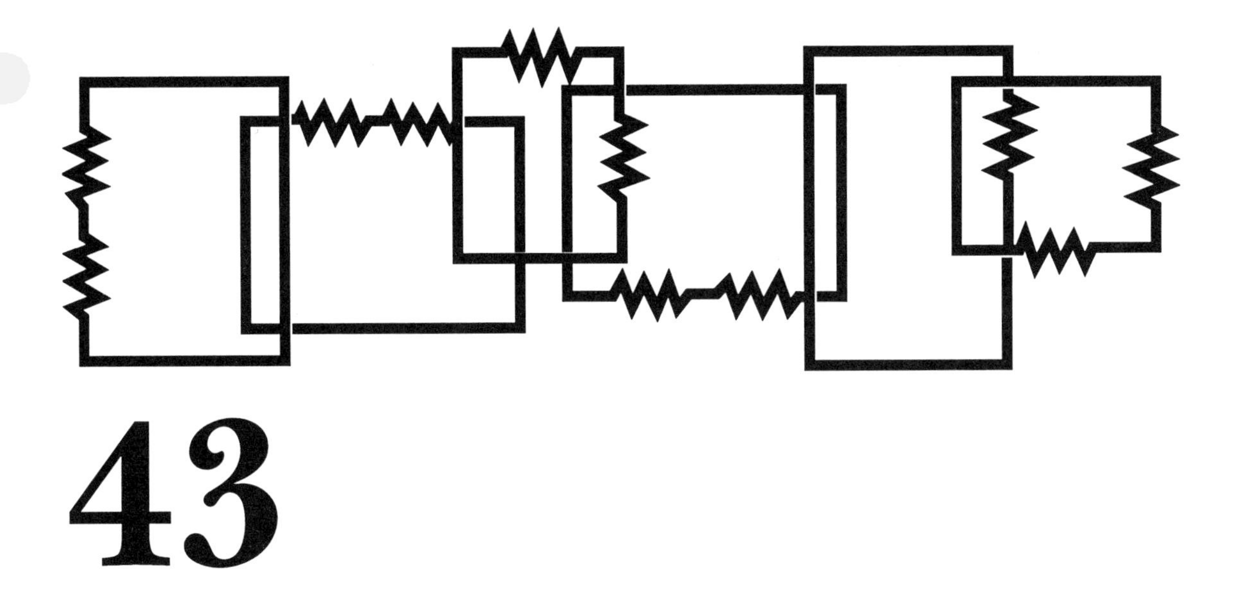

43

Resonance—Characteristics

Objectives

After completing this experiment, you will be able to:

1. Describe the characteristics of resonant circuits.
2. Evaluate a series and parallel resonant circuit.

Introduction

It is the purpose of this experiment to examine the characteristics of resonant circuits, both series and parallel. To perform this experiment, you will need:

A function generator
Two 15Ω resistors
A 120V:12.6V transformer
A 1μF capacitor
An oscilloscope

Procedure

1. Remove the activity sheet for this experiment and then construct the circuit shown in figure 43–1.
2. Set the oscilloscope to operate in the differential mode. (A-B)
3. Place the scope probes across the complete circuit to monitor the generator output. Set the function generator to:
   Sinusoidal output
   500Hz
   8V peak to peak
4. While looking at the scope display, vary the generator's output frequency from 500Hz up to about 20KHz. At some frequency within this range, the voltage magnitude displayed will decrease to a minimum. Rock the frequency dial on the generator back and forth around this frequency to make sure that you have found the "dip." This is the circuit's resonant frequency.
5. Connect the scope probes across R1 and adjust the scope controls for the best display. Again, rock the frequency control on the function generator above and below the resonant frequency. How does the scope display change? Measure the frequency at resonance. $f = \frac{1}{T}$
6. Construct the circuit shown in figure 43–2.

FIGURE 43–1 Resonant series circuit

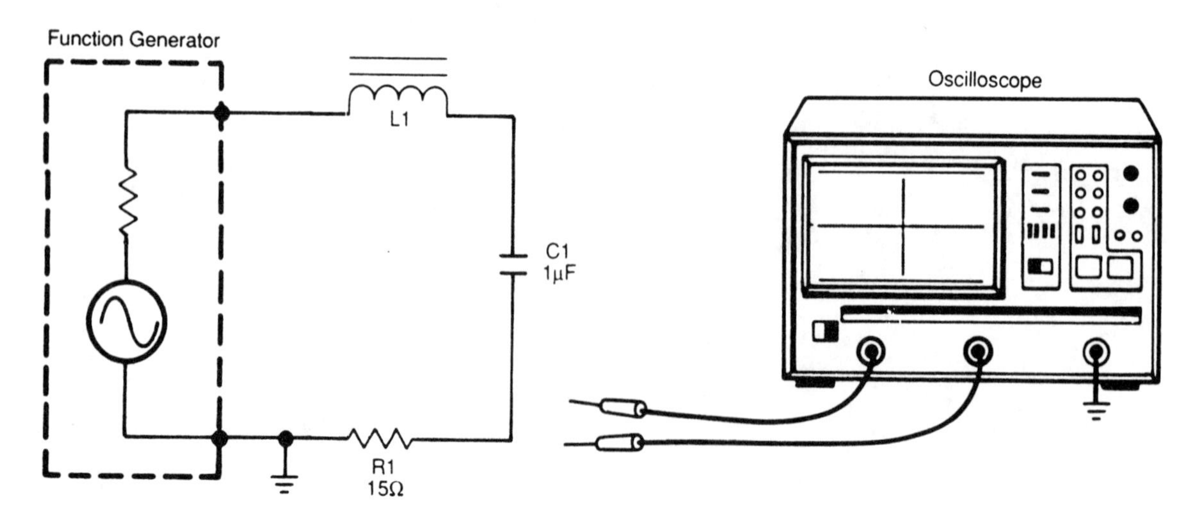

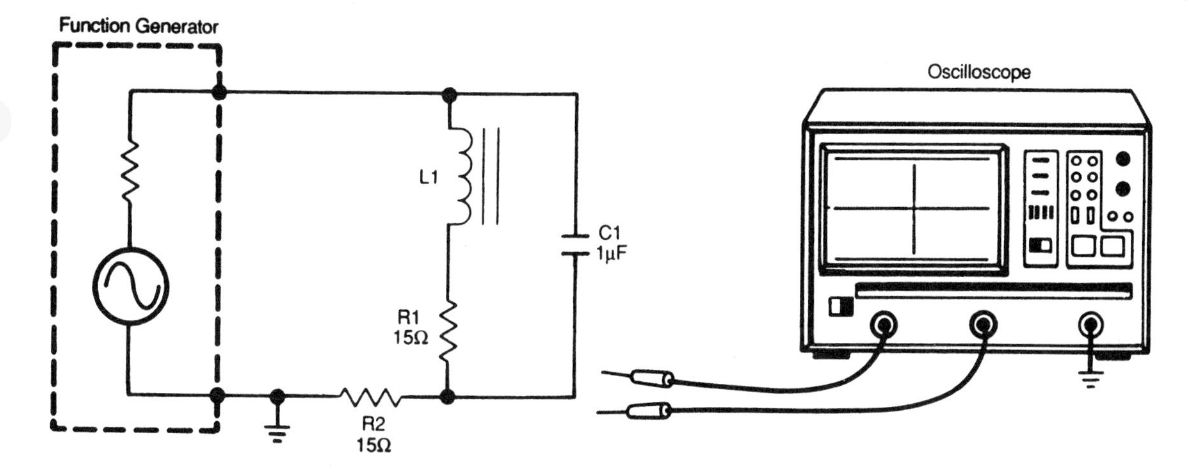

FIGURE 43–2 Resonant parallel circuit

7. Place the scope probes across the circuit to monitor the generator's output and then adjust the generator's frequency while observing the display voltage. What do you observe on the scope at the resonant frequency of the circuit?
8. Measure the frequency of the generator at resonance. Is it the same as the series connected circuit?
9. Connect the scope probes across R1. Measure and record the voltage at resonance.
10. Connect the scope probes across R2. Measure the voltage at resonance. Why are the two voltages so different?

Review

You've just tested two resonant circuits made from the same components. One was series connected and the other parallel connected. You found that both have resonant conditions which are opposite. By measurement you observed that the series circuit, at resonance, permits maximum current flow and therefore has its least opposition to current flow at the resonant frequency.

When parallel connected you discovered that, at resonance, the circuit presented maximum opposition to current flow. This was indicated by the voltage drop across R2 which is proportional to source current. At the same frequency the voltage drop across R1 was considerably greater, proving that the circulating current is many times greater than the source current.

Resonant circuits are very useful, and important, to the broad field of electronics. You will encounter them numerous times in the future as you proceed onto more advanced applications.

ACTIVITY SHEET EXPERIMENT 43

NAME ______________________________

DATE ______________________________

Step 4

A. Once you have located resonance how does the scope display change?

____

____

Step 5

B. How does the voltage observed across R1 appear at the resonant frequency?

____

____

By measurement, fr = ____________________ Hz fr = $\frac{1}{T}$

Step 7

C. At the resonant frequency of the circuit what does the scope display show you?______________________________

____

Step 8

D. By measurement, fr = ____________________ Hz. Are the two resonant frequencies equal?______________________________

Steps 9–10

E. Measure and record the voltage drops.

V_{R1} ____________________ V V_{R2} ____________________ V

Compare the two voltages. Why are they so much different?____

____

Could these voltages be used to determine the "Q" or current magnification factor of the circuit?____________________

Explain: ______________________________

____

____

____

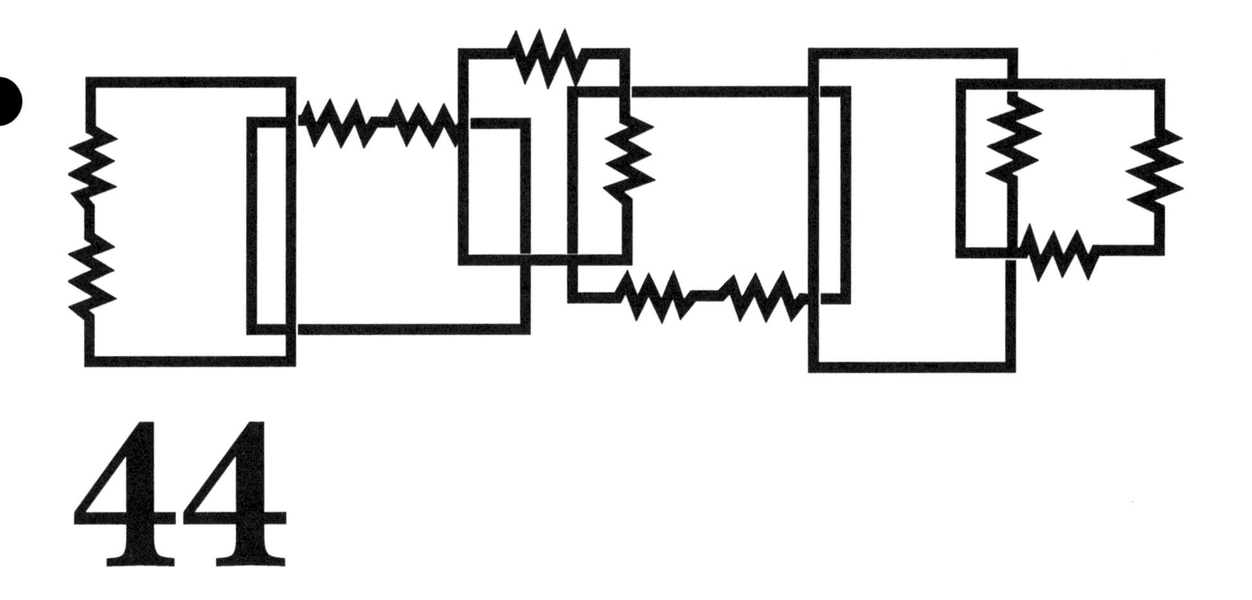

44

Quality

Objectives

After completing this experiment, you will be able to:

1. Evaluate the quality of a resonant circuit.
2. Demonstrate how the Q of a resonant circuit determines the damping of oscillations.

Introduction

This experiment will introduce you to a simple, and perhaps the original, method of evaluating the quality of a resonant circuit. To perform this experiment you will need:

A function generator
A 120V:12.6V transformer
One 1.0μF capacitor
One 1kΩ resistor
An oscilloscope

Procedure

1. Construct the circuit shown in figure 44–1 and remove the activity sheet for this experiment.
2. Set the function generator to squarewave and turn the amplitude output to maximum.
3. Set up the scope as a differential voltmeter. Connect the probes across the LC circuit.
4. Adjust the function generator's frequency to about 100Hz. The exact frequency is not critical as long as it is not too high.
5. Adjust the scope for the best display. On the oscilloscope display you should see a series of damped oscillations. These are the result of the low frequency square wave shock exciting the resonant circuit. What is the frequency of the oscillation? How many cycles can you count before the oscillation dissipates?
6. Connect the 1kΩ resistor in parallel with the resonant circuit. How does the scope display change? How many cycles can you count now?

Review

Shock exciting a resonant circuit is the electrical equivalent of ringing a bell or striking a tuning fork. A jolt of energy is applied to the circuit and it breaks into oscillation at its natural frequency. Since

FIGURE 44–1 Resonant circuit quality damping resistor in dashed symbol

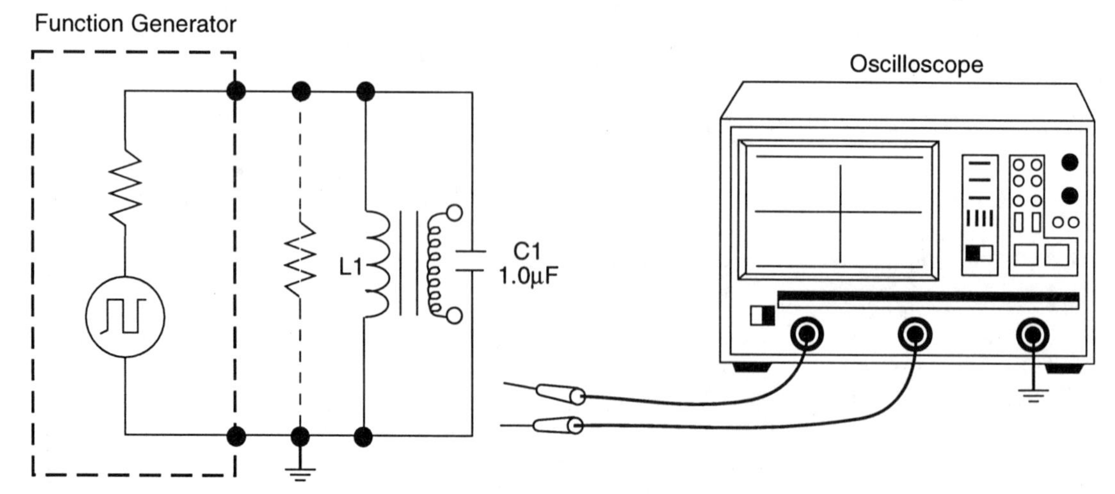

the input energy is not continuously applied to the circuit, the oscillations quickly die out or are "damped."

A high quality circuit will oscillate longer than a low quality circuit. Your first circuit had a fairly high quality and tended to oscillate for several cycles. When the damping resistor was added to the circuit the oscillations died out more quickly. That was because the resistor took power from the oscillating circuit.

In future experiments you will learn more sophisticated ways to evaluate circuit "Q" and bandwidth.

FIGURE 44–2 Oscillating waves

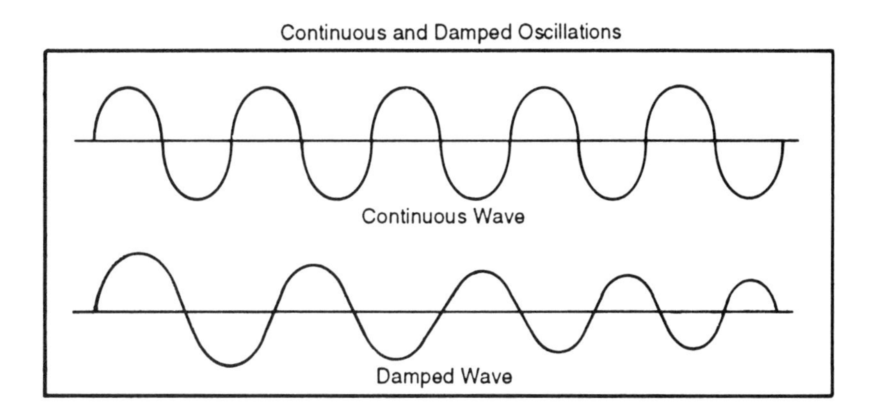

ACTIVITY SHEET EXPERIMENT 44

NAME ______________________________

DATE ______________________________

Step 5

A. How many cycles of oscillation do you count from the scope display?

____

What is causing the circuit to oscillate? ____________________

____

____

Why do the oscillations die out? ____________________

____

____

By measurement, fr = ____________________ Hz

Step 6

B. With the damping resistor added to the circuit, how many cycles of oscillation do you count? ____________________ Why are there fewer oscillations now? ____________________

____

Has the oscillating frequency changed? ____________________

____

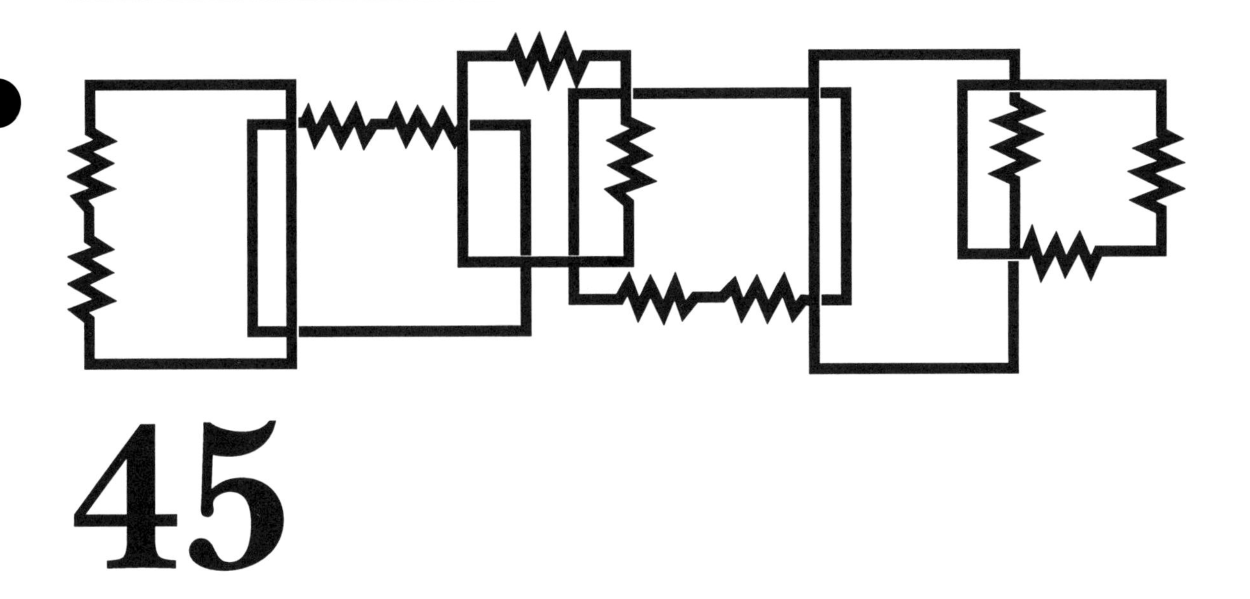

45

Semiconductor Diodes

Objectives

After completing this experiment, you will be able to:

1. Verify the low voltage electrical properties of the diode.
2. Compare the properties of diodes made from different semiconductor materials.

Introduction

As you carry out this experiment you will verify the important low voltage electrical properties of the diode. You will compare the properties of diodes made from different semiconductor materials to observe their similarities, and their differences. You will need:

1 Digital Multimeter & test leads (With diode test function)
1 Variable dc Power Supply
1 Resistor, 470 Ohms, $\frac{1}{2}$ watt or more
4 Alligator Clip test leads
1 Germanium Diode, 1N34 or equivalent
1 Silicon Diode, 1N4002 or equivalent
1 Zener Diode, 5.1 Volts, 1 watt, 1N4733A or equivalent
1 Schottky Diode, 1N5819 or equivalent
1 Light Emitting Diode, any visible color, with leads

Procedure

1. Assemble the test circuit as depicted in figure 45–1. With the Digital Voltmeter connected across the diode insertion test points, TP1 and TP2, adjust the power supply output for 6.0 volts. The voltmeter will remain across the diode test points as you test each of the diodes.
2. Pull the activity sheet for this experiment. Note how it is laid out to enable you to record the measurements as you proceed.
3. Connect the first diode across the diode test points, observing polarity to assure that it will be forward biased. Measure and record the forward voltage. Then turn the diode around so that it will be reverse biased. Measure and record the reverse voltage across it.
4. Continue to test each of the diodes, one at a time. For each diode you will measure and record the forward voltage, then the reverse voltage.
5. After testing each of the diodes, analyze the data you have collected. Do you find that each type of semiconductor material has its own unique characteristics?
6. You may disassemble the diode test circuit. The measurements which follow will be made using the Diode Test mode of your digital multimeter.
7. Set your digital multimeter to the Diode Test mode and individually test each of the diodes in both the forward and reverse directions. The smaller reading for each diode will be its forward characteristic. Record the meter reading for each diode in its appropriate space on the activity sheet.
8. After you have finished testing all of the diodes, compare the data collected by the two measurement techniques. Do the measurements made using the test circuit agree with the measurements made with the digital meter alone?

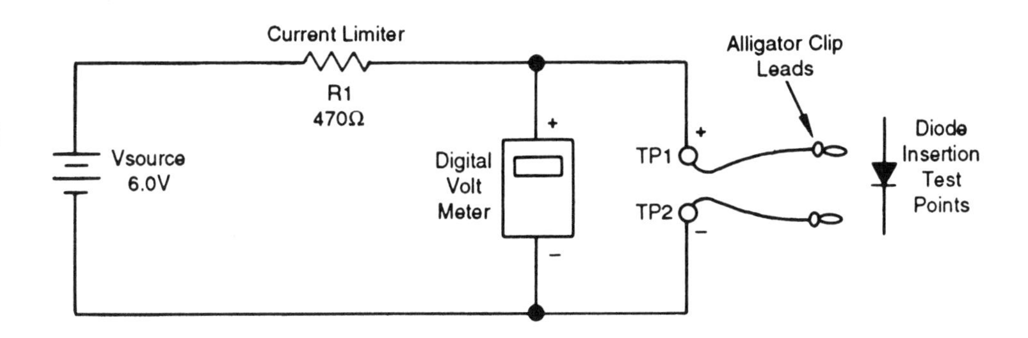

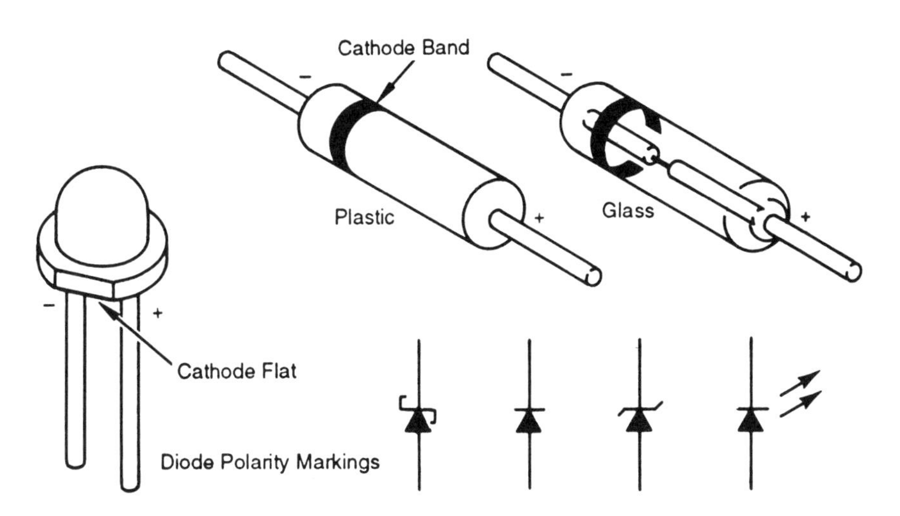

FIGURE 45–1 Diode test circuit and diode polarity markings

Review

In the first part of this experiment you measured the forward and reverse voltages for each type of diode, and recorded your measurements for comparison. You proved that each of the diode types does have a different forward voltage, except for the silicon diode and the zener diode. Those diodes have essentially the same forward characteristic because they are both made from the same semiconductor material. You also observed that the LED produced light when forward biased.

As for the reverse voltages, you determined that six volts of applied emf is not enough to cause the diodes to "breakdown", except for one. The one reverse voltage that was different from the others was the zener diode. In that case the applied six volts was enough to cause it to "break down" with sufficient current flow to develop a voltage less than the circuit source voltage. Did the reverse voltage measured across the zener diode surprise you?

Then, in the second part of this experiment, you used the digital multimeter as a stand alone diode test instrument. When the digital meter is placed in the diode test mode, an internal circuit is activated which employs one of two kinds of measurement principles. In one case the meter is similar to an OHMMETER, and the readings developed as diodes are tested will be qualitative. For a good diode you will obtain a low forward reading, and a very high backward reading, usually indicated as "OL" which means "too great to measure."

In the other case the diode test circuit is similar to a VOLTMETER, and the readings developed will be quantitative. When a good diode is tested the forward reading will be the normal forward voltage characteristic of the diode. The backward reading will be indicated as "OL" or "out of range" if the diode is good.

The readings you gathered while testing diodes using the digital multimeter Diode Test mode, whether relative or absolute, are effective for determining whether a diode has normal LOW VOLTAGE junction characteristics. In future laboratory exercises you will learn how to evaluate semiconductor diodes more completely.

9. Return all materials to their proper places.

ACTIVITY SHEET EXPERIMENT 45

NAME ______________________________

DATE ______________________________

Step 1 **A.** Adjust power supply output to 6.0 volts:

Measured power supply output ____________________ V

Steps 3–4 **B.** Measure and record diode voltages:

	Forward	**Reverse**
Germanium	V	V
Silicon	V	V
Zener	V	V
Schottky	V	V
LED	V	V

Step 7 **C.** Measure and record meter readings:

	Forward	**Reverse**
Germanium		
Silicon		
Zener		
Schottky		
LED		

Step 8 **D.** Do the results of both types of diode test verify normal diode operation? ______________________________

Is one method better than the other? ____________________

____

____

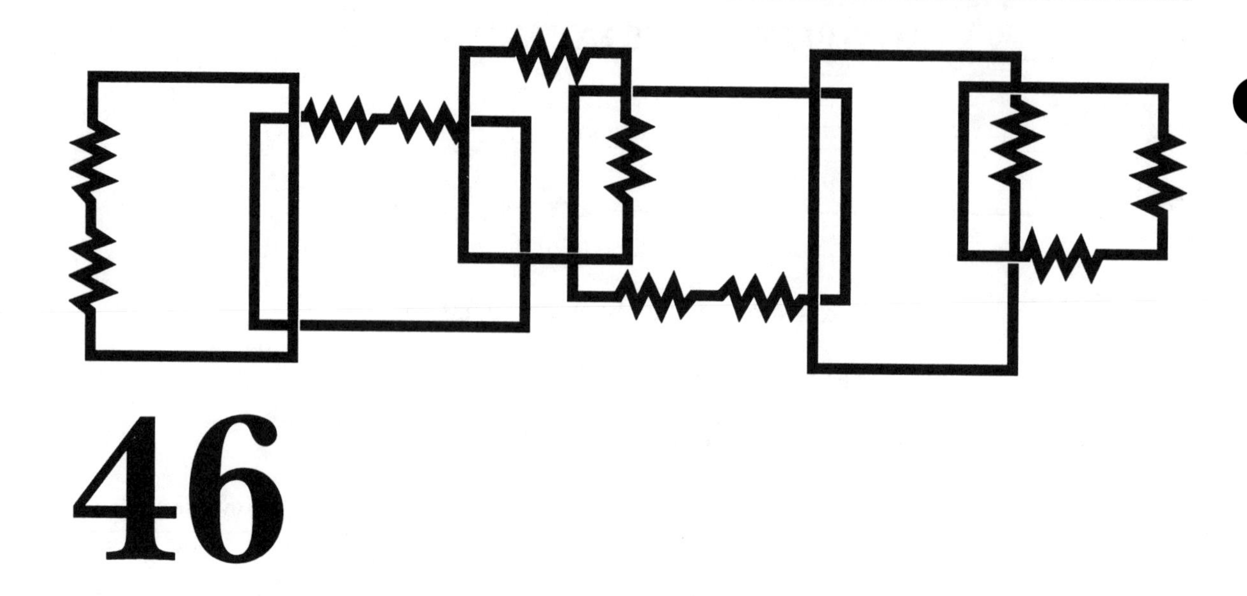

46

Semiconductor Stress

Objectives

After completing this experiment, you will be able to:

1. Observe the harmful effects of stress on a semiconductor diode.
2. Compare how two diodes made from different semiconductor materials handle an excess forward current flow.
3. Determine the type of failure caused when a diode is stressed.

Introduction

This experiment is about the harmful effects of stress, as applied to a semiconductor diode—actually, to two diodes made from different semiconductor materials. We will deliberately stress each diode by allowing excess forward current flow, then determine the type of failure it caused. You will need the following:

- 1 Semiconductor Reference Manual
- 1 Variable dc power supply
- 1 Digital Multimeter with leads
- 1 Resistor, 47 ohms, $\frac{1}{2}$ watt minimum
- 4 Alligator clip test leads
- 1 Silicon diode, 1N914 or equivalent
- 1 Red LED (T $1\frac{3}{4}$ style)

Procedure

1. Pull the activity sheet for this experiment and assemble the circuit shown in figure 46–1. Set the power supply output voltage to minimum. Leave the digital voltmeter connected across the diode connection points.
2. Connect the silicon diode into the circuit while observing polarity to assure forward current will flow when the circuit is energized.
3. Slowly increase the power supply voltage while monitoring the voltage across the diode. Continue to slowly increase the source voltage until there is an abrupt change in the voltage across the diode, then quickly disconnect the diode from the circuit.

CAUTION

The diode may be very hot. Do not handle it directly.

   Leave the power supply voltage as it is. The diode has reacted to the excess stress!

4. Record the power supply output voltage, as indicated on the digital voltmeter. Using this voltage and the resistance value of the series resistor R1, calculate the current flow at the time of diode failure.

FIGURE 46–1 Diode stress circuit

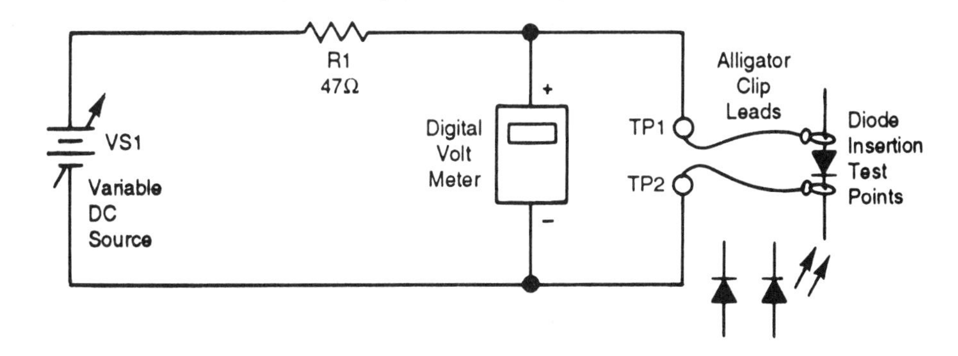

5. Turn the power supply voltage down to minimum and connect the LED into the circuit. Observe polarity to assure forward current flow when the circuit is again energized.
6. Slowly increase the power supply output voltage while monitoring the voltage across the LED. Also visually monitor the intensity of photon output of the LED as you continue to slowly increase the power supply voltage.
7. As soon as you detect an abrupt change in the voltage across the LED, or sudden change in the light produced, or both, quickly disconnect the LED from the circuit.

CAUTION

The LED may be very hot.

Record the power supply voltage as displayed on the digital voltmeter.

8. Using the power supply voltage and the series resistance value, calculate the current flow through the LED when it became defective.
9. Use the Semiconductor Reference Manual to determine the maximum forward current rating of the silicon diode. Also determine the maximum recommended forward current for the red LED. If you are unable to find data for the LED, use 50 milliamperes. This is typically the maximum forward current for most LEDs.
10. Compare the listed forward current ratings of the diodes with the calculated current at the time of failure. Were the diodes able to tolerate a significant amount of excess current flow before they were destroyed?
11. Disassemble the test circuit. With the digital multimeter set to Diode Test, evaluate the two stressed diodes. Record your findings.
12. Set the DMM to measure resistance, then measure and record the resistance of each diode in both directions. Record your findings.
13. How would you classify the failures of the diodes? Did the stress of excess current flow burn them out? Or do your measurements indicate another form of abnormality?

Review

Every kind of electrical or electronic component is susceptible to failure due to excess current flow. Some devices are very tolerant and forgiving in that they are able to withstand much more than their rated maximum current before they fail. Others are fragile and less tolerant, not able to withstand much more than their rated maximum current flow. The diodes used in this experiment were of the latter variety, not able to stand up to current flow much greater than their rated maximum.

As you subjected the diodes to increasing stress, you reached a point where the stress was too great for them, and an abrupt change in the voltage measured across them occurred. The excess current flow caused excess heat buildup in the semiconductor material, causing it to lose its normal properties. As a result, the diodes may have been converted to low ohm resistors. Their measured resistances would be a couple of ohms or less, with no difference in forward or reverse values. Or the semiconductor material may have opened up, similar to a blown light bulb, its internal current path destroyed. Or even some other possibility revealed by the measurements could have occurred.

In the case of the LED you may have seen a relatively rare phenomenon. Sometimes instead of burning out abruptly as a light bulb does, the LED will continue to illuminate with an increasing forward voltage drop, up to 5 or more volts, and perhaps changing the color of light produced just before burnout.

The important point of this experience was to demonstrate that this kind of electrical stress is damaging, and that a damaged diode can be tested to prove that it is no longer useful. It has been destroyed, though physically it may not always appear to be bad at all.

14. Return all materials to their proper places.

ACTIVITY SHEET EXPERIMENT 46

NAME ______________________________

DATE ______________________________

Step 3 **A.** When the diode voltage changed abruptly, what was the direction of change?______________________________

Step 4 **B.** Calculate diode current at instant of diode failure:

Power supply voltage ____________________ V

Diode forward voltage ____________________ V

$$I_{fail} = \frac{V_{power\ supply} - V_{diode}}{R_{series}}$$

I_{fail} = ____________________ mA

Steps 7–8 **C.** When the diode voltage changed abruptly, what was the direction of change?____________________

Calculate diode current at instant of diode failure:

Power supply voltage ____________________ V

Diode forward voltage ____________________ V

$$I_{fail} = \frac{V_{power\ supply} - V_{diode}}{R_{series}}$$

I_{fail} = ____________________ mA

Step 12 **D.** Measure and record diode resistances:

	Forward	**Reverse**
Silicon	Ω	Ω
LED	Ω	Ω

Step 13

E. Based on your resistance measurements, what is wrong with the diodes?

Silicon:____

LED: ____

Are there any visual indications that the diodes are bad?

Silicon:____

LED: ____

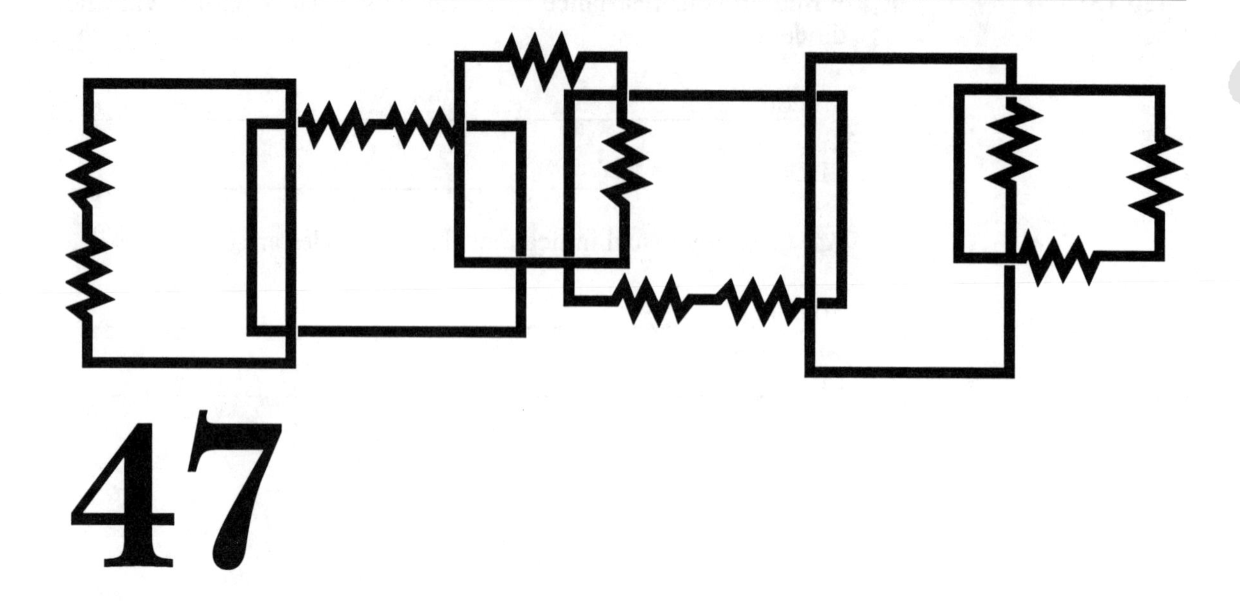

47

Rectification

Objectives

After completing this experiment, you will be able to:

1. Demonstrate how diodes can be used to rectify.
2. Construct and evaluate two simple rectifier circuits.
3. Evaluate the resultant dc output from the rectifiers.

Introduction

Rectification is what diodes are good at. Almost every type of electronic equipment in use today, whether in the home, in the automobile or aboard an airplane, needs dc power in order to operate. Electrical power that is most readily produced by the power companies, the automobile, or an aircraft, is nearly always ac power. Therefore an efficient, inexpensive means of converting ac to dc is in great demand. In this experiment you will construct two simple rectifier circuits to evaluate the process of rectification, and to investigate the characteristics of the resultant dc. You will need:

1 Step down transformer (120 VAC to 12.6 VAC, 500 mA min.)
1 Incandescent lamp (14V, 100 mA) Type 7373 or equivalent
4 Rectifier diodes, 1N4001 or equivalent
1 Digital Multimeter with leads
1 Oscilloscope with probe

Procedure

1. Pull the activity sheet for this experiment, then construct the circuit shown in figure 47–1.
2. Energize the circuit and connect the incandescent lamp to the step down transformer low voltage secondary, across TP1 to common. With the digital voltmeter, measure the ac voltage across the lamp.
3. Reconnect the lamp in the circuit so that it is in series with the first rectifier diode, from TP2 to common. With the digital multimeter measure the dc voltage across the lamp. Did you observe a change in the brightness of the lamp?
4. Reconnect the lamp again so that it is in series with the second rectifier diode, from TP3 to common. Measure the dc voltage across the lamp with the digital voltmeter. What has changed?
5. Now use the oscilloscope to view the waveforms of the voltages directly across the transformer secondary, then across the lamp when it is in series with each of the two rectifier diodes. Measure the peak voltages as accurately as possible with the scope. Then de-energize the circuit.
6. Modify the circuit as shown in figure 47–2, by adding the two additional diodes and test points. Now re-energize the circuit.

FIGURE 47–1 Half-wave rectifier test circuit

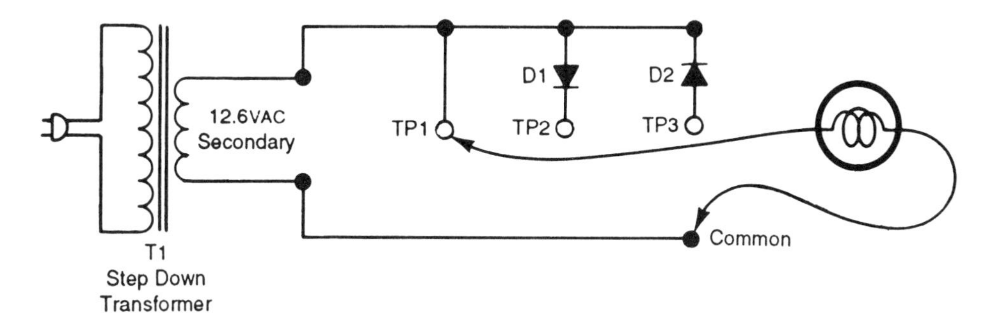

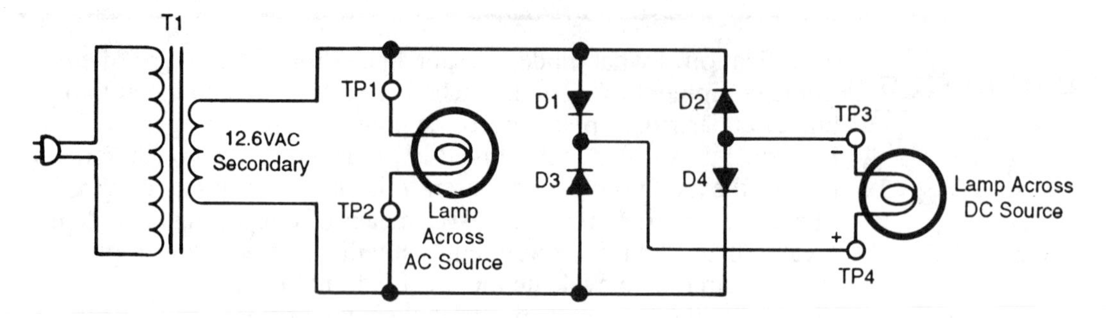

FIGURE 47–2 Bridge rectifier test circuit

7. Connect the lamp directly to the transformer secondary at TP1 and TP2, then measure the ac voltage across it with the digital voltmeter. Record the voltage.
8. Reconnect the lamp so that it is connected across the output terminals of the diode array, TP3 and TP4. Then measure and record the dc voltage across the lamp with the digital voltmeter. Did you observe a change in brightness of the lamp after you switched from ac power to dc power?
9. Again use the oscilloscope to view the voltage waveforms across the transformer secondary at TP1 and TP2. Then view the voltage waveform across the lamp connected to TP3 and TP4. Measure the voltage peaks as accurately as possible from each waveform.

Review

With the first test circuit you powered an incandescent lamp with ac, then with pulsating dc in both directions of current flow. You found that when the lamp was switched from ac to dc there was a considerable lessening of its brightness. But when powered with pulsating dc of either polarity the lamp was equally bright for both. When you viewed the circuit voltage waveforms with the oscilloscope you found that the ac waveform consisted of two equal magnitude, opposite polarity half cycles. The dc waveforms consisted of unipolar pulsations which were half of the ac waveform. In one case the dc pulsations were completely positive with the negative half cycles clipped off. In the other case you saw negative half cycles with the positive half cycles clipped off. In both cases however, this is referred to as half wave rectification.

When you measured circuit voltages with the digital voltmeter, you found that the dc voltages were much less than the ac source voltage. That was also apparent from the brightness of the lamp in each case. Pulsating half wave dc does not convey the same amount of power that is contained in the ac wave.

With the additional diodes of the modified test circuit you found that the brightness of the lamp when powered by ac or pulsating dc was nearly the same. You found that the Digital Multimeter measured values of the ac voltage and the dc voltage were much closer too. In this circuit the brightness of the lamp revealed that the ac power and the dc power were almost equal. When the voltage waveforms were viewed on the scope you found the ac wave to be the same as before, but the dc wave was different form those previously viewed. The dc produced by this circuit was unipolar pulsations similar to the previous circuit, but both half cycles were present. In this case it is referred to as full wave rectification.

10. Return all materials to their proper places.

ACTIVITY SHEET EXPERIMENT 47

NAME ______________________________

DATE ______________________________

Step 2 **A.** Measure and record the ac voltage across the lamp:

V_{AC} ____________________ V

Step 3 **B.** Measure and record the dc voltage across the lamp:

V_{DC} ____________________ V Polarity ____________________

Step 4 **C.** Measure and record the dc voltage across the lamp:

V_{DC} ____________________ V Polarity ____________________

Step 5 **D.** Measure and record peak voltages of circuit. On a separate sheet of paper draw the waveforms viewed on the oscilloscope. Label each drawing to indicate what is being viewed.

V_{AC} ____________________ V_{PK}

V_{DC} ____________________ V_{PK} Polarity ____________________

V_{DC} ____________________ V_{PK} Polarity ____________________

Step 7 **E.** Measure and record the ac voltage across the lamp:

V_{AC} ____________________ V

Step 8 **F.** Measure and record the dc voltage across the lamp:

V_{DC} ____________________ V

Step 9 **G.** Measure and record peak voltages:

V_{AC} ____________________ V_{PK}

V_{DC} ____________________ V_{PK}

H. While performing steps 2, 3 and 4, why was there such a great difference between the ac voltage and the two dc voltages?

____

____

____

____

____

Was there a direct correlation between the brightness of the lamp

and the RMS or average dc voltages? ____________________

Explain: ____

____

____

____

____

Explain the difference between half wave and full wave rectification as it affects average dc voltages and average power:

____

____

____

____

____

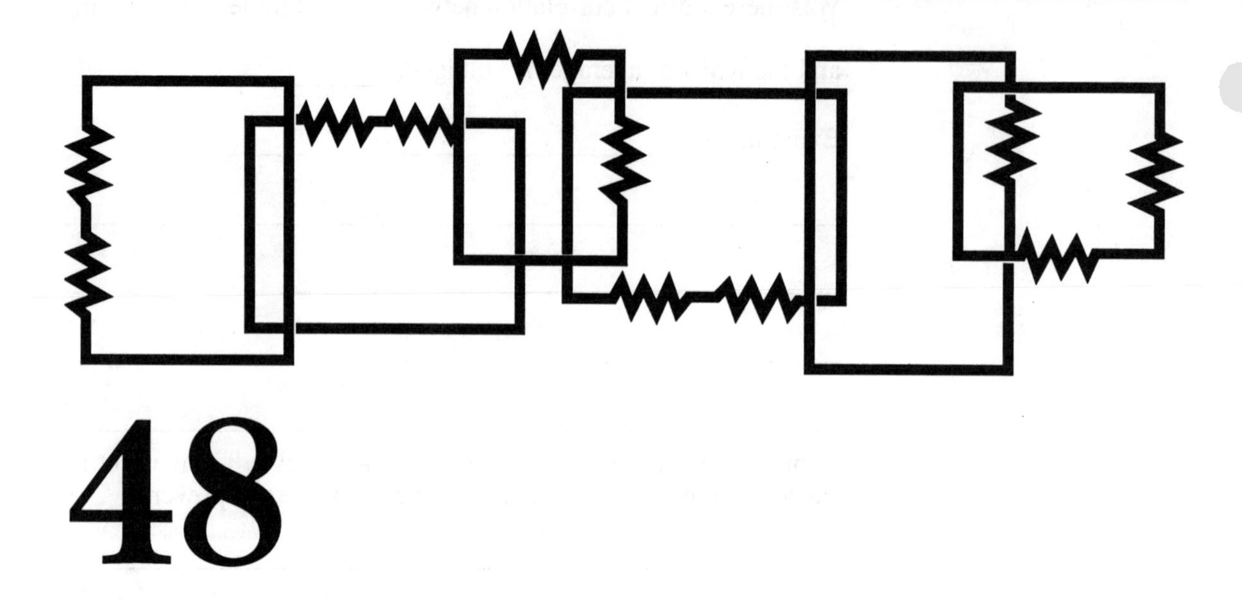

48

The Reference Diode

Objectives

After completing this experiment, you will be able to:

1. Demonstrate how a zener diode can be used to provide a stable dc reference voltage.
2. Evaluate the electrical characteristics of a zener diode.

Introduction

The reference diode, or more commonly, the zener diode, is a diode designed especially to provide a stable dc reference voltage where needed in electronic circuitry. In this experiment you will closely examine the electrical characteristics of one of these diodes. The materials you will need are:

1 Zener diode, (9.1 volts, 1 watt), type 1N4739A or equal
1 Variable dc power supply
1 Digital multimeter with leads
1 Oscilloscope, dual trace, with probes
1 Capacitor, 0.1 microfarad, 20 Volts min.
1 Resistor, 150 ohms, $\frac{1}{2}$ watt
4 Alligator clip leads

Procedure

1. Pull the activity sheet for this experiment. Assemble the circuit as shown in Figure 48–1. Do not energize the power supply yet.
2. Set the scope up for dual trace operation and connect the scope probes to the two test points shown. Set the two vertical channels to dc input. Set both input sensitivities to the same range (2 volts per centimeter). Set sweep speed to 200 microseconds/ cm.
3. With both probe inputs grounded, adjust the vertical position of the two traces so that they are precisely superimposed and appear as only one line at the bottom of the graticule. Since the voltages to be displayed will be positive, causing an upward deflection on the scope, the zero volt reference will be positioned at the bottom of the display graticule. This will allow measurement of positive dc voltages from zero to the full height of the graticule. Unground the probes.
4. Insert the diode to be tested into the test circuit. Assure that the diode polarity with relation to the circuit power supply will cause the diode to be reverse biased. Connect the dc Digital

FIGURE 48–1 Zener diode test circuit

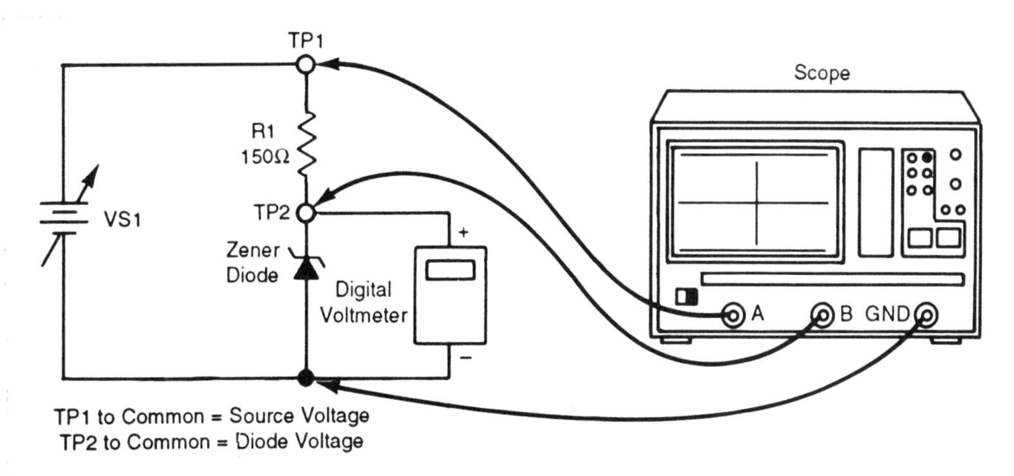

Voltmeter across the diode. Set the power supply dc output to minimum and turn the power supply on.

5. Slowly increase the output of the power supply while watching the scope display. When the voltage level is reached which causes the diode to begin avalanche, the superimposed scope traces will just begin to separate. At that point, read the diode voltage from the digital voltmeter and record it on your sheet.
6. Now as you continue to slowly increase the power supply voltage, the two traces on the scope will further separate. The upper trace, which corresponds to the power supply voltage will move the most. The lower trace, which corresponds to the diode voltage, will move only slightly and then stabilize at its rated reference voltage. When you reach this point, read and record the diode voltage from the digital voltmeter.
7. Continue to increase the power supply voltage until its trace is at the top of the scope graticule (about 16 volts.) Read and record the diode voltage from the digital voltmeter. Turn the power supply output down to minimum and turn the power supply off.

Review

In the first part of this experiment you made a test circuit which utilized two pieces of test equipment to quickly locate the voltage at which diode avalanche just begins, and to precisely measure it. The scope was used to indicate when the diode first began to conduct current flow in the reverse direction. Though you could easily see this on the scope when the traces separated, it would not have been possible to read the voltage value precisely from the scope graticule. For the accurate voltage reading you used the digital voltmeter. An analog device was used to detect an event, and a digital device was used to accurately measure it. You then went on to evaluate avalanche, looking for the voltage at which avalanche stabilized, and finally the diode voltage at approximately optimum zener current, which was maximum current for this circuit.

Now we will closely scrutinize avalanche magnified to see what sort of electronic noise is produced, and whether we can "turn" it for maximum amplitude.

8. Set the scope up to display small ac signals. The only trace on the scope display needed for this exercise is the one which displays diode voltage. The other channel may be disabled and the scope set to operate in single channel mode. Set input coupling to AC, and sensitivity to 2 millivolts per centimeter. Ground the vertical input and move the trace (zero volt baseline) to the center of the scope graticule to allow deflection both above and below the zero volt reference. Unground the scope input.
9. Check that the dc power supply output is set to minimum then turn the power supply on. Leave the digital dc voltmeter connected to the diode in order to measure the voltage at which maximum noise appears. While watching the scope display, slowly increase the power supply voltage until you see noise appear. Then rock the voltage in and around this level until you see the maximum amount of noise. Readjust scope sweep and

sensitivity as necessary to achieve best display. When you've achieved the maximum noise condition, read and record the dc diode voltage from the digital voltmeter. From the oscilloscope display measure and record the maximum obtainable noise voltage as a peak to peak ac value.

Review

You've just verified that the zener diode can be an electronic noise source of surprising amplitude. The noise displayed on the scope, also known as "grass", or "white noise", is a very complex signal consisting of many, many different frequencies. If you were to amplify this noise and apply it to a loudspeaker, you would find that it reminds you of something. For most people, listening to this noise has a soothing effect.

As you proceed onto more advanced studies, you will encounter noise again. You will learn that it has some valuable applications in spite of the fact that this sort of electronic noise is generally considered to be a nuisance. Remember that it is very easy to construct a small signal noise source, whenever you need one.

10. Now take the 0.1 microfarad capacitor and place it parallel to the diode in the circuit. Look at the scope display to verify that the noise is gone. Check the digital dc voltmeter to see whether the dc diode voltage has changed. Why has the noise disappeared?

11. De-energize the circuit and return all materials to their proper places.

ACTIVITY SHEET EXPERIMENT 48

NAME ______________________________

DATE ______________________________

Step 5 **A.** Measure and record voltage when the superimposed traces begin to separate:

V_{diode} ____________________ V

Step 6 **B.** Measure and record the diode voltage when it begins to stabilize:

V_{diode} ____________________ V

Step 7 **C.** Measure and record the diode voltage when the dc source is about 16V:

V_{diode} ____________________ V

What is the difference in potential between the well stabilized diode voltage and the "just beginning to avalanche" voltage?

V_{diff} ____________________ V

Step 9 **D.** When maximum noise is attained, the dc diode voltage is:

V_{DC} ____________________ V

E. The maximum noise peak to peak ac voltage from the scope is:

V_{noise} ____________________ V_{pp}

Step 10 **F.** After placing the capacitor parallel to the noisy diode, why does the noise disappear? ______________________________

____

____

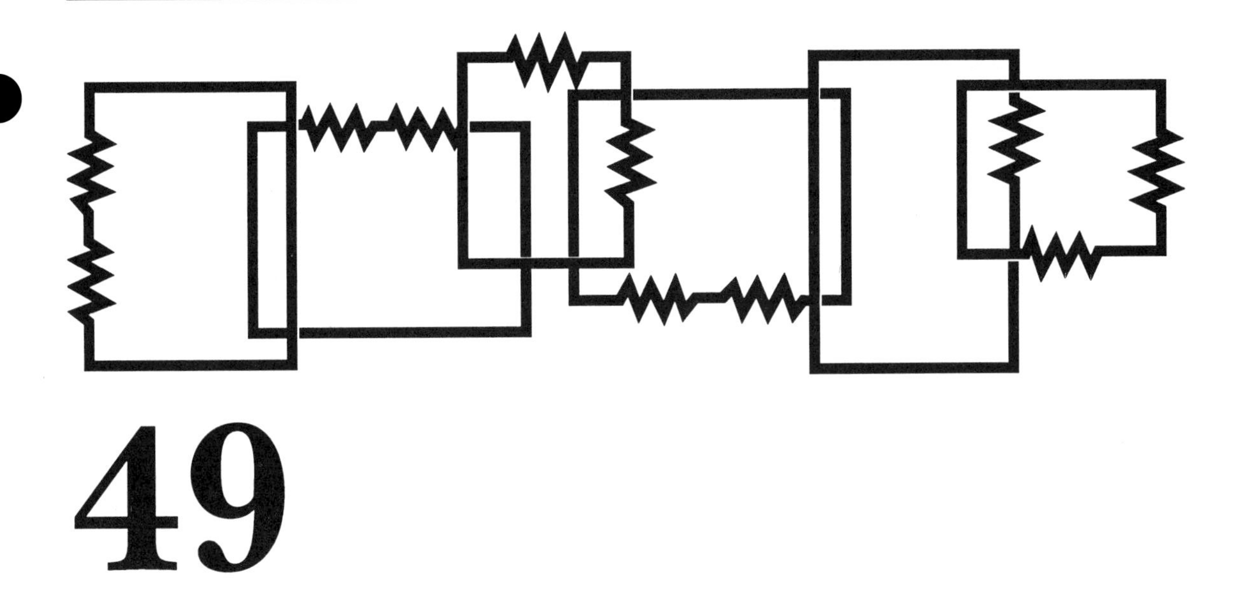

49

The Filtered Power Supply

Objectives

After completing this experiment, you will be able to:

1. Construct a simple dc power supply.
2. Analyze the operation of the power supply when loaded.
3. Describe some of the limitations of a dc power supply.

Introduction

This experiment will carry your examination of ac to dc power conversion several steps forward. By construction of a simple dc power supply and analysis of its operation when loaded, you will see first hand some of its limitations. The materials you will need are:

- 1 12.6 Volts ac power source, 500mA minimum
- 1 Rectifier Diode, Type 1N4001 or equivalent
- 2 20 Microfarad Electrolytic Capacitors, 25 Volts min.
- 1 100 Microfarad Electrolytic Capacitor, 25 Volts min.
- 1 1000 Microfarad Electrolytic Capacitor, 25 Volts min.
- 1 10 Ohm Resistor, 1 Watt
- 1 Incandescent Lamp, Type 7382 (14V, 80mA)
- 1 Incandescent Lamp, Type 7373 (14V, 100mA)
- 1 Small Loudspeaker, 8 ohms to 40 ohms, 100mW
- 1 Oscilloscope, Dual Trace with probes
- 1 Digital Multimeter
- 6 Alligator clip test leads

Procedure

1. Pull the activity sheet from this experiment. Construct the circuit shown in figure 49–1. Use a 20 microfarad capacitor as C1. Do not apply power to the circuit yet.
2. Double check that the polarities of CR1, C1 and C2 are correct. When satisfied that there are no errors, apply power to the circuit. Listen intently to the loudspeaker. Is there any sound?

FIGURE 49–1 Power supply test circuit

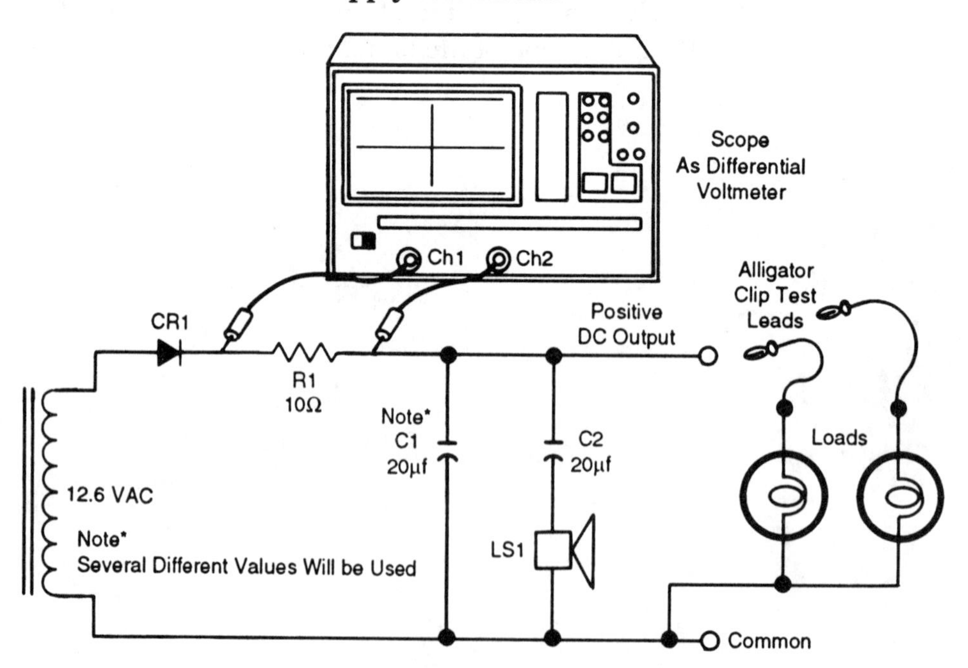

3. Connect the 7382 lamp across the output terminals of the dc power supply circuit. Is there any change in the output of the speaker?
4. Connect the 7373 lamp across the dc output, paralleling the 7382 lamp. Does this affect the brightness of the first lamp? Is there a change in the output of the speaker? Disconnect and reconnect the 7373 lamp several times to verify.
5. Set the oscilloscope up to operate as a differential voltmeter with both inputs set to the same sensitivity. Start with 200 mV per centimeter, dc signal sense. Connect the scope probes across resistor R1 with the channel A input probe to the end of R1 connected to CR1. Adjust scope controls as necessary to obtain best presentation showing two to four cycles of waveform deflecting in the positive direction on the screen. Leave the probes connected for the remainder of the experiment.
6. Measure and record the time duration of one dc pulse at its zero volt baseline. Then for comparison, calculate and record the time duration of one half cycle at 60 hertz. Considering that the rectifier is half wave, and the frequency of the ac input is 60 Hz, why is the measured time duration of the pulse not equal to the time duration of a half cycle at 60 Hz? What do the pulses represent?

Review

You've constructed a simple capacitor filtered dc power supply in order to evaluate its performance under varying conditions. Resistor 1, although not essential to circuit operation, is sometimes included in the circuit to function as a surge limiter. The value of 10 ohms is larger than it would ordinarily be in order to exaggerate the effect it produces, and to allow measurement of waveforms across it. The capacitor/loudspeaker leg in the circuit is there to allow you to hear the sound of ripple when it is present.

In this first part of the experiment you have evaluated the operation of lightly filtered supply with no load, then a moderate load, and finally a heavier load. The speaker, the lamps and the oscilloscope together allow you to see and hear indicators of its performance.

7. De-energize the ac input to the supply and disconnect both lamps from the output of the supply. Then remove the 20 microfarad capacitor C1 and replace it with the 100 microfarad capacitor.
8. Re-energize the supply and repeat the steps of evaluation. Listen for the presence of hum when the supply is unloaded, moderately loaded by connection of the 7382 lamp to its output, and more heavily loaded with both the 7382 lamp and the 7373 lamp connected. Then monitor the waveform developed across R1. Record your observations in the appropriate spaces on the activity sheet.
9. De-energize the ac input to the supply and disconnect both lamps from its output. Remove the 100 microfarad filter capacitor C1 and replace it with the 1000 microfarad capacitor.
10. Re-energize the supply and repeat the steps of evaluation. Record your observations on the activity sheet.

Review

The remainder of the experiment was spent evaluating the supply with different values of filter capacitor. You discovered that as the size of the filter capacitor was increased, the less ripple was evident at the loudspeaker as you loaded the supply. There was also less of a difference in lamp brilliance as both lamps were used as loads. If you have concluded that more capacitance enhances the stability of the dc output of the supply, you are right. By observing the waveforms across R1 in each case, and comparing pulse widths, you have probably concluded that as filter capacitance is increased, the time duration of the charge replenishment current pulses is decreased. The larger filter capacitor charge is not severely depleted while it is supplying current to the load between charge replenishment pulses.

The capacitor/loudspeaker branch of the circuit is not part of the power supply. It was included to enable you to hear the sound of power supply ripple when it is present. Capacitor C2 operates as a coupling capacitor in this case, blocking the flow of dc current through the speaker yet allowing the ripple component to be coupled and heard.

11. De-energize and disassemble the circuit. Return all materials to their proper places.

ACTIVITY SHEET EXPERIMENT 49

NAME ______________________________

DATE ______________________________

Steps 1–2

A. Construct the test circuit using the 20μF capacitor as the first value for C1. Double check circuit to be sure all polarized components are properly hooked up. Energize the circuit while listening to the loudspeaker. What do you hear?

____

____

Step 3

B. Connect a load across the output of the supply, the 7382 lamp.

What do you hear at the loudspeaker? ____________________

____

Step 4

C. Increase the load on the power supply by attaching the 7373 lamp in parallel with the 7382. Describe what you hear now.

____

____

Disconnect and reconnect the 7373 lamp several times. What happens? ______________________________

____

Steps 5–6

D. Configure the scope to operate as a differential voltmeter and connect the probes across R1. Adjust scope as necessary for best display. Measure the time duration of one pulse at its zero volt baseline.

t ____________________ ms

Calculate period of one half cycle at 60 Hz.

t ____________________ ms

Why is the duration of the pulse less than a half cycle at 60 Hz?

____

____

____

What is causing the pulses across R1?___________________

____

____

Step 7

E. Safely replace C1 with the 100μF capacitor.

Steps 7–10

F. Evaluate each filter capacitor.

C1 value	Single lamp load hum	Dual lamp load hum	Pulse across R1 voltage
100 μF			
1000 μF			
	Describe hum loudness		Measure pulse

Why does the hum level change when the capacitance of C1 is increased?___________________

____

____

Why do the pulses across R1 change as the value of C1 is changed? ___________________

____

____

How does increasing the filter capacitor capacitance affect the pulses across R1? ___________________

____

____

Why is this so?___________________

____

____

____

____

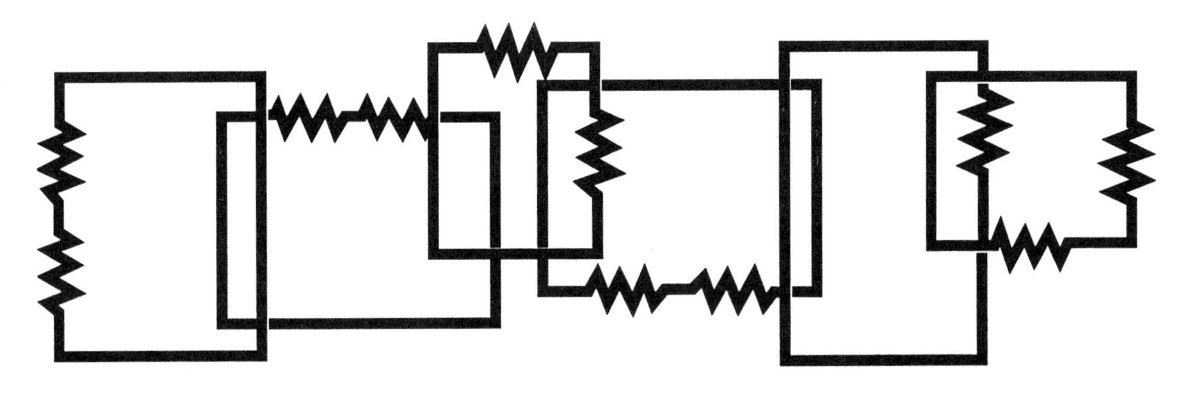

50

Full Wave, Dual Polarity, and Doubling Supplies

Objectives

After completing this experiment, you will be able to:

1. Evaluate two dc power supply circuits with capacitive filters, namely:
   a. A full wave bridge supply
   b. Two half wave opposite polarity supplies, or a single voltage doubler supply.
2. Observe how the loaded supply voltages from a dc power supply differ from the unloaded supply voltages.

Introduction

This experiment will be sent evaluating two dc power supply circuits which utilize capacitor filters. The first will be a full wave bridge supply and the second will be two half wave opposite polarity supplies, or a single voltage doubler supply. The materials you will need are:

1 12.6 volt ac power source, 500 mA min.
4 Rectifier diodes, 1N4001 or equivalent
2 1000 microfarad capacitors, 25 volts or better
1 Digital multimeter with leads
6 Alligator clip test leads
1 Oscilloscope with test probes
1 60 watt, 120 volt lamp with test leads attached

Procedure

1. Pull the activity sheet for this experiment. Assemble the circuit shown in figure 50–1.
2. Energize the circuit. Set the digital multimeter to measure dc volts. Set the scope to measure ac input. Connect both across the dc power supply circuit output terminals. Set the range selectors on the dc voltmeter and the scope appropriately to monitor the supply dc output voltage and the ac ripple component.
3. Using an oscilloscope, measure and record the dc output voltage and the ac peak to peak ripple voltage.
4. Now connect the 60 watt incandescent lamp to the power supply output. Again measure and record the dc output voltage and the ac peak to peak ripple voltage. Adjust the scope as necessary to obtain best display and accuracy.
5. Disconnect the digital multimeter from the circuit and set it up to measure dc current flow. Insert the ammeter in series with the incandescent lamp dummy load. Measure and record the load current flow.
6. De-energize the circuit. Perform calculations called for on activity sheet.

FIGURE 50–1 Full wave supply

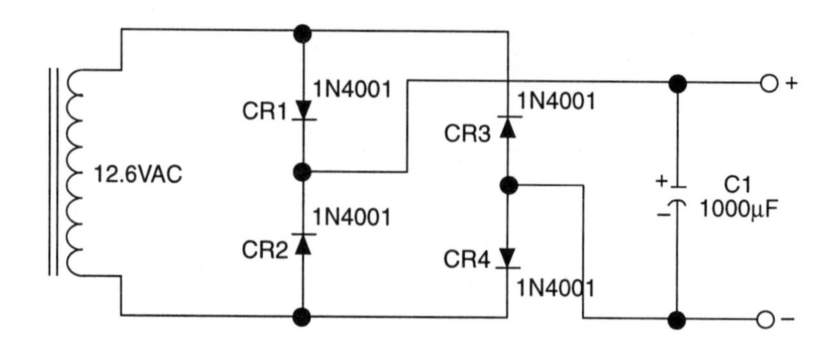

You have constructed and evaluated a full wave bridge, capacitive filtered power supply circuit. You first measured its unloaded dc ouput and ac ripple voltages. Then you attached a dummy load to draw several watts of power from the supply and measured the supply voltages again. The loaded supply voltages were somewhat different from the unloaded voltages, one of which changed a little and the other changed quite a lot. This is a normal characteristic of simple, unregulated power supply circuits such as this.

7. Modify the power supply test circuit as shown in figure 50–2(a).

8. Using the digital multimeter and the oscilloscope, measure and record the unloaded dc output and ac ripple voltages for each of the two half wave outputs.

9. Attach the dummy load to each of the supply outputs in turn, and measure and record the dc output and ac ripple voltages.

10. Remove the dummy load from the supply. Reconfigure the common side of the test circuit as shown in figure 50–2(b). Notice that the circuit does not need extensive rewiring, but just an adjustment of perspective. By relocating the common side of the circuit output, which eliminates the negative output, a single positive output with respect to common is produced. This single positive output, with respect to common, is the sum of the voltages stored in the two series connected filter capacitors.

11. Using the digital multimeter and the oscilloscope, measure and record the unloaded dc output voltage and ac ripple voltage.

12. Attach the dummy load to the test circuit output terminals and again measure the supply output voltages, dc and ac ripple.

FIGURE 50–2

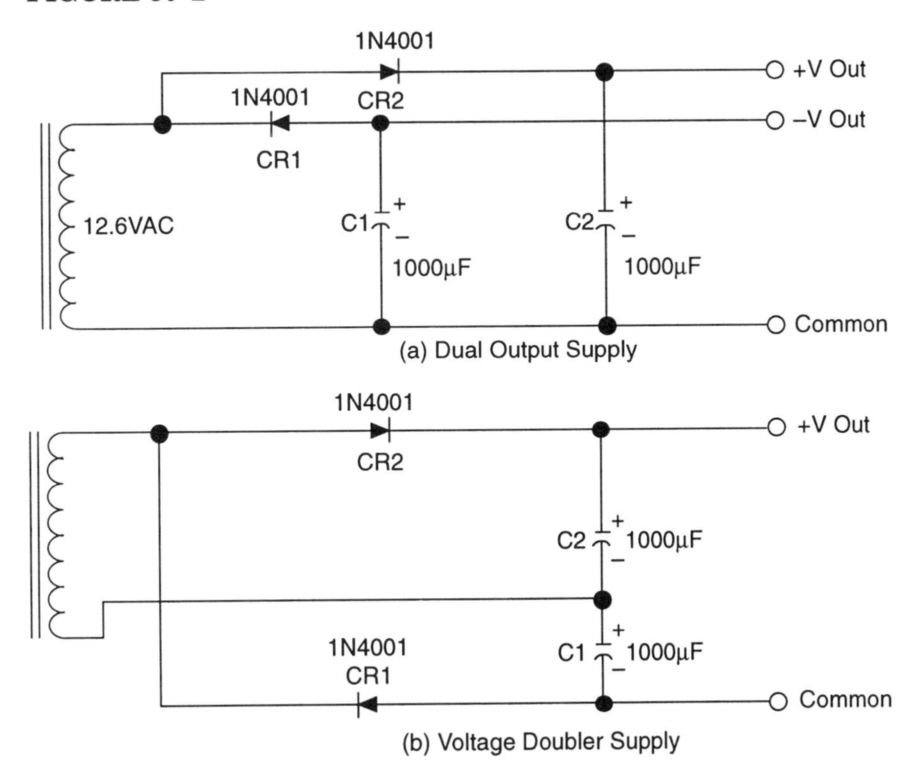

13. Disconnect the digital multimeter from the test circuit and set it up to measure dc current flow. Then connect the digital ammeter in series with the dummy load to measure and record load current. When finished, de-energize the test circuit and perform calculations on the activity sheet.

Review

In the second part of the experiment you evaluated the performance of two half wave, opposite polarity power supply circuits derived from a single ac transformer secondary winding. Then you rearranged connections to the circuit to obtain an output voltage of twice the half wave outputs, and evaluated the performance of the doubler. You found that in all cases, the loaded supply voltages were different from the unloaded voltages. This variation is characteristic of all unregulated dc power supplies. You discovered that simple ac to dc conversion need not require a complex circuit. Whether a half wave or full wave rectifier configuration is employed depends on how heavily the supply is to be loaded. The amount of filter capacitance used is determined by the amount of load current expected to be drawn from the supply. To keep ripple to an acceptably small value, large values of capacitance are frequently used. One rule of thumb states that filter capacitance should be no less than 2000 microfarads for each ampere of load current.

14. Disassemble the test circuitry and return all materials to their proper places.

ACTIVITY SHEET EXPERIMENT 50

NAME ______________________________

DATE ______________________________

Step 2 — **A.** Connect the digital voltmeter and the oscilloscope to the circuit output dc output and ac ripple voltages.

Steps 3–4 — **B.** Full wave power supply outputs

	dc volts	Ripple volts pk-pk
Unloaded	V	$V_{p\text{-}p}$
Loaded	V	$V_{p\text{-}p}$

Step 5 — **C.** Measured load current: ____________________ mA

Step 6 — **D.** Using measured load voltage and current calculate

load resistance ____________________ Ω

load dissipation ____________________ W

Which of the power supply voltages increased when the supply was loaded? ____________________ Which voltage decreased? ____________________ Explain why each voltage changed.

____

____

____

____

____

Steps 8–9 — **E.** Measure and record the unloaded and loaded outputs of the two supplies.

	Positive Supply		Negative Supply	
	dc output	Ripple	dc output	Ripple
Unloaded	V	$V_{p\text{-}p}$	V	$V_{p\text{-}p}$
Loaded	V	$V_{p\text{-}p}$	V	$V_{p\text{-}p}$

Steps 11–12

F. Measure and record the unloaded and loaded outputs of the voltage doubler.

	Voltage Doubler Supply	
	dc output	**Ripple**
Unloaded	V	V_{p-p}
Loaded	V	V_{p-p}

Step 13

G. Measure load current ____________________ mA

Using measured load voltage and current calculate

Load resistance ____________________ Ω

Load dissipation ____________________ W

Consider the performance of the loaded doubler. Was the supply loaded beyond its ability by the 60W lamp? ________________

Explain: ____

____

____

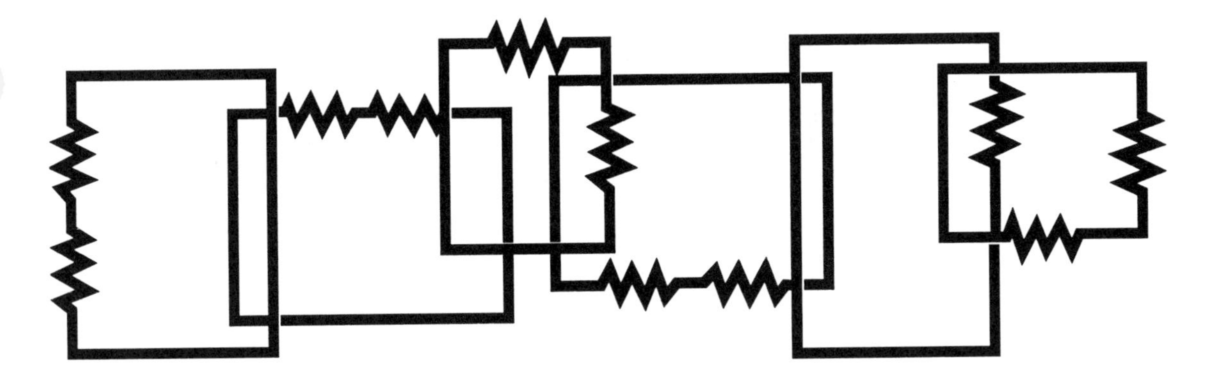

51

Transistor Operation—Cutoff and Saturation

Objectives

After completing this experiment, you will be able to:

1. Examine how a transistor operates.
2. Demonstrate how a transistor can be made to operate in its extreme conditions of cut-off and saturation, and in the linear mode between the extremes.
3. Verify that the transistor can be made to operate as a voltage controlled switch or voltage controlled variable resistor.

Introduction

This experiment will enable you to look into how a transistor operates. You will examine the extreme conditions of operation, cutoff and saturation, as well as the linear mode which occupies the region between the extremes. You will find that the transistor is really just a special kind of resistor, one that can be controlled electrically. To conduct the experiment you will need:

1 Variable dc power supply
1 Function generator or variable audio signal source
1 Transistor, NPN, 2N3904 or equivalent
1 Incandescent lamp, 14V, 80mA, type 7382 or equal
2 Resistors, 10.0 kilohms, $\frac{1}{4}$ watt, 2% tolerance
1 Potentiometer, 1.0 megohm
1 Capacitor, 20 microfarads, 25 volts
1 Digital multimeter with leads
6 Alligator clip test leads

Procedure

1. Pull the activity sheet for this experiment. Assemble the test circuit shown in figure 51–1.
2. Energize the dc power source and measure its output voltage with the digital multimeter set to operate as a voltmeter. Adjust the power supply output to 12.0 volts.
3. Rotate the shaft of the 1 megohm rheostat connected potentiometer in both directions. While rotating the shaft you should see the lamp brightness vary. Turn the shaft fully in the direction that makes the lamp become dimmer. The lamp should be extinguished.
4. With your digital multimeter set to measure voltage, measure and record the voltage across R2 from TP1 to TP2. Use the measured voltage to calculate current flow through the resistor. Is there any base current flow?
5. Measure and record the voltage from TP3 to common, which is from collector to emitter of Q1. Does this voltage indicate current flow through the transistor and through the lamp?

FIGURE 51–1 Transistor test circuit

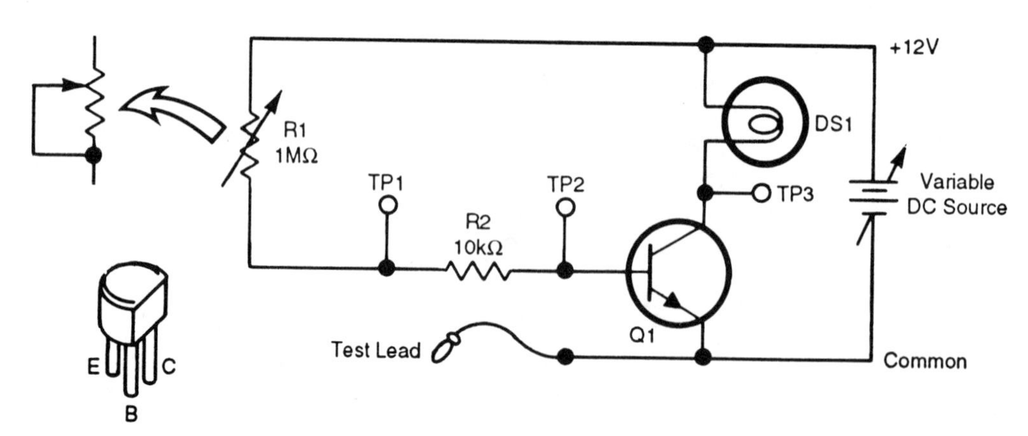

6. With the digital voltmeter connected from TP3 to common, which is across Q1 from collector to emitter, adjust the variable resistor R1 until the measured voltage is exactly 6.0 volts. Then connect the digital voltmeter across R2 from TP1 to TP2. Measure and record the voltage across R2. Use the voltage across R2 to calculate current flow through it. How much base current is there? What is the internal resistance of Q1 from collector to emitter?
7. Connect the digital voltmeter from TP3 to common, across Q1 collector to emitter. While monitoring this voltage use an alligator clip test lead to connect TP2 to common. This will place a short circuit across Q1 base-emitter junction. What happens to the lamp? Disconnect and reconnect the short circuit from TP2 to common several times. With the short circuit in place, what is the internal collector to emitter resistance of Q1? Has the transistor been damaged by placing the short circuit from its base to its emitter?

Review

So far in this experiment you have constructed a simple transistor circuit with a lamp in the collector circuit. To enable base current flow the base circuit has two series resistors, R1 and R2. The variable resistor R1 allows you to increase or decrease base current while R2 assures that the base circuit resistance will never decrease below 10 kilohms. This limits the base current to a safe value when R1 is decreased to minimum resistance. You saw that a tiny base current, (microamperes), causes a much larger collector current, (milliamperes), as evidenced by the lamp illumination. You found that when base current is continuously adjustable, the collector current is also continuously variable, you were able to smoothly control the brilliance of the lamp in the collector circuit.

Then when you placed a short circuit across the transistor base-emitter junction, you shunted current flow in the base supply circuit (R1 and R2) away from the transistor. The base-emitter junction thereby lost its forward bias and base current, which caused collector current flow to cease. The lamp went out while the short circuit was in place. When the short circuit was disconnected, the transistor resumed normal operation. The short across the base-emitter junction was harmless to the transistor and proved to be an easy way to cut the transistor off. No base current means no collector current.

8. Turn variable resistor R1 fully in the direction which increases the illumination of DS1. Measure the voltage drop across R2, from TP1 to TP2 with the digital voltmeter. Then measure the voltage from TP3 to common. Record the voltages and determine Q1 base current. With the digital voltmeter connected from TP3 to common, slowly increase the resistance of R1 until the voltage indicated on the digital voltmeter just begins to increase. Measure the voltage across R2 from TP1 to TP2 and again calculate base current. Lamp DS1 should be brightly illuminated throughout.

By decreasing the resistance of the base supply circuit to minimum, you allowed maximum base current to flow, which in turn caused maximum collector current flow. In this case virtually all of the source voltage is dropped across the lamp DS1 while Q1 emitter to collector voltage is very small. The transistor is effectively a short circuit internally from emitter to collector. This is saturation.

Then you slowly increased the base circuit resistance, which gradually decreased base current, until the transistor just began to drop out of saturation. This was where the collector to emitter voltage just began to increase. The base current calculated for this condition was the minimum base current capable of saturating the transistor. Any further increase in base current, while possible, does not further increase the collector current. The lamp intensity remained unchanged until base current was decreased sufficiently to exit saturation and enter the linear mode. In linear mode, any further decreases in base current caused corresponding decreases in collector current, as evidenced by the lamp dimming.

9. Modify the transistor test circuit by replacing DS1 with a fixed resistor and adding C1, as shown in figure 51–2. The output of the audio source is applied to the base circuit via C1. Energize the circuitry and set the output of the audio generator to zero. With the digital voltmeter connected from TP3 to common, adjust variable resistor R1 until the meter indicates exactly 6.0 volts.

10. Set the frequency of the audio generator to approximately 5.0 kilohertz. Set the scope up to view dc voltage and connect the scope input probe to TP3. Connect the scope ground lead to circuit common. Leave the digital meter (still set for dc voltage) connected as well, to simultaneously provide a scope presentation and a digital measurement.

11. Turn the audio output of the signal generator up until the scope displays a wave of 3.0 volts peak to peak. Adjust the scope as necessary for best display, showing two to four cycles of waveform. Measure and record the dc collector voltage with the digital voltmeter.

FIGURE 51–2 Modified transistor test circuit

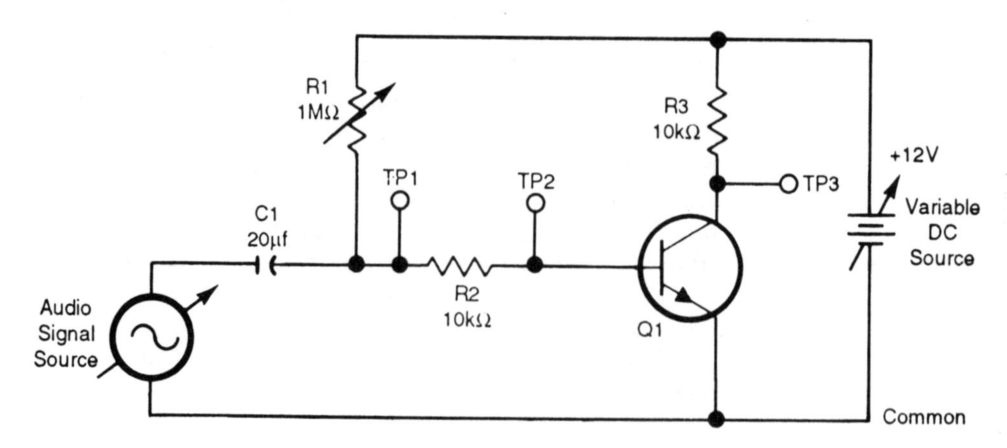

12. Continue to slowly increase the output of the signal generator while observing both the scope and the digital voltmeter. When you reach the point where the signal displayed on the scope begins to distort by flattening or clipping the top or bottom of the wave, or both, decrease generator output voltage. Adjust the generator voltage carefully to attain maximum peak to peak undistorted collector output on the scope. Measure and record both the collector dc voltage with the digital voltmeter, and the peak to peak amplitude of the signal from the scope display. Then move the scope probe from TP3 to the output of the audio generator at TP1. Adjust the scope as necessary for best display, then measure and record the peak to peak amplitude of the generator output. Perform calculations on activity sheet.
13. Place the scope probe back to TP3 and continue to increase the generator output. Adjust the scope as necessary to maintain best display. The waveform should now be severely distorted with considerable clipping at the tops and bottoms. Which part of the scope display represents cutoff? Which part saturation? Which part linear mode?

Review

The final phase of the experiment was spent evaluating amplifier operation. You adjusted circuit bias so that the internal collector to emitter resistance of Q1 was dropping half the dc supply voltage. As you applied signal to the circuit from the audio generator, you observed an amplified waveform on the scope. Oddly, the voltage indicated on the digital voltmeter remained constant. This is a characteristic of linear circuit operation. The collector voltage is dc varying at an ac rate, as seen on the scope, but its average value is still 6.0 volts DC.

The linear operation of the transistor gave way to non-linear operation when the generator output signal was increased to a magnitude sufficient to drive Q1 into saturation and cutoff. The wave distortion produced by overdriving Q1 was easily discerned on the scope. The flattened positive top of the signal corresponds to cutoff, where there is no collector current flow, and the collector voltage for that part of the waveform is equal to 12 volts. The flattened bottom of the collector waveform is the saturated part. Here collector current flow is maximum because the internal collector to emitter resistance of Q1 is essentially a short circuit. The signal voltage at the collector for the saturated part of the waveform is extremely small, about zero volts; the vertical parts of the waveform, between the flattened tops, represent the linear region of transistor operation.

14. Disassemble the test circuit and return all materials to their proper places.

ACTIVITY SHEET EXPERIMENT 51

NAME ______________________________

DATE ______________________________

Step 2 **A.** Adjust dc power supply for 12.0 volts output.

Step 3 **B.** Adjust R1 fully in the direction which extinguishes DS1.

Step 4 **C.** Measure voltage drop across R2. Calculate current flow.

V_{R2} ________________ V I_{RZ} ____________________ μA

Step 5 **D.** Measure collector to emitter voltage of Q1.

$VQ1_{C\text{-}E}$ ____________________ V

Based upon measurements just made, is there base current flow in Q1?____________________ Is there collector current flow? ____________________ Explain why lamp DS1 is not illuminated. ____

____

____

Step 6 **E.** Adjust base current control R1 until Q1 collector to emitter voltage is 6.0 volts.

Measure R2 voltage drop. V_{R2} ____________________ V

Calculate Q1 base current. I_{R2} ____________________ μA

Determine Q1 internal resistance collector to emitter.

____________________ Ω

Step 7 **F.** Measure voltage from Q1 collector to emitter when Q1 base-emitter is shorted.

____________________ V

Has the transistor been damaged by the base-emitter short? ____________________ What is your proof?____________________

____

____

Step 8 **G.** Turn base current control R1 fully in the direction which illuminates lamp DS1.

Measure voltage across R2. V_{R2} ________________ V

Calculate Q1 base current. I_B ________________ μA

Measure Q1 collector-emitter voltage.

V_{CE} ________________ V

Decrease Q1 base current until voltage from collector to emitter just begins to increase.

Measure voltage across R2. V_{R2} ________________ V

Calculate Q1 base current. I_B ________________ μA

Step 9 **H.** Modify transistor circuit and set Q1 collector to emitter voltage to 6.0 volts by adjusting R1.

Steps 10–11 **I.** Set audio generator to 5 kmz, and adjust audio generator output until scope displays waveform of 3.0 volts peak to peak at Q1 collector. Measure Q1 collector voltage with digital voltmeter.

$VQ1_{C\text{-}E}$ ________________ V

Step 12 **J.** View Q1 collector signal while adjusting audio signal level to obtain maximum undistorted output. Measure Q1 collector voltage.

Collector signal voltage peak to peak ________________ $V_{p\text{-}p}$

Collector dc voltage ________________ V

Measure generator output voltage peak to peak

V_{GEN} ________________ $V_{p\text{-}p}$

Calculate signal gain of Q1 ________________

Step 13 **K.** View Q1 collector signal while increasing audio generator output until signal is severely distorted at top and bottom. Which parts of the displayed scope waveform represent:

Saturation ________________________________

Cutoff ________________________________

Linear region ________________________________

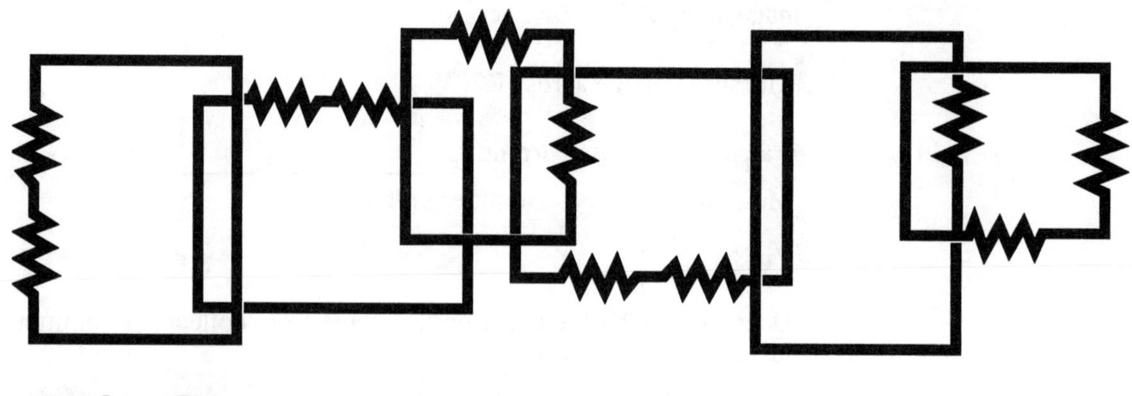

52

Thyristors—AC and DC Power Control

Objectives

After completing this experiment, you will be able to:

1. Construct several test circuits to evaluate the operational characteristics of the SCR and TRIAC.
2. Evaluate the differences and similarities between the SCR and TRIAC in both ac and dc power control.

Introduction

The purpose of this experiment is to introduce you to the two most prominent members of the thyristor family, the Silicon Controlled Rectifier and the TRIAC. You will construct several test circuits to evaluate the operational characteristics of the devices as electronic switches, and then discern their similarities and differences when used in both ac and dc power control. To accomplish the tests of the thyristors you will need:

- 1 Oscilloscope with probes
- 1 12.6VAC source, 500 mA rating
- 2 Rectifier diodes, 1N4001 or equal
- 1 SCR, S401E or equal, 400V, 1A
- 1 TRIAC, Q401E3 or equal, 400V, 1A
- 1 Electrolytic capacitor, 100 microfarad, 25V
- 1 Incandescent lamp, 14V, 0.1A, 7373 or equivalent
- 1 Resistor, 2200 ohms, $\frac{1}{2}$ watt
- 1 Potentiometer, 5000 ohms, 1 watt
- 6 Alligator clip test leads

Procedure

1. Pull the activity sheets for this experiment, then construct the circuit shown in figure 52–1. Connect test lead 1 to the gate of SCR1. SCR1 is wired into the circuit to switch current flow from an ac source through load DS1.
2. Energize the circuit. Vary the resistance of variable resistor R1 through its complete range of adjustment several times. What effect does this adjustment have on the brilliance of the lamp? Remove test lead 1 from the gate of SCR1 briefly, then reconnect it. What happened? Is the SCR a "normally on" device, or a "normally off" device? What seems to be needed to make it work?
3. Set the oscilloscope up to view dc voltage, then connect the oscilloscope input across lamp DS1. Set variable resistor R1 to the center of its range and adjust the scope for best display of the voltage waveform across the lamp. Again vary resistor R1 from end to end. How does the adjustment of R1 affect the waveform across the lamp? Is there any correlation between the waveshape and the brilliance of the lamp?

FIGURE 52–1 AC operation of SCR

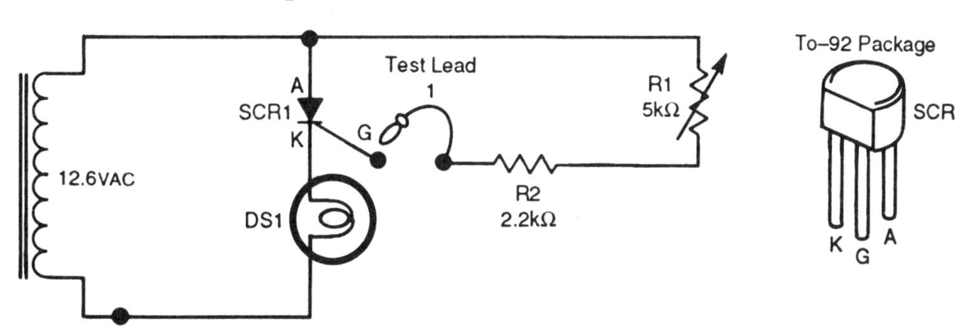

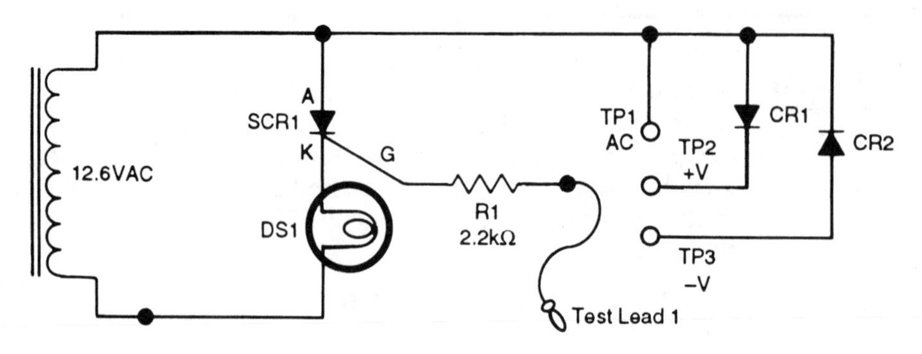

FIGURE 52–2 AC operation of SCR with AC, +V and -V gate inputs

4. De-energize the circuit and reconstruct it as shown in figure 52–2. Connect the scope input across lamp DS1.
5. Energize the circuit and note the status of lamp DS1. Now briefly touch test lead 1 to TP1 to apply alternating gate voltage to SCR1. Do this several times. Is the gate input needed to turn the SCR on? Now clip test lead 1 to TP1. What sort of waveform is developed across the lamp?
6. Move the scope input probe from across DS1, and place it from the gate of SCR1 to its cathode, to view the gate voltage waveform. What is the peak voltage at the gate of SCR1 in the positive direction? What is the peak voltage in the negative direction? What does this waveform tell you about the gate characteristic of the SCR? Place the scope input probe across DS1 again.
7. Remove test lead 1 from TP1 and briefly touch it to TP2 several times. This will apply a positive voltage to the gate of SCR1. What affect does this have on the brilliance of the lamp? Clip test lead 2 to the gate of SCR1. Has the waveform across DS1 changed?
8. Remove test lead 1 from TP2 and briefly touch it to TP3 several times. This will apply a negative voltage to the gate. Does the negative gate voltage cause the same circuit response as the positive gate voltage did? Why, or why not?

Review

You have constructed two test circuits to evaluate the operation of the SCR when connected to switch alternating current through a load. In the first circuit you applied an alternating voltage to the SCR gate by means of a current limiting variable resistance connection. You found that by varying the resistance of R1 you were able to vary the brilliance of lamp DS1. You saw that the current flow through the lamp was pulsating dc where the duration of SCR conduction could be varied to some extent, operating somewhat like a lamp dimmer. In this first circuit the SCR functions as a controlled half wave rectifier. There is no load current until the SCR is turned on, and its "turn on" can be controlled with a variable resistance in the gate circuit.

The second circuit allowed you to apply three different voltages to the gate of the SCR; alternating, positive dc and negative dc. The alternating gate input resulted in load current flow that was pulsating

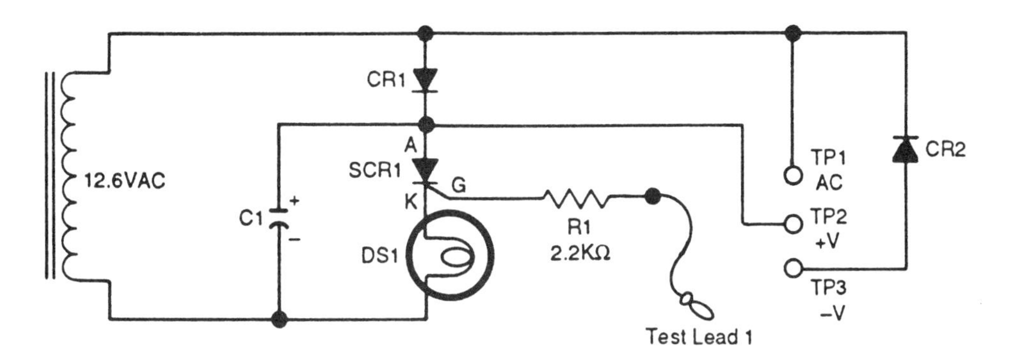

FIGURE 52–3 DC operation of SCR with AC, +V and –V gate inputs.

dc. The gate voltage viewed on the scope showed that gate current flows only during the positive alternation. The positive dc gate input produced essentially the same effect as the ac voltage. Load current was pulsating dc. When the negative voltage was applied to the SCR gate you found that an entirely different condition resulted.

You verified that the SCR functions as a controllable half wave rectifier in an ac circuit, and that the polarity of the gate voltage is quite important. Next you will evaluate how the SCR switches dc.

9. De-energize the circuit and reconstruct it as shown in figure 52–3. In this circuit the SCR is connected to switch direct current through lamp DS1.
10. Energize the circuit. Note the state of lamp DS1. Now touch test lead 1 briefly to TP1. What happened?
11. De-energize the circuit until lamp DS1 is extinguished, then reenergize it. Touch test lead 1 briefly to TP2. What happened?
12. Again de-energize the circuit until lamp DS1 is extinguished, then re-energize it. Now touch test lead 1 briefly to TP3. What happened?

Review

The SCR operates differently in a circuit where the current to be switched is dc. You discovered that when the SCR is gated on with a gate voltage of the proper polarity, it latches, and load current flows even when gate voltage is removed from the SCR. Once the SCR latches on, it remains in conduction until the source of load current is disabled. In the test circuit you interrupted load current by turning the supply off. Load current flow can also be interrupted by opening the current path with a switch. Next you will investigate the TRIAC.

13. Construct the circuit shown in figure 52–4. Connect the oscilloscope input across lamp DS1. Energize the circuit and note the status of load DS1.
14. Touch test lead 1 to the gate of TRIAC1 several times. What happens? Clip test lead 1 to the gate of TRIAC1 and vary the resistance of variable resistor R1 from end to end. What effect does this have on lamp DS1?

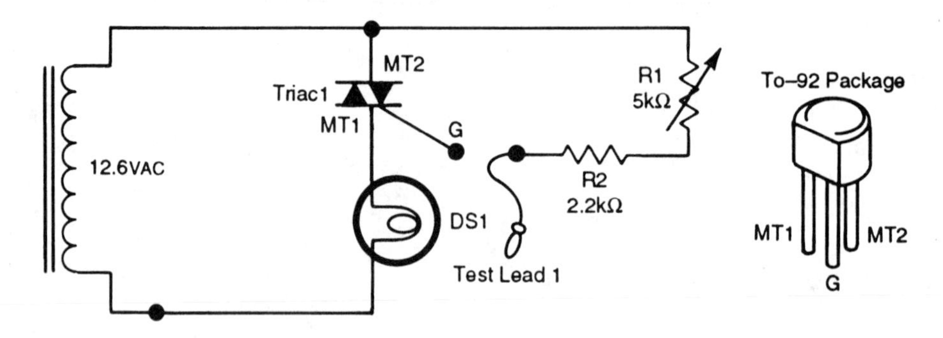

Figure 52–4 AC operation of TRIAC

15. View the voltage waveform developed across lamp DS1. Adjust the scope as necessary for best display. Again vary the resistance of R1 while watching the waveform. What does the waveform tell you about the operation of the TRIAC?
16. De-energize the circuit and reconstruct it as shown in figure 52–5. Connect the scope probe across lamp DS1. Energize the circuit and note the status of the lamp.
17. While observing both the lamp and the waveform across the load, touch test lead 1 to TP1 several times. What happens when the gate is momentarily touched? Clip test 1 to TP1 and analyze the waveform across the load. How is the waveform different from the SCR waveform when the SCR was used to switch an ac current?
18. Move the scope probe from the lamp to the TRIAC to view the gate waveform from the gate to MT1. Adjust the scope as necessary for best display of the gate waveform. What does the waveform tell you about the gate characteristics of the TRIAC? Put the scope probe back across the load lamp DS1. How is the TRIAC different from the SCR when used to switch alternating current?
19. Remove test lead 1 from TP1 and perform the same tests using TP2 and TP3 in succession. Note the state of the lamp with the test lead touching and not touching the positive and negative gate voltage points. Analyze the waveforms developed across the load when test lead 1 is connected to each of the gate voltage test points. How is the polarity of the gate voltage related to the polarity of the TRIAC output?

FIGURE 52–5 AC operation of TRIAC with AC, +V and –V gate inputs

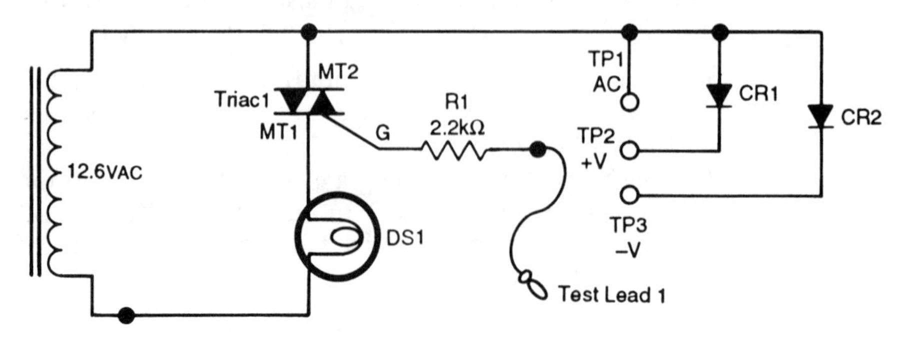

In the just completed series of tests you subjected the TRIAC to the same conditions you had done previously with the SCR. You found that in ac switching applications the TRIAC is bidirectional, it is able to be turned on in both directions of current flow. Like the SCR, its conduction duration can be varied when alternating gate current is applied through a variable resistance. But with the TRIAC it is controllable during both half cycles. The TRIAC doesn't rectify the load current, but rather, permits load current flow in either direction, as long as the gate voltage is also alternating.

Then when you used a dc voltage applied to the gate you discovered that the TRIAC is selective. It only passes the half cycles that are of the same polarity as the gate voltage. Positive gate voltage results in positive output, while negative gate voltage results in negative output. In this case it is like the SCR in that half wave rectification is accomplished, but the polarity of the rectification is selectable. Next you will evaluate TRIAC operation in dc switching applications.

20. Construct the circuit shown in figure 52–6. Energize the circuit and note the status of the lamp. Touch test lead 1 to TP1 making only momentary contact. What happens? Unclip one end of test lead 2 to extinguish the lamp, then re-attach it.

21. Perform the same test with test lead 1 to test points 2 and 3 in succession. Momentarily touching them to apply first positive gate voltage and then negative gate voltage. Unclip test lead 2 as necessary to extinguish the lamp when TRIAC1 latches. Is the TRIAC different from the SCR in this application? De-energize the circuit.

22. Re-configure the test circuit as shown in figure 52–7. Perform the same sequence of tests as done with the previous circuit. Momentarily touch each of the three test points 1, 2 and 3 with test lead 1 to apply ac, +dc and –dc to the gate. Detach test lead 2 as necessary to extinguish load DS1. How did this dc operated TRIAC circuit operate as compared to the previous one? Is gate polarity important when the TRIAC is used to switch dc current?

FIGURE 52–6 DC operation of TRIAC with AC, +V and –V gate inputs

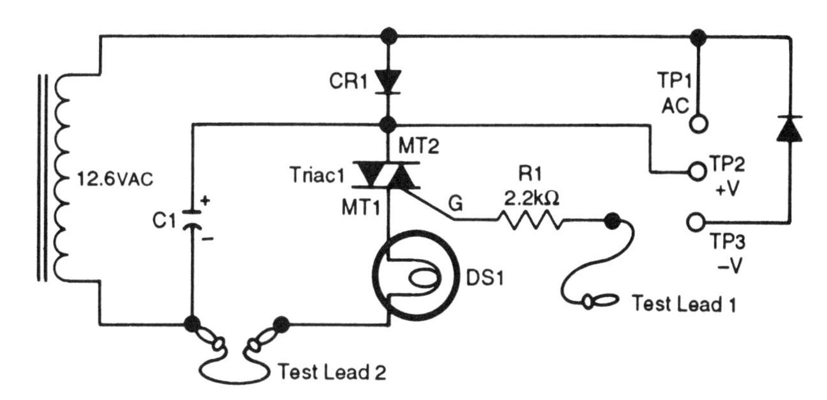

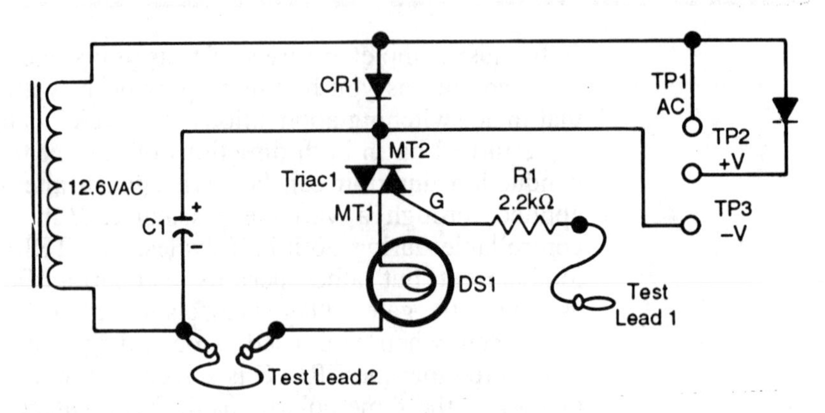

FIGURE 52-7 DC operation of TRIAC with AC, +V and -V gate inputs

Review

The last set of tests you performed revealed that the TRIAC is very similar to the SCR when used to switch dc current. In fact the TRIAC and the SCR could be used interchangeably in this application. The only difference is that the TRIAC is capable of being gated on with either positive or negative gate voltage that corresponds to the polarity of Main Terminal 2. A single TRIAC can be used to switch either positive or negative current. Like the SCR, the TRIAC also latches when gated on in the dc circuit. In order for the TRIAC to turn off, load current must be interrupted. This was accomplished by temporarily disconnecting an end of test lead 2 from the circuit.

You have completed a long and tedious, but necessary, series of electrical tests to examine the low voltage characteristics of two widely used thyristors. In advanced studies of the two you will explore high voltage, high current applications, where their switching differences will be more pronounced.

23. Return all materials to their proper places.

ACTIVITY SHEET EXPERIMENT 52

NAME ______________________________

DATE ______________________________

Thyristors—SCR

Steps 1–2

A. With test lead 1 connected to the gate of SCR1, and the circuit energized, rotate variable resistor R1 from end to end several times. As R1 is adjusted, what happens to the lamp?

____

With the lamp dimly illuminated, remove test lead 1 from the gate of SCR1. Reconnect it. Is the SCR dependent upon gate current in order to switch?

____

Step 3

B. With lamp DS1 illuminated, view the waveform across it with the oscilloscope. Is the displayed wave ac or dc? ___________

How do you know? ________________________________

____

____

While viewing the waveform vary R1 from end to end several times. Is the frequency, the amplitude, or the "on" period of the wave being altered?

____

What correlation do you see between the waveform and the brightness of DS1? ________________________________

____

____

Step 5

C. With test lead 1 disconnected from all of the test points, what is the state of DS1? ____________________ With test lead 1 connected to TP1, is the waveform developed across DS1 ac or dc? ____________________ Is the wave complete or incomplete when compared to the circuit ac input? ____________________

Step 6 **D.** View the waveform at the gate of SCR1 with respect to its cathode, while test lead 1 is connected to TP1. Study the wave carefully. Which part of the wave is clipped? ______________

Why? ____

____

Why is the opposite part of the wave not clipped?

____

____

Is there current flow in the gate circuit? ______________

How can you prove it? ______________________________

____

Step 7 **E.** When test lead 1 is touched to TP2, is the lamp lit? __________

Is it as bright as when test lead 1 is touched to TP1?__________

Why? ____

____

Step 8 **F.** Touch test lead 1 to TP3 several times. Is there any gate current flow? ____________________

Why? ____

____

Step 10 **G.** When you touch test lead 1 to TP1, what happens to DS1?

____________________ Why doesn't it turn off?

____

Step 11 **H.** After extinguishing DS1, touch test lead 1 to TP2. What happens to DS1?____________________ In this circuit, why does DS1 shine more brightly than in the previous circuits?____________

____

____

Steps 13–14 **I.** Without gate current, does TRIAC1 permit load current flow to DS1?____________________ When lamp DS1 is illuminated, is the waveform across it as viewed on the scope ac or dc?

____

Step 15 **J.** As R1 is varied from end to end, what aspects of the wave viewed across DS1 change?

Frequency?________________ Amplitude?________________

Pulse width?________________ Polarity?________________

How does the TRIAC differ from the SCR with regard to load current?____

____

____

Steps 16–17 **K.** Does the TRIAC seem to switch ac exactly like the SCR?

____________________ What seems to be different about the illumination of DS1? ____________________________________

____

Step 18 **L.** Observe the gate to MT1 waveform of the TRIAC. What part of the waveform is clipped? ______________________________

Why? ____

____

Does gate current flow during each half cycle?______________

Step 19 **M.** How does the gate input from TP2 affect the TRIAC?

______________________ Does DS1 light as brightly as it did with ac input to the gate?______________________ Does gate input from TP3 affect the TRIAC in a different way?

______________________ Does DS1 light as brightly as when gate input was from TP2?____________________

View the load waveform while gating the TRIAC from TP2, then TP3. What is different about the waveforms? ______________

____

Are they ac or dc? __________________ Are their polarities different?__________________

Steps 20–21

N. When switching dc current, will ac gate input activate the TRIAC?__________________ Why? ______________

____

What is the purpose of disconnecting test lead 2 temporarily?

____

Does +V gate input from TP2 switch the TRIAC?

____

When test lead 1 is touched to TP3, does gate current flow?

____________________ Why doesn't TRIAC 1 switch on?

____

____

Step 22

O. The polarity of the load current source has been reversed. Will the TRIAC permit load current flow in this case?

____________________ Why? ______________________

Which of the three gate inputs to TRIAC1 switch it on?

TP1__________ TP2__________ TP3__________

When the TRIAC is used to switch dc, why is the polarity of the gate signal important?

____

____

____

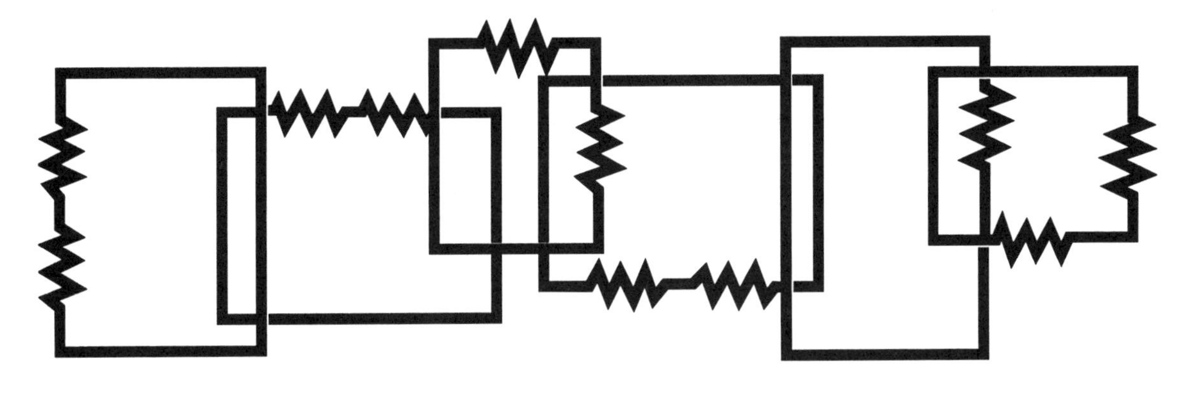

53

The Transistor as an Amplifier

Objectives

After completing this experiment, you will be able to:

1. Demonstrate the amplification characteristics of the transistor and transistorized amplifier circuits.
2. Perform tests on the transistor to determine its current gain and its voltage gain.
3. Construct amplifier circuits to evaluate small signal amplification and large signal amplification.

Introduction

This experiment will be a study of the amplification characteristics of the transistor and transistorized amplifier circuits. You will perform tests on a transistor to determine its current gain and its voltage gain. Then you will construct amplifier circuits to evaluate small signal amplification and large signal amplification. To accomplish these tests you will need:

1 Variable dc power supply
1 Function generator or variable audio source
1 Digital multimeter with test leads
1 Oscilloscope with test probe
2 NPN transistors, 2N3904 or equal
1 N channel JFET, type MPF102 or equivalent
1 One megohm potentiometer, 1 watt
1 10 megohm, $\frac{1}{2}$ watt resistor
1 100 ohm resistor, $\frac{1}{2}$ watt
1 220 ohm, $\frac{1}{2}$ watt resistor
1 470Ω resistor, $\frac{1}{2}$ watt
2 2200 ohm resistors, $\frac{1}{2}$ watt
1 1500 ohm resistor, $\frac{1}{2}$ watt
1 10,000 ohm resistor, $\frac{1}{2}$ watt
2 1000 ohm resistors, $\frac{1}{2}$ watt
2 4700 ohm resistors, $\frac{1}{2}$ watt
1 0.10 microfarad ceramic capacitor
4 10 microfarad electrolytic capacitors, 25 volts
6 Alligator clip test leads

Procedure

1. Pull the activity sheet for this experiment. Then construct the circuit shown in figure 53–1a. Energize the circuit and adjust the output of the dc power supply to 7 volts.
2. Set the digital multimeter up to measure dc volts. Connect the dc voltmeter across transistor Q1 from collector to emitter. Adjust variable resistor R1 until the measured voltage across Q1 is 3.0 volts.
3. With the dc voltmeter, measure the voltmeter, measure the voltage drop across base resistor R3. Then measure the voltage drop across collector resistor R2.
4. Using the measured voltages and the resistance values of the two resistors, calculate Q1 base current and collector current.
5. Now that you have determined the transistor input (base) current, and the transistor output (collector) current, calculate the current gain of the transistor.

Explanation

The attenuator shown in figure 53–1b, consisting of RA1 and RA2, is positioned between the signal generator and the test circuit. It is a voltage divider circuit which decreases, or attenuates, the signal level from the signal generator. Sometimes, when dealing with small signal amplitudes, the signal generator by itself is not able to output directly

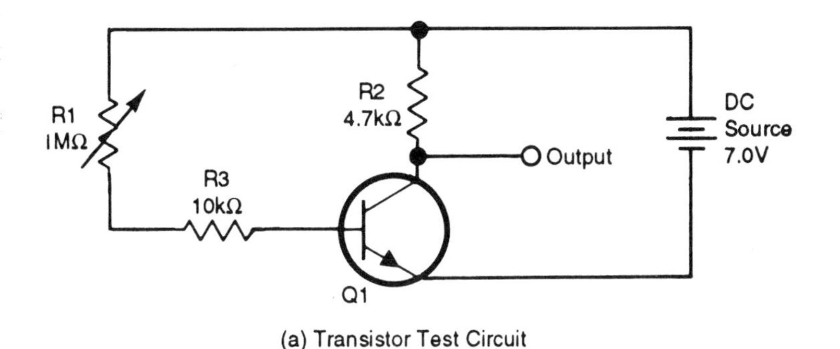

(a) Transistor Test Circuit

Signal Generator and Attenuator

(b) Small Signal Amplifier Circuit with input from Attenuated Signal Generator.

FIGURE 53–1 Transistor circuits

a signal as small as is needed. The external attenuator circuit is an easy way to assure that the signal can be made very tiny.

6. Modify the circuit by adding the coupling capacitor and voltage divider components shown in figure 53–1b. Set the oscilloscope up to measure ac voltages and connect the scope probe across Q1 from collector to emitter.
7. Energize the ac signal source and set its output frequency to approximately 5 kilohertz. Whole monitoring the scope display, adjust the output voltage of the ac source until a signal of 500 millivolts peak to peak is attained at the collector of Q1.
8. Now place the scope probe across the base-emitter junction of Q1 and measure the signal voltage present at the base. Adjust the scope as necessary for best display.
9. Using the measured input (base), and output (collector) signal voltages, calculate the voltage gain of Q1. Then de-energize the circuit.

Review

You have just constructed a simple amplifier circuit and used it to determine two characteristics of the transistor, dc current gain and ac voltage gain. You may be wondering "Well what about ac current gain and dc voltage gain?" The answer is "yes, there are those characteristics also!" It's just that the characteristics you have derived are somewhat easier to find, and fortunately, are virtually the same as their counterparts.

You found that the current gain is a very respectable quantity. Typically it is between 80 and 120 for a general purpose transistor such as yours, sometimes even as much as 200. You also found the voltage gain to be very high as well, typically more than 300. These are the qualities which make the transistor such a useful device; it is capable of a lot of amplification. A very tiny current at the base causes a very much larger current at the collector.

10. Leave your first test circuit intact and construct the circuit shown in figure 53–2a alongside it. The two circuits will be joined together later.
11. Set the output voltage of the signal generator to minimum. Energize the dc power supply (still set to 7 volts), then measure

FIGURE 53–2 Amplifier circuit

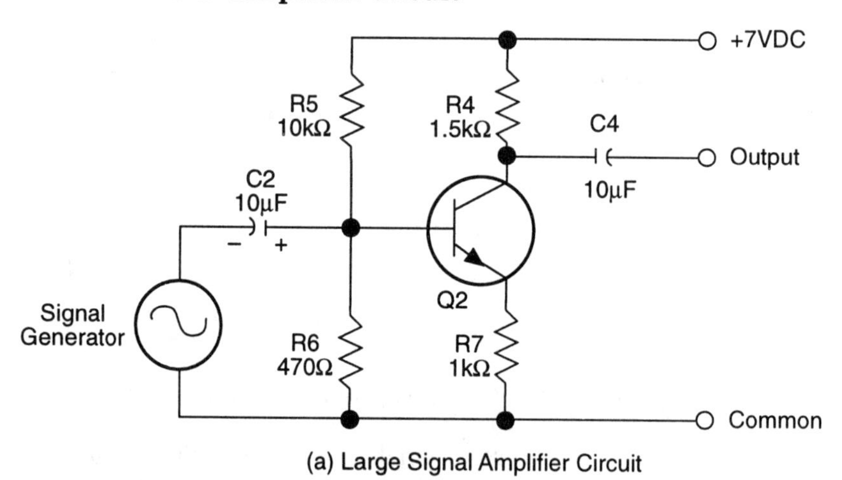

(a) Large Signal Amplifier Circuit

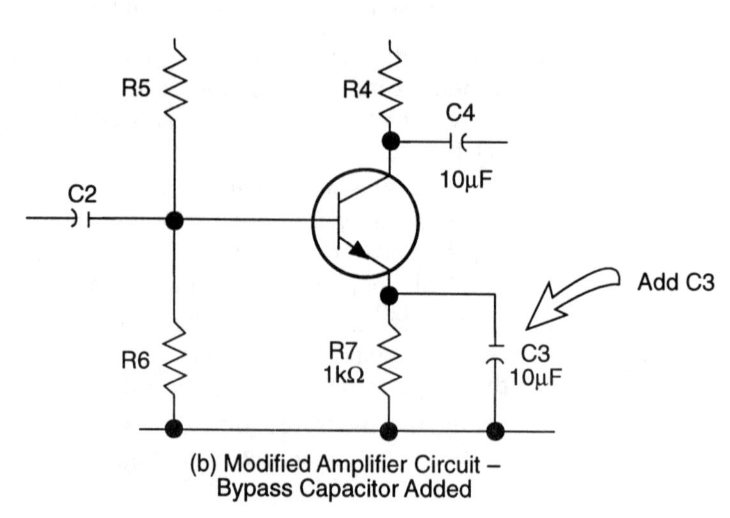

(b) Modified Amplifier Circuit – Bypass Capacitor Added

the dc voltage from Q2 collector to common with the digital voltmeter. It should be about four volts.

12. Connect the scope from the collector of Q2 to common. Increase the output of the signal generator until you see 2 volts peak to peak at the output of the circuit.
13. Move the scope probe to the base of Q2 to measure the signal input voltage to the circuit. Then calculate the voltage gain of the circuit. Is there something wrong with the circuit? Why is the voltage gain so small?
14. De-energize the circuit and modify it as shown in figure 53–2b by adding C3. Then re-energize the circuit.
15. With the scope attached to the collector of Q2, view the output waveform. If the signal output of the signal generator has not been changed you will see a severely distorted output. What has happened?
16. Re-adjust the signal output of the signal generator until the circuit output at the collector of Q2 is once again a clean, non-distorted two volts peak to peak. Then place the scope probe at the base of Q2 to measure the circuit input voltage. Re-adjust the scope as necessary for best display.
17. With circuit input and output voltages, calculate the circuit gain. Why is circuit gain so much higher now that the circuit has been modified?

Review

The second circuit you've constructed is a large signal amplifier and it has some added resistors to stabilize the quiescent current flow. When you first measured the circuit voltage gain, it was surprisingly small. In fact the output signal voltage was not a whole lot larger than the input. The circuit component most responsible for the poor voltage gain is resistor R7, in the emitter leg of Q2. This loss of signal gain is the cost of adding the emitter resistor, whose function is to eliminate quiescent current drift due to transistor thermal characteristics.

In the modified circuit you saw that the circuit voltage gain was much higher. The addition of bypass capacitor C3 boosted circuit voltage gain a lot, so that it was necessary to turn the signal level at the input down in order to return the output to 2 volts peak to peak.

18. De-energize the dc power supply and join the two amplifier circuits as shown in figure 53–3. Turn the signal generator output down to minimum.
19. Connect the scope to the circuit output at the collector of Q2. While monitoring the output waveform, increase the signal generator output voltage until a 2 volt peak to peak signal is obtained. Then move the scope probe to the base of Q1 to measure the circuit signal input. Re-adjust the scope as necessary for best display.
20. Using the measured signal input and output voltages, calculate the voltage gain of the two stage amplifier circuit.

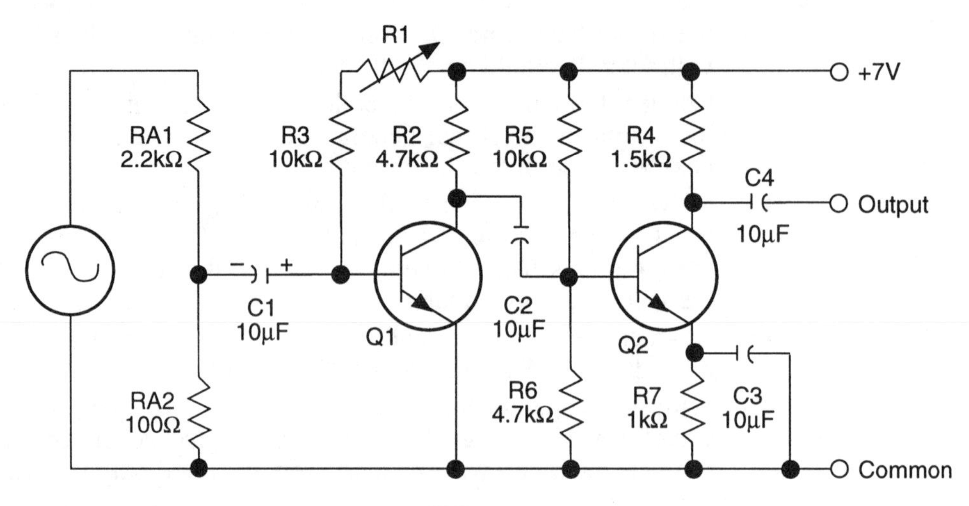

FIGURE 53–3 Two stage amplifier

Review

To create the two stage test circuit you cascaded the small signal amplifier and the large signal amplifier. When you measured and calculated the voltage gain of the combined circuit, you may have thought it to be smaller than it should. This is because when the two circuits are joined, the second circuit loads the first circuit and reduces its gain. Therefore the overall gain isn't as great as you might have expected when considering the individual circuit gains calculated previously.

21. Measure the input and output signal levels of the first amplifier stage and re-calculate its voltage gain. This will confirm that the gain has actually decreased in the cascaded circuit because the second circuit input loads the first circuit output. Next you will add a new front end to see if you can regain some of the lost voltage gain.

22. Modify the circuit as shown in figure 53–4, by adding a pre-amplifier stage. Energize the circuit. Place the scope probe at the drain of Q3 and measure its output. Then move the probe to the gate of Q3 to measure its input.

FIGURE 53–4 Three stage amplifier, pre-amplifier added

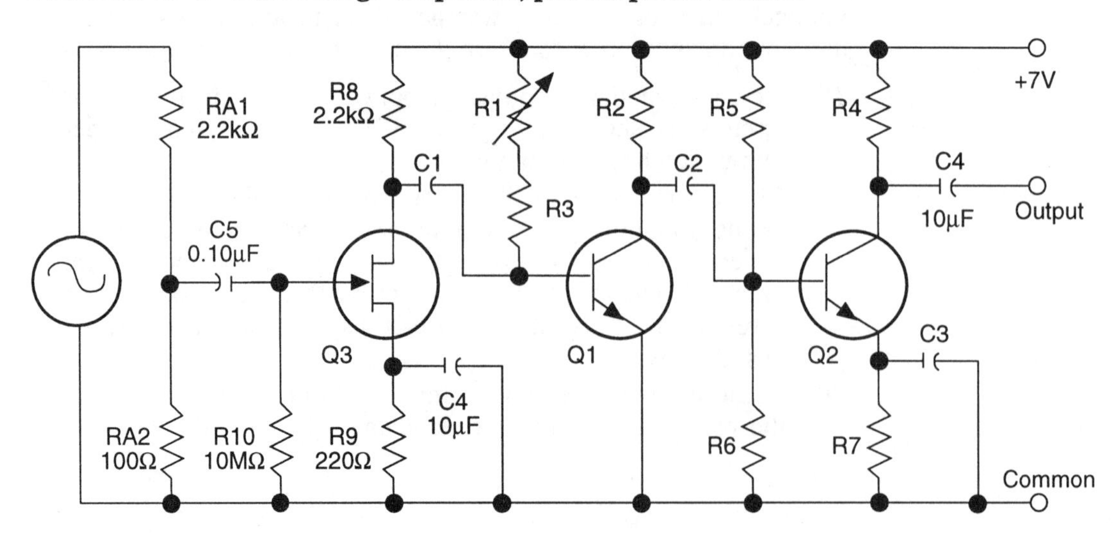

23. With the measured input and output voltages of the preamplifier stage, calculate its voltage gain.
24. Move the scope probe to the collector of Q2, the circuit output, and adjust the signal generator to attain a signal of two volts peak to peak.
25. Place the scope probe at the gate of Q3 to measure the circuit input signal voltage. Using the measured input and output voltages, calculate the three stage circuit voltage gain. Has the lost gain been restored?

Review

The purpose of putting another transistor into the circuit was to try to recapture some of the gain that was lost by loading the output of Q1 with the input of Q2. Transistor Q3, an N channel JFET, is the new "front end" of the amplifier. Although the JFET isn't used to its best advantage in this case, it should perform well. Field Effect Transistors have qualities which make them especially good candidates for small signal pre-amplifiers. Their amplification characteristics are very quiet, meaning that they don't introduce as much noise into the amplification process as most junction transistors. Their input impedance is very high, and they are voltage operated. Variations in gate voltage (input), produced corresponding changes in the drain current (output), without any gate current flow. There are other very important characteristics of the JFET, which you will study in advanced courses, which make them particularly well suited for circuit front ends.

26. Return all materials to their proper places.

ACTIVITY SHEET EXPERIMENT 53

NAME ______________________________

DATE ______________________________

Step 1 **A.** Adjust dc power supply output to 7 volts.

Measured dc power supply ____________________ V

Step 2 **B.** Adjust R1 for 3.0 volt drop across Q1 collector to emitter.

Measured Q1 V_{CE} ____________________ V

Step 3 **C.** Measured voltage drop across R3 ____________________ V

Measured voltage drop across R2 ____________________ V

Steps 4–5 **D.** Using measured voltages and resistance values, calculate Q1 base and collector currents.

Q1 I_B ____________________ μA

Q1 I_C ____________________ mA

Calculate Q1 current gain.

$$\beta = \frac{I_C}{I_B}$$

Q1β ____________________

Step 7 **E.** Set signal generator to 5 KHz and adjust output level until signal viewed at Q1 collector = 500m$V_{p\text{-}p}$.

Measured Q1 output signal ____________________ V_{pp}

Step 8 **F.** Measure Q1 input signal at Q1 base.

Measured Q1 input signal ____________________ V_{pp}

Step 9 **G.** Using measured input and output voltages, calculate Q1 voltage gain.

$$\text{Voltage gain} = \frac{V_{OUT}}{V_{IN}}$$

Av ____________________

Step 11 **H.** With power supply output voltage unchanged, measure Q2 drop from collector to emitter.

V_{CE} ____________________ V

Step 12 **I.** With signal generator output frequency unchanged, set signal output level to attain 2 V_{pp} at Q2 collector.

Measured Q2 output signal ____________________ V

Step 13 **J.** Measure Q2 input signal at base of Q2.

Measured Q2 input signal ____________________ V

Using the measured output and input signal voltages, calculate circuit voltage gain. Av ____________________

Step 15 **K.** After adding emitter bypass capacitor, with circuit input voltage unchanged, view Q2 collector signal. Why is the signal clipped?

____

Step 16 **L.** While viewing Q2 collector signal, decrease audio input level to circuit until 2V_{pp} is again attained.

Measured Q2 output ____________________ V

Measure circuit input signal at Q2 base.

Measured Q2 output ____________________ V

Step 17 **M.** With the improved Q2 circuit input and output signal voltages, calculate circuit voltage gain. Av ____________________

Step 19 **N.** While viewing the combined circuit output at Q2 collector, adjust signal generator input to Q1 base until the output signal is 2.0V_{pp}.

Measured Q2 output ____________________ V

Now measure circuit input signal voltage at Q1 base.

Measured Q1 input ____________________ V

Step 20 **O.** Calculate the voltage gain of the two stage amplifier circuit.

Av ____________________

Is the voltage gain of the cascaded circuit smaller than expected?

____________________ Why is this so? ____________________

Step 21 **P.** Measure the input and output signals of the first stage at Q1 base and collector.

input signal ____________________ V

output signal ____________________ V

Calculate first stage loaded voltage gain.

Av_{loaded} ____________________

Step 22 **Q.** Measure and record the output and input signal voltages at Q3.

Q3 output voltage at drain ____________________ V_{pp}

Q3 input voltage at gate ____________________ V_{pp}

Step 23 **R.** Calculate the voltage gain of Q3.

$$Av = \frac{V_{OUT}}{V_{IN}} = \text{\_\_\_\_\_\_\_\_\_\_\_\_\_\_\_\_\_\_\_\_}$$

Steps 24–25 **S.** Adjust signal generator for $2.0V_{pp}$ at Q2 collector.

Amplifier output voltage ____________________ V_{pp}

Measure amplifier input voltage at Q3 gate.

Amplifier input voltage ____________________ V_{pp}

Calculate the amplifier voltage gain.

$$\text{Overall amplifier gain} = \frac{V_{OUT}}{V_{IN}} = \text{\_\_\_\_\_\_\_\_\_\_\_\_\_\_\_\_\_\_\_\_}$$

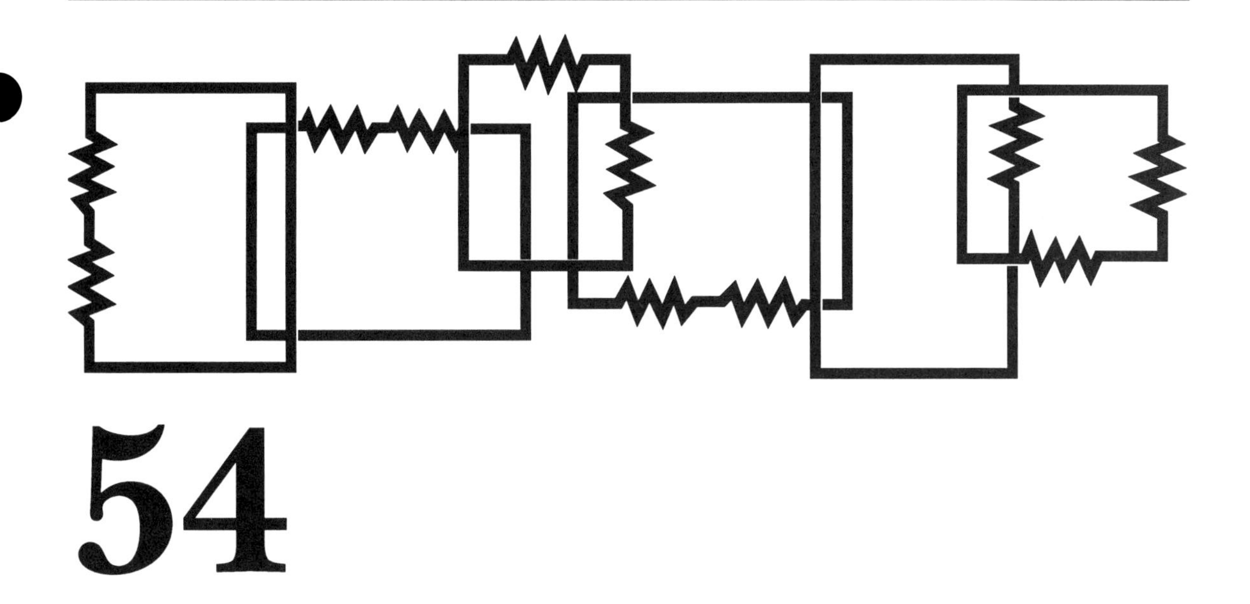

54

Amplifier Applications

Objectives

After completing this experiment, you will be able to:

1. Construct and evaluate two electronic circuits which rely upon amplifiers to make them work, namely:
   a. An RC network which provides regenerative feedback to an amplifier which in turn oscillates.
   b. An error correction amplifier which employs dc feedback to maintain a steady dc output.
2. Observe the relationship of frequency to resistance in the first circuit and the second circuit's susceptibility to failure when stressed.

Introduction

In this experiment you will construct and evaluate two electronic circuits which rely upon amplifiers to make them work, The first circuit makes use of a resistor-capacitor network to provide regenerative ac signal feedback to an amplifier which subsequently oscillates. The second circuit employs dc feedback to an error correction amplifier circuit which helps it maintain a steady dc output and therefore regulates. To construct these circuits you will need:

1 Variable output dc power supply
1 Digital multimeter with test leads
1 Oscilloscope with test probe
3 0.01 microfarad ceramic capacitors
3 10 microfarad, 25 volt capacitors
1 47 ohm, $\frac{1}{2}$ watt resistor
1 470 ohm, $\frac{1}{2}$ watt resistor
3 2.2 kilohm, $\frac{1}{2}$ watt resistors
2 4.7 kilohm, $\frac{1}{2}$ watt resistors
2 10 kilohm, $\frac{1}{2}$ watt resistors
1 47 kilohm, $\frac{1}{2}$ watt resistor
1 470 kilohm, $\frac{1}{2}$ watt resistor
2 2N3904 NPN transistors, or equivalent
1 Mini-loudspeaker, 8 ohms to 40 ohms
6 Alligator clip test leads
1 12.6 volt ac source, 500 mA or better
1 Rectifier diode, 1N4001 or equivalent
1 1N4733A 5.1 volt, 1 watt zener diode, or equivalent
1 100 microfarad, 25 volt, capacitor
2 1 kilohm, $\frac{1}{2}$ watt resistors
1 5 kilohm potentiometer
1 incandescent lamp, 14V, 100 mA, type 7373 or equal
1 incandescent lamp, 6V, 200 mA, type 7328 or equal

Procedure

1. Adjust the dc power supply output to 7.0 volts. Construct the circuit shown in figure 54–1. Pull the activity sheet for this experiment.
2. Energize the circuit. Monitor the output of the circuit with the oscilloscope. View the waveform at Q1 collector.
3. Adjust the scope for best display of the signal and measure the period of the signal. From this measurement calculate the frequency of the wave. Is it within the audio range?
4. Analyze the wave for signs of distortion. Do you see any evidence of wave distortion? What do you see?
5. Measure the signal amplitude at the circuit output from the scope display. Then move the scope probe to the base of Q1. Adjust the scope for best display, then measure the signal amplitude at the base. With these values determine the voltage gain of the circuit.

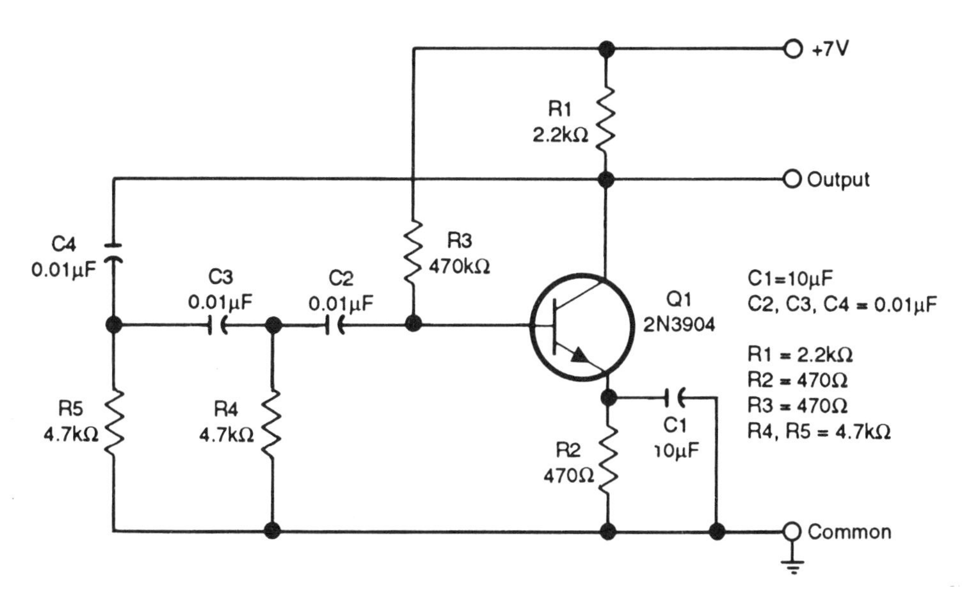

FIGURE 54–1 Oscillator circuit

6. De-energize the circuit and add to it the components shown in figure 54–2. Then re-energize the circuit.
7. Are you able to hear the output of the circuit? With the scope, measure the signal amplitude at the base and emitter of Q2. What is the voltage gain of this circuit? What aspect of the input signal is Q2 amplifying? Voltage, current or power?
8. Attach the scope probe to Q1 collector and measure the signal amplitude. Compare this measurement with the value measured previously. Did connection of the second stage of the circuit load Q1 output appreciably? Based on this comparison, would you say that the input impedance of the Q2 circuit is quite high?
9. De-energize the circuit, and replace resistors R4 and R5 with the 2.2 kilohm resistors. Re-energize the circuit. Has the output frequency increased or decreased? What electrical parameters of the collector to base feedback circuit have changed?
10. De-energize the circuit, and replace resistors R4 and R5 with the 10 kilohm resistors. Re-energize the circuit. What change in signal frequency do you observe now?

FIGURE 54–2 Loudspeaker driver

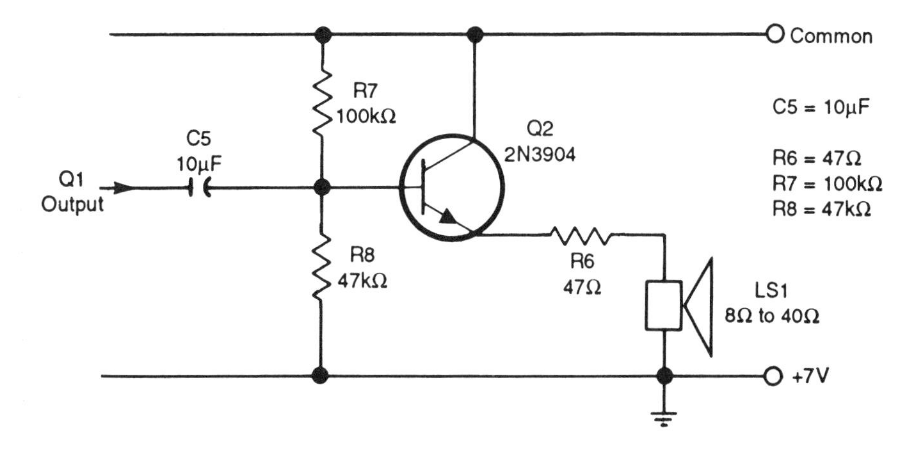

Review

The first circuit you constructed is a phase shift oscillator. The frequency determining components in the circuit are the feedback network comprised of capacitors C2, C3, C4, resistors R4, R5, and the input impedance of Q1. The feedback network introduces a 180 degree phase shift, from collector to base of Q1, at a frequency determined by the time constant of the network. Q1 also introduces a 180 degree shift internally from base to collector, making the total shift 360 degrees, or zero degrees. This is known as constructive or regenerative feedback, and its purpose is to support oscillation.

The circuit which was added on to the first is an emitter follower. This circuit delivers a voltage gain of less than unity, but produces a large current gain. So although there is no voltage increase, the large current gain makes for a power gain. The power output of Q2 is applied to the small loudspeaker to make the oscillator signal audible. The 47 ohm resistor in series with the loudspeaker dissipates some of the power to keep the speaker from making too much noise.

In order to demonstrate the frequency determining capability of the RC feedback network, you changed the values of the two resistors. First you decreased the values of the resistors, then you increased them. Did you note a direct relationship of frequency to resistance, or an inverse relationship? The frequency could have been changed by replacing the three capacitors in the feedback network as well.

The phase shift oscillator circuit is one which has good frequency stability. The frequency of oscillation doesn't drift much while it is in operation.

11. Dismantle the oscillator circuit and construct the circuit shown in figure 54–3.

12. Energize the circuit and attach the digital voltmeter to the unregulated input at the collector of Q1. Measure and record the unregulated dc input voltage. Then attach the digital voltmeter to the regulated output. Adjust potentiometer R3 from end to end and measure the voltage extremes that are developed at the regulated output.

FIGURE 54–3 Voltage regulator

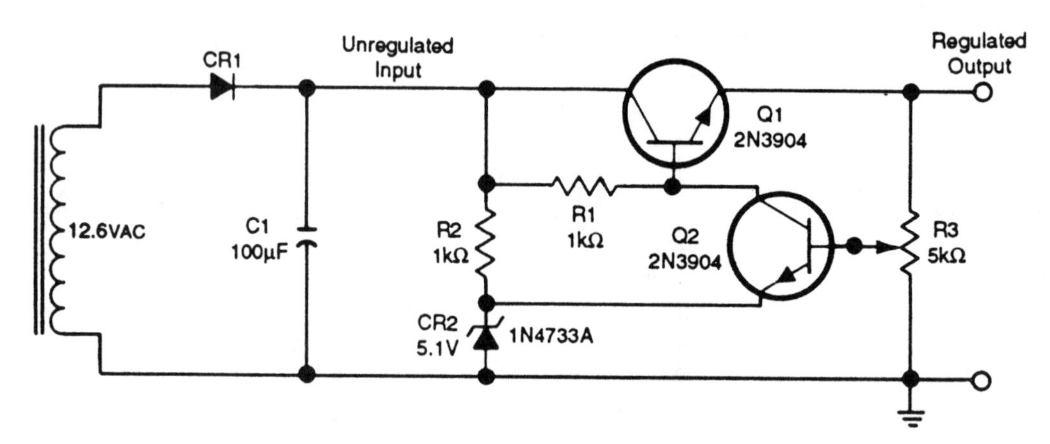

13. Load the regulator circuit by connecting the 14 volt lamp to the circuit output. Again adjust R3 from end to end and measure the maximum and minimum voltage that the loaded regulator is capable of.
14. Adjust the output voltage adjustment, R3, for a measured output of 12.0 volts, and leave the digital voltmeter connected to the regulator output. The lamp should be brightly illuminated.
15. With the oscilloscope view the ripple waveform, superimposed upon the dc voltage, at the unregulated input. Adjust the scope as necessary for best display, then measure the peak to peak amplitude of the ripple voltage.
16. Move the scope probe to the regulated output of the circuit and measure the ripple component of the loaded dc output. Then, while viewing the ripple waveform, disconnect the lamp momentarily. Does the ripple voltage increase or decrease?
17. While monitoring the digital voltmeter (still connected to the regulated output of the circuit) voltage, disconnect the lamp from the regulator output once again. Does the dc output voltage of the regulator change when the load is removed? Reconnect and disconnect the load a few times to be sure.
18. Remove the 14 volt lamp from the regulator output. Adjust output voltage potentiometer R3 until the regulated voltage indicated on the digital voltmeter is 6.0 volts.

Review

Thus far you've constructed and evaluated the performance of a voltage regulator circuit. Transistor Q1, which is called the pass transistor, is configured as an emitter follower. Transistor Q2, which functions as the error correction amplifier, is configured as a common emitter amplifier. Note that Q2 emitter is connected to the cathode of zener diode CR2. CR2, which is reverse biased, provides a stable reference voltage for Q2. The base input for Q2, provided by potentiometer R3, is a portion of the regulated output voltage which appears at the emitter of Q1. Q2 compares the base input voltage from R3 to its stable emitter voltage developed by CR2, and produces a voltage at its collector to control pass transistor Q1.

You measured the unregulated input to the pass transistor Q1, and while adjusting R3, measured the regulated output voltage capability of the circuit. You found that the circuit has limitations, which are characteristic of all regulators. The regulated output has definite upper and lower limits, and there is always some amount of ripple present there. Also, when the regulator is loaded, its dc output voltage decreases slightly, while the ac ripple voltage increases.

This test regulator was purposely put together with two weaknesses, to demonstrate the major problems associated with these circuits. First, filter capacitor C1 is much smaller in value than it would normally be. A small value was used to exaggerate the amount of ripple in both the input and output of the circuit. A voltage regulator circuit is capable of keeping ripple at the unregulated input from appearing in the regulated output, as long as the ripple is not severe at the unregulated input. Whenever you find excessive ripple in a regulated power supply output, suspect the filter capacitor at the unregulated input. Second, a transistor with relatively small current handling capability was used as the pass transistor. Normally, Q1 would be a power transistor mounted on a heat sink. Even at relatively small current levels, around a hundred milliamps or so, a lot of power is dissipated by the pass transistor. Especially when the regulated output voltage is less than half the unregulated input voltage. The power dissipated produces heat which can damage the pass transistor. To assure that the pass transistor would be heavily stressed by heat buildup, a small plastic transistor was used. You will now observe the effects of transistor stress.

CAUTION

As you proceed with the experiment, transistor Q1 will develop a great deal of heat. DO NOT TOUCH the transistor until it has cooled for several minutes after it self-destructs.

19. While monitoring the voltage indicated on the digital voltmeter, quickly connect and disconnect the 6 volt lamp to the regulator output. Keep the lamp connected to the regulator output only long enough for the digital meter display to stabilize, then disconnect the lamp. Record the measured loaded voltage. If all is well, the digital voltmeter will continue to display 6.0 volts when the lamp is disconnected.

20. Now connect the 6 volt lamp to the regulated output and let it remain connected until the pass transistor exceeds its power handling capability. This will only take a few seconds. It is probable that the 6 volt lamp will burn out as well.

21. After the regulator pass transistor has expired, de-energize the circuit and allow the overheated transistor to cool. Describe the visual and olfactory indications of regulator burnout on the activity sheet.

22. After the defective pass transistor has cooled, remove it from the circuit and test it using the digital multimeter in the diode test mode. Record your observations on the activity sheet.

Review

Although the regulator test circuit you have used was particularly susceptible to failure by design, it demonstrates a possibility that all regulators are subject to. Whenever a regulator circuit is loaded beyond its abilities, something will be stressed. The component in the series regulator which is most likely to fail is the pass transistor. When the pass transistor fails because of excessive heat buildup, it generally shorts from collector to emitter. This in turn subjects the load to the full unregulated input voltage, and it too is often destroyed. In advanced studies of regulated power supplies you will learn that protective features can be built into the regulator circuit to prevent pass transistor burnout. There are also features that can be included in regulator design to avoid subjecting the load to hazardous overvoltage and possible destruction.

23. Return all materials to their proper places.

ACTIVITY SHEET EXPERIMENT 54

NAME ______________________________

DATE ______________________________

Steps 1–3

A. View signal waveform at Q1 collector. Measure the period of the wave.

Period of one cycle t = ____________________ S

Frequency $= \frac{1}{t} =$ ____________________ Hz

Step 4

B. Analyze the wave for any signs of distortion. What kind of waveform do you see? ______________________________

____

Is the wave visibly distorted? ______________________________

Is the amplifier operating as a linear amplifier? ______________

____

Step 5

C. Measure the output and input signal voltages Q1.

Output voltage ____________________ V_{pp}

Input voltage ____________________ V_{pp}

Calculated voltage gain $= \frac{V_{OUT}}{V_{IN}} =$ ____________________

Step 7

D. Does the added circuitry produce an audible output? __________

Measure the input and output voltages at Q2.

Input voltage at Q2 base ____________________ V_{pp}

Output voltage at Q2 emitter ____________________ V_{pp}

Q2 voltage gain ____________________

Explain what Q2 amplifies:

Voltage? ________ Current? ________ Power? ________

____

Step 8 **E.** Measure signal voltage at Q1 collector.

Q1 output voltage ______________ V_{pp}

Compare this output voltage with the output previously measured at C (step 5).

Explain loading: ______________

____

____

____

Step 9 **F.** Measure period and determine frequency of new output signal.

t = ______________ S

f = ______________ Hz

What aspect of the feedback network have you changed?

____

Step 10 **G.** Determine period and frequency of new signal.

t = ______________ S

f = ______________ Hz

Step 12 **H.** Measure the unregulated dc input to the circuit at Q1 collector.

______________ V

Measure the output of the regulator circuit with R3 set for minimum and maximum.

V_{out} min. ______________ V

V_{out} max. ______________ V

Step 13 **I.** Load the regulator output with the incandescent lamp, and again measure the circuit minimum and maximum outputs.

V_{out} loaded min. ______________ V

V_{out} loaded max. ______________ V

Steps 14–16 **J.** Adjust R3 until the loaded regulator output is 12.0 volts as shown on digital voltmeter.

Loaded dc output ______________ V

Measure ripple voltage at unregulated input and regulated output of the circuit.

Ripple at input ____________________ V_{pp}

Ripple at output ____________________ V_{pp}

When the load is momentarily disconnected from the regulator, does the amount of ripple seen in the dc output change? ______ ______________ Explain why. ________________________

____

____

Step 17

K. When the load is removed from the regulator output, does the dc output voltage change? ____________________

Explain. ____

____

____

Step 18

L. With the load removed, adjust the regulator output to 6.0 volts.

V_{out} unloaded ____________________ V

Step 19

M. Connect the incandescent lamp to the regulator output only long enough to read the loaded output voltage.

V_{out} loaded ____________________ V

Has connection of the load to the regulator caused the output voltage to change?____________________ Explain. ________

____

____

____

Steps 21–22

N. While the pass transistor Q1 was being stressed, what did you see and smell?

____

____

Test the stressed transistor and describe your findings.

____

____

____

____

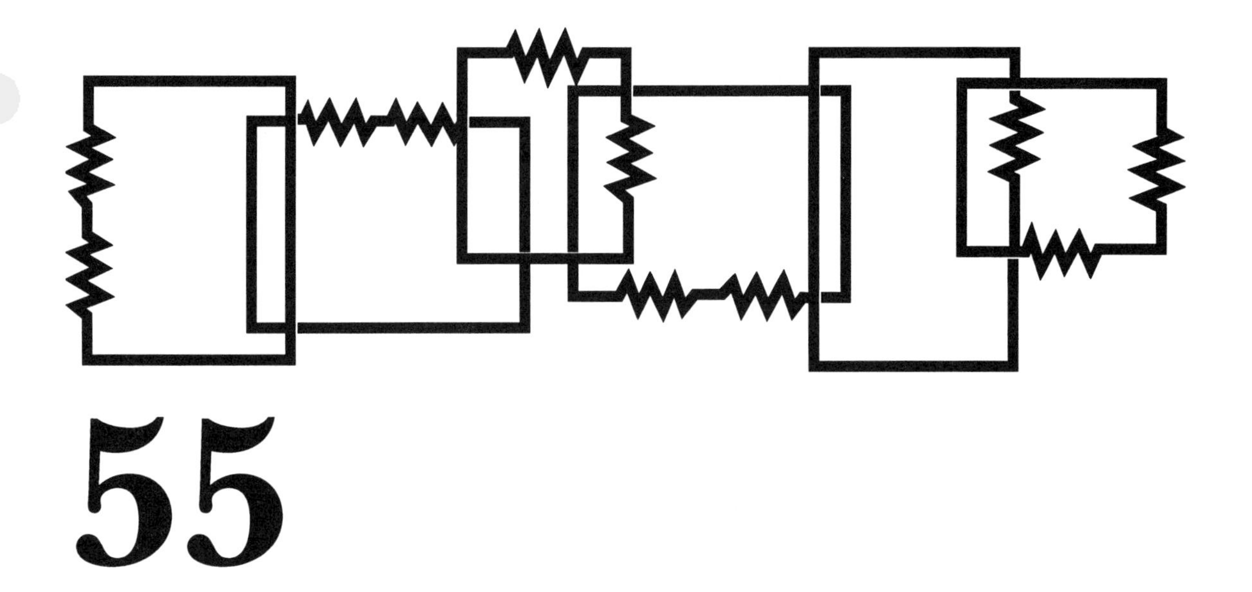

55

Operational Amplifiers

Objectives

After completing this experiment, you will be able to:

1. Construct a two stage circuit to demonstrate the versatility of the operational amplifier.
2. Demonstrate how a two stage amplifier can be used to amplify AM radio signals and light source signals.

Introduction

In this experiment you will construct a two stage circuit to demonstrate the versatility of the operational amplifier. The Op Amp is one of the marvels of modern semiconductor electronics. It is a complex multitransistor circuit, built upon a small chip of silicon, which displays the most desirable characteristics of amplifier circuits. And it comes in a small, easy to use package. To perform this experiment you will need:

- 1 Digital Multimeter with probes
- 1 Oscilloscope with test probe
- 1 Dual output dc power supply
- 2 LM741CN op amp chips, or equivalent
- 1 1 megohm potentiometer
- 1 10 kilohm, $\frac{1}{2}$ watt resistor
- 1 4700 ohm, $\frac{1}{2}$ watt resistor
- 1 47 ohm, $\frac{1}{2}$ watt resistor
- 1 0.10 microfarad ceramic capacitor
- 1 1N34 germanium diode, or equal
- 1 Photoresistor, 10 kilohms "dark resistance" or better
- 1 length of hookup wire, #22 to #26, 10 to 20 feet long
- 1 mini-loudspeaker, 8 to 40 ohms
- 6 alligator clip test leads

Procedure

1. Set the outputs of the dual voltage power supply to positive 10 volts and negative 10 volts, both referenced to common or

FIGURE 55–1 High gain amplifier

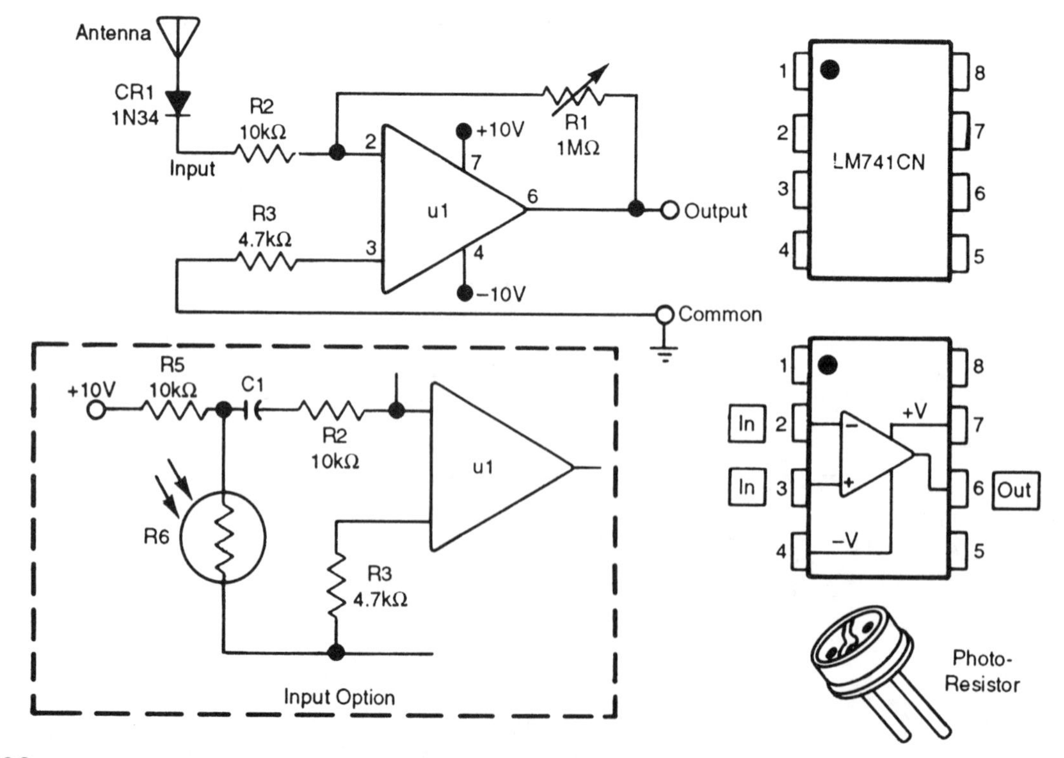

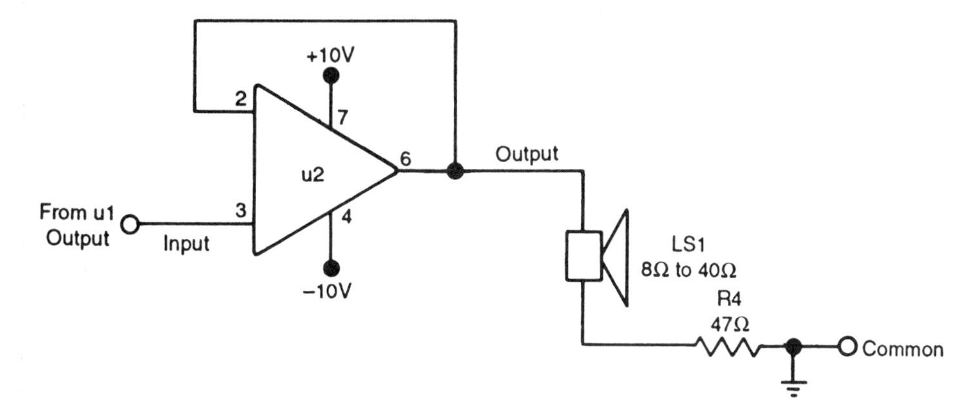

FIGURE 55–2 Speaker driver

ground. Construct the circuit shown in figure 55–1. The antenna for the circuit is a ten to twenty foot length of hookup wire suspended from the ceiling if possible. It can also be draped horizontally across any convenient object at hand.

2. Pull the activity sheet for this experiment.
3. Energize the circuit. Place the scope probe at the circuit output, and while viewing the scope screen, adjust variable resistor R1 slowly from end to end. Adjust the scope controls as necessary for best display. Set R1 to produce maximum undistorted output and measure the peak to peak amplitude of the signal.
4. Move the scope probe to the connection point of CR1 and R2, the input of the circuit. Adjust the controls of the scope as necessary for best display, then measure the amplitude of the signal. Using the measured input and output signals of the circuit, calculate its signal gain.
5. De-energize the circuit and add to it the components shown in figure 55–2. Then re-energize the circuit.
6. You should hear sounds from the loudspeaker. Adjust variable resistor R1. How does it affect the sound?
7. Adjust R1 until the level of audio output is just audible at the speaker. With the scope, measure signal amplitude at the input to U2, and at U2 output. Calculate the voltage gain of U2. Would you describe the function of U2 as being a voltage, current or power amplifier?
8. Refer back to figure 55–1, and alter the input of U1 as shown in the input option diagram. Clip the photoresistor into the circuit with a pair of alligator clip leads so that you can point it at various light sources. Adjust R1 as necessary to maintain an appropriate volume level from the loudspeaker. Point the photoresistor in the direction of available electrically operated light sources and describe the sound produced. Based on what you hear, would you say that the light sources are producing a steady light output, or a pulsating light output?
9. View the output of the two stage amplifier and describe the signal waveform. Measure the period of the waveform and calculate the signal frequency. Is there a discrepancy in the measurement?

Review

The circuit you have constructed consists of two stages of circuitry. The first stage operates as a moderately high impedance, high gain audio amplifier. If you were able to measure input and output signals, you found that it has a voltage gain of 100 or less, as determined by the adjustment of feedback resistor R1. The input to the circuit, a wire antenna and diode detector CR1, should allow you to receive and detect radio signals from local AM radio stations. Depending on how many local radio stations there are in your locale, and what their output power is, you should be able to measure a small signal at the input to the amplifier, and a larger signal at the output. From the scope display, you should be able to recognize audio signals varying in amplitude.

When you attach the second part of the circuit, with the loudspeaker, you should be able to hear the audio signal from one or more AM radio stations. When variable resistor R1 is varied, the volume of the speaker is increased and decreased. When you determine the voltage gain of the speaker driver, U2, you find it is unity, or 1. Therefore the circuit, which operates as a voltage follower (no voltage increase but large current increase), functions as an impedance matching device to drive the speaker efficiently. If the signal amplitude from the antenna/detector is too weak to hear well, resistor R4 in series with the loudspeaker, can be removed or bypassed with a short circuit.

The second input to the amplifier circuit, constructed from the photoresistor, fixed resistor and capacitor, enables you to "hear" light sources. With the photoresistor pointed at an ac operated incandescent or fluorescent lamp, you should hear a distinctive sound. When you measure its frequency, you may wonder why it is the second harmonic of the power line frequency.

10. Return all materials to their proper places.

ACTIVITY SHEET EXPERIMENT 55

NAME ______________________________

DATE ______________________________

Step 3 **A.** With the test circuit hooked up and energized, adjust R1 for maximum undistorted signal at the output of U1.

$V_{signal\ out\ max.}$ ____________________ V_{pp}

Step 4 **B.** Measure the signal amplitude at the input of U1.

$V_{signal\ in}$ ____________________ V_{pp}

Determine gain of U1.

Av ____________________

Step 6 **C.** Describe the sound you hear from the loudspeaker.

____

____

____

How does adjustment of R1 affect the sound?

____

Step 7 **D.** Measure the input and output signals of U1.

V_{in} ____________________ V_{pp}

V_{out} ____________________ V_{pp}

Av ____________________

Step 8 **E.** Point the photoresistor at various light sources. Describe what you hear. ____

____

Are the light sources producing a steady output? ____________

Explain. ____

____

____

Step 9

F. From the scope display of the signal produced by a fluorescent lamp, measure the period of the wave.

t ____________________ s

Calculate its frequency.

f ____________________ Hz

Explain the apparent discrepancy.

____

____

____

____

____

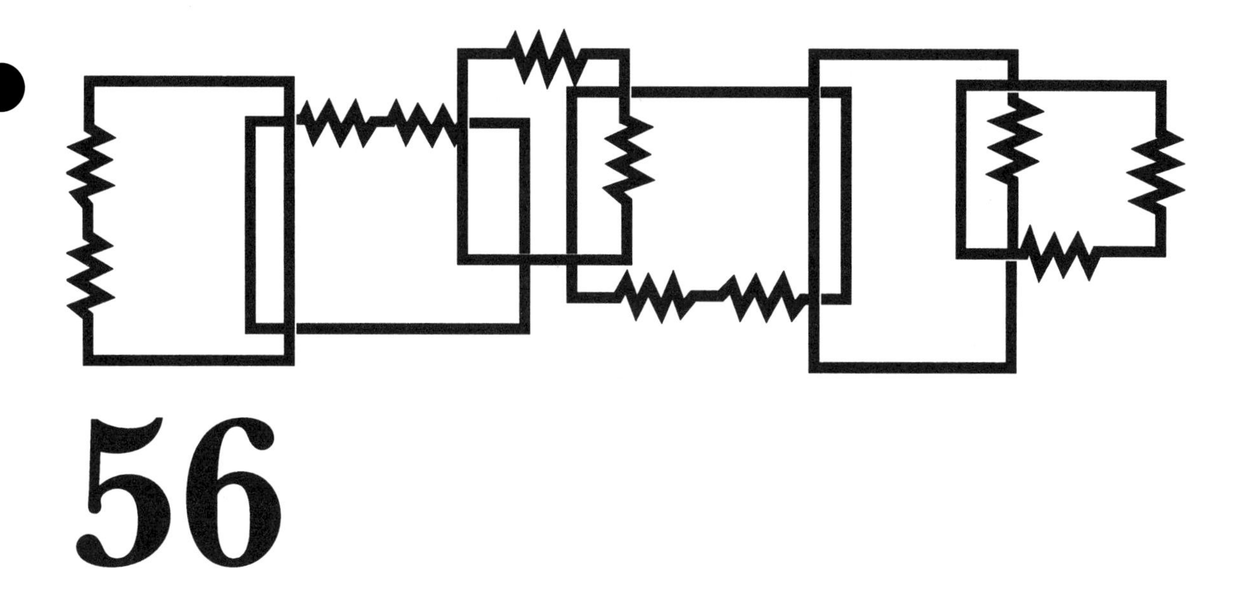

56

Op Amps as Comparators

Objectives

After completing this experiment, you will be able to:

1. Construct and evaluate two op-amp comparator circuits.
2. Compare a simple single reference comparator circuit to a dual reference comparator circuit.

Introduction

This experiment is about op amp circuits which are used to compare voltages. The first circuit is a simple comparator, which shows you by visual indication, when a variable voltage is either too low or too high when compared to a fixed reference voltage. But, as you will see, the circuit has a weakness. The second circuit will be a more complex kind of comparator which will compare a variable voltage to two predetermined references, a low limit and a high limit. It will also show you visually the status of the variable voltage. To accomplish your evaluation of these circuits you will need:

1 Dual output dc power supply
1 Digital multimeter with test leads
1 Zener diode, 5.1 volts, 1 watt, type 1N4733A or equal
1 Light emitting diode, color red
1 Light emitting diode, color yellow
1 Light emitting diode, color green
2 LM741CN op amps, or equivalent
1 5.0 kilohm, 1 watt, potentiometer
2 10.0 kilohm, $\frac{1}{2}$ watt, resistors
3 1.0 kilohm, $\frac{1}{2}$ watt, resistors
2 470 ohm, $\frac{1}{2}$ watt, resistors

Procedure

1. Pull the activity sheet for this experiment. Construct the circuit shown in figure 56–1. Adjust the dual dc power supply outputs to positive 10 volts and negative 10 volts, both referenced to common or ground.
2. With the digital voltmeter, measure and record the voltage developed by CR1, at TP1. This is the reference voltage input to the circuit.
3. Connect the voltmeter to TP2 to monitor the adjustable voltage input to the circuit. Vary potentiometer R3 from end to end, while observing LED 1 and LED 2.

FIGURE 56–1 Comparator with high-low indicators

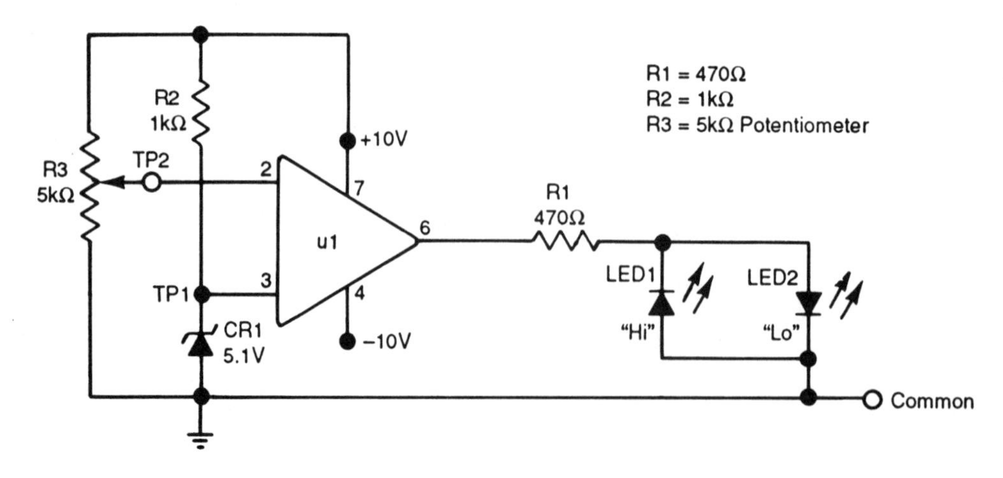

4. Rotate R3 to increase the variable voltage input from minimum to maximum. Measure and record the voltage at which the leads change states. Repeat several times to obtain an accurate measurement.
5. Rotate R3 to decrease the variable voltage from maximum to minimum. Measure and record the voltage at which the leds change states. Repeat several times to obtain an accurate measurement.

Review

The circuit you've just evaluated is a simple comparator. Op amp U1 is being used in the open loop configuration where there is no gain limiting feedback. This causes the amplifier to operate with maximum gain, so that it "over responds" to any input. Non linear operation is the result, and this mode is characterized by the output of the op amp being driven into one of two states, positive saturation and negative saturation.

As you rotated R3 back and forth, you found that the leds changed states near the middle of their range. When you measured the variable voltage input, you found that the leds changed states when the variable voltage was exactly equal to the reference voltage. The only information the leds can convey in this circuit, is that the variable voltage is either greater than, or less than, the reference voltage. You found that one led always meant "greater than", and the other led always meant "less than" the reference voltage. The next circuit will tell you more.

6. Disassemble the first test circuit and replace it with the circuit shown in figure 56–2.
7. With the digital voltmeter, measure and record the voltages at TP1 and TP2. These two voltages represent the upper and lower limits against which the variable voltage input will be compared.

FIGURE 56–2 Window comparator with indicators

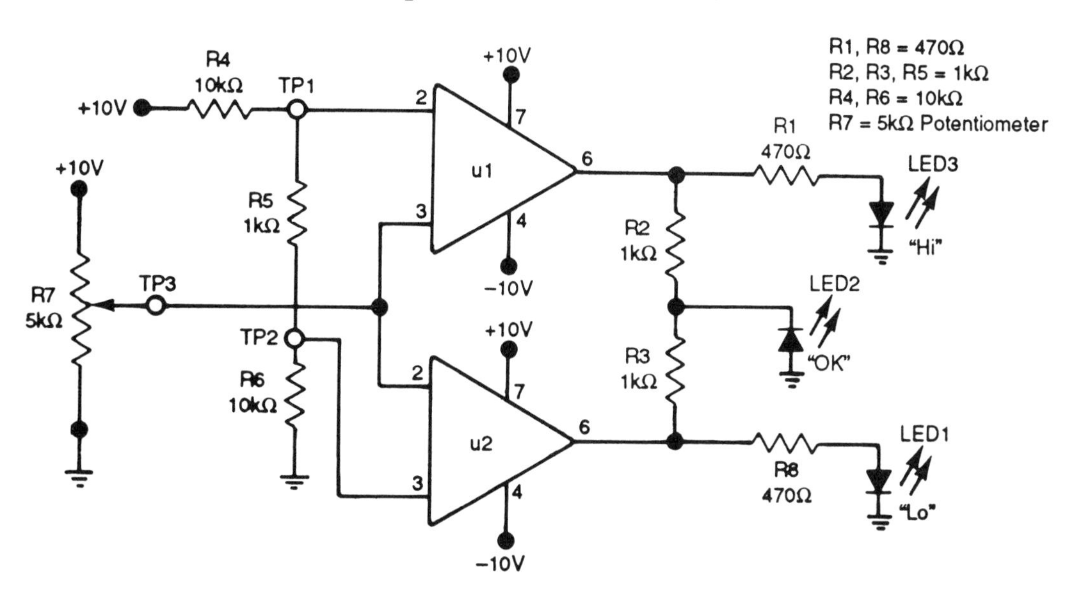

8. Connect the digital voltmeter to TP3, the variable voltage input, and rotate R7 from end to end several times. Observe the action of the leds as you do this.
9. While monitoring the variable input voltage at TP3, rotate R7 slowly. Measure and record the voltages at which the leds change states. You should see only one of the leds illuminated at any time. As you adjust R7 from minimum voltage to maximum voltage, the three leds should illuminate in sequence. Then when you adjust R7 from maximum voltage to minimum voltage, the illumination sequence of the leds will be reversed. Record the voltages at which the illuminated led changes.
10. Analyze the data you have just collected. What does it mean when LED 1 is illuminated? LED 2? LED 3?

Review

The circuit you've just evaluated is known as the Window Comparator. It has three leds at its output, any one of which can be illuminated at any time, but only one at a time. When the variable voltage input was varied, increasing or decreasing beyond one of the two reference limits, the states of the leds would change. The fixed voltage inputs, taken from the resistive voltage divider, are two different voltages that define a "window". Whenever the variable voltage falls within this window, a certain led will illuminate. As the variable voltage is varied above and below this window, one led will always tell you when it is "high", and another led will always tell you when it is "low".

With a little improvisation the circuit could be used to monitor the output of a variable power supply. The state of the leds would reveal to you when the output voltage of the supply is too high, too low, or right on, as programmed by the upper and lower limits of the window. A sort of electronic meter.

11. Put all materials in their proper places.

ACTIVITY SHEET EXPERIMENT 56

NAME ______________________________

DATE ______________________________

Step 2 **A.** Measure the voltage across CR1 at TP1.

$V_{reference}$ ____________________ V

Steps 3–5 **B.** Monitor the voltage at TP2. While rotating R3 from minimum to maximum voltage, measure voltage where leds change states.

V_{change} ____________________ V

Rotate R3 from maximum to minimum voltage. Measure voltage where leds change states.

V_{change} ____________________ V

Step 7 **C.** Measure the circuit reference voltages at TP1 and TP2.

V_{TP1} high reference ____________________ V

V_{TP2} low reference ____________________ V

Steps 8–10 **D.** Rotate R7 slowly from minimum to maximum voltage while monitoring the voltage at TP3. Measure the "on" range of each LED.

LED1 "on" from 0.0 V to ________________ V

LED2 "on" from ______________ V to ______________ V

LED3 "on" from ______________ V to 10.0 V

Calculate the width of the window represented by LED2 in volts

LED2 window ____________________ V

Could a circuit such as this be useful in monitoring an automotive electrical system? ______________________________

What could it tell you? ______________________________

____

____

____

Going beyond:

Modify the values of resistors used to establish the high and low limits of the window comparator. Customize it to fit another voltage range.

How about negative voltages? Will the window comparator sense negative voltages as well?

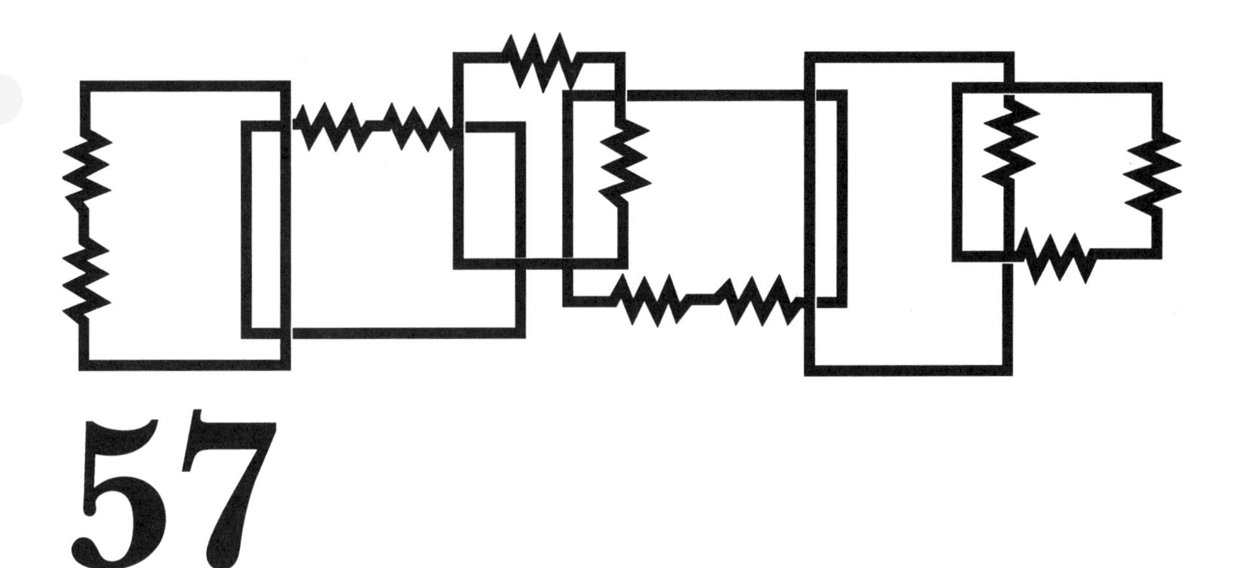

57

Series Circuit Measurements

Objectives

After completing this experiment, you will be able to:

1. Evaluate voltage distribution in a series circuit.
2. Determine actual resistance of individual resistors.
3. Confirm the principles of Ohm's Law.

Introduction

In Experiment 57 you will use a computer and software program to construct a series resistive circuit and explore the electrical characteristics of that circuit. You will be conducting tests to measure resistance, voltage, and current. The test equipment you will use in the computer application program is very similar to what you will use on the lab workbench or out in a maintenance work shop.

Procedure

1. Open Electronics Workbench (EWB)/MultiSim. See your instructor for details on how to access and open the application.
2. Insert the CD from the back of the lab manual into the CD drive.
3. Place three resistors from the parts bin onto the work surface.
4. Configure the resistors as shown in Figure 57–1.

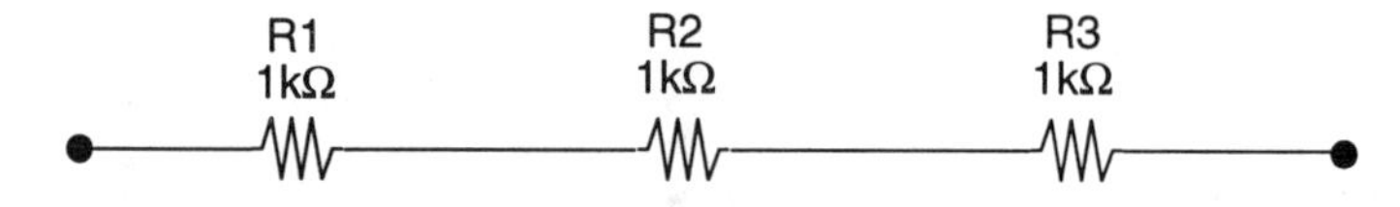

FIGURE 57-1

5. Move the multimeter from the tool bar onto the work surface and set it up to function as an ohmmeter.
6. Measure the value of each resistor by connecting the meter leads so that the meter is in parallel with the component to be tested.
7. Record the measured value of each resistor in the experiment activity sheet.
8. Disconnect the meter from the circuit.
9. Add a 10 volt battery to the circuit as shown in Figure 57–2.

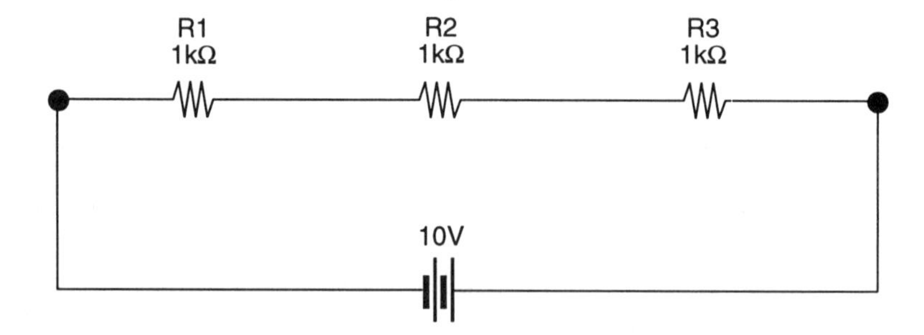

FIGURE 57-2

10. Select the multimeter and set it up to measure voltage.
11. Measure the voltage drop across each resistor by connecting the meter leads so that the meter is in parallel with the component to be tested. Record the results in the experiment activity sheet.
12. Set the multimeter to measure current.
13. Measure the current through each component by connecting the meter in series with each component. See Figure 57–3 for an illustration of how to connect the meter.

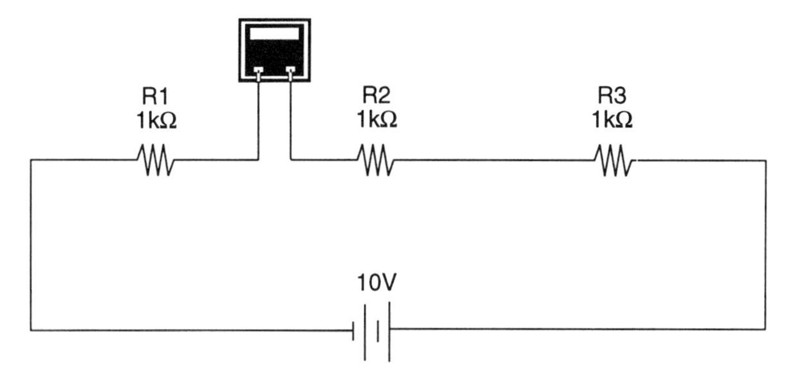

FIGURE 57–3

14. Record the results in the experiment activity sheet.

15. Verify test results by calculating the values of voltage and current for the circuit in Figure 57–2.

ACTIVITY SHEET EXPERIMENT 57

NAME ______________________________

DATE ______________________________

Step 7 R1= ____________________ R2= ____________________

R3= ____________________

Step 11 E_{batt}= ____________________ V_{R1}= ____________________

V_{R2}= ____________________ V_{R3}= ____________________

Step 14 I_{R1}= ____________________ I_{R2}= ____________________

I_T= ____________________

Step 15 Using the information from steps 7, 11, and 14, in the space below, show the Ohm's Law formulas and calculations to verify the current and voltage values measured in steps 11 and 14.

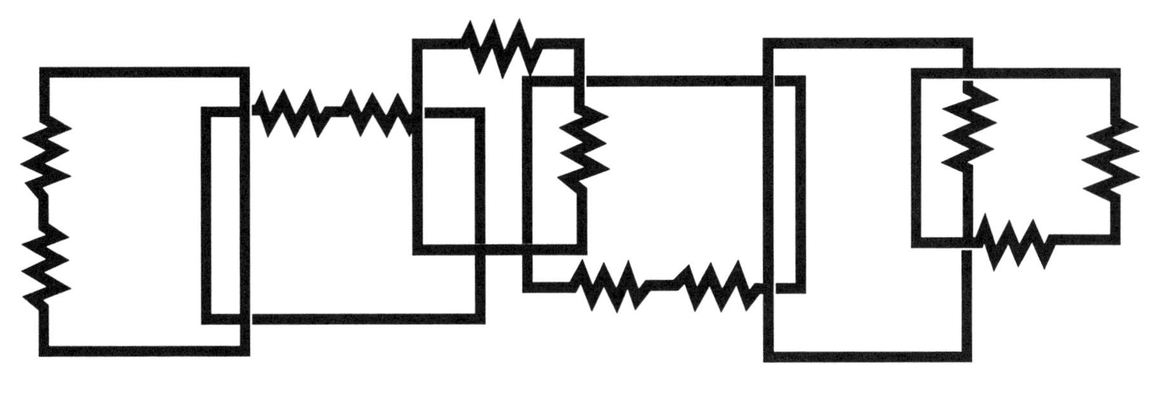

58

Troubleshooting Series Circuits 1

Objectives

After completing this experiment, you will be able to:

1. Recognize malfunctions in a series circuit.
2. Locate a failed component in a series circuit.

Introduction

The purpose of this experiment is to develop and practice troubleshooting skills. You will be using Electronics Workbench (EWB) and a prebuilt, prefaulted circuit. You will use the test equipment in the EWB application to make voltage, current, and resistance checks to locate and identify a failed component.

Procedure

1. Open Electronics Workbench (EWB)/MultiSim. See your instructor for details on how to access and open the application.
2. Insert the student CD provided with the lab manual into the CD drive.
3. Select *Open* from the *File* menu.
4. Change to the CD drive and double click on CX581.ewb.
5. Complete the activity sheet for Experiment 58.

ACTIVITY SHEET EXPERIMENT 58

NAME ______________________________

DATE ______________________________

Step 5 Using the information from the schematic diagram and Ohm's Law, complete Table 58–1.

	Ebatt (V)	**R1**	**R2**	**R3**	**DS1**	**Totals**
Resistance						
Voltage						
Current						

TABLE 58–1

Using the multimeter in EWB, make circuit and component tests necessary to complete Table 58–2.

	Ebatt (V)	**R1**	**R2**	**R3**	**DS1**	**Totals**
Resistance						
Voltage						
Current						

TABLE 58–2

Compare the data from Table 58–1 to data from Table 58–2. Where does the data differ?

____

____

____

Based on the data from the two tables, determine which circuit component is bad. List the component below. Obtain your instructor's concurrence before proceeding to the next step.

____

____

Explain why the failure of this component caused the symptoms noted in Table 58–2.

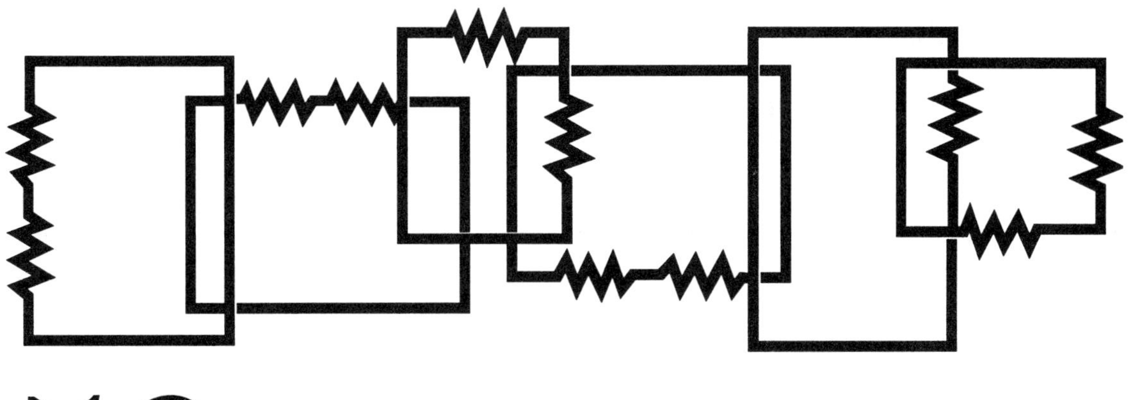

59

Troubleshooting Series Circuits 2

Objectives

After completing this experiment, you will be able to:

1. Recognize malfunctions in a series circuit.
2. Locate a failed component in a series circuit.

Introduction

The purpose of this experiment is to develop and practice troubleshooting skills. You will be using Electronics Workbench (EWB) and a prebuilt, prefaulted circuit. You will use the test equipment in the EWB application to make voltage, current, and resistance checks to locate and identify a failed component.

Procedure

1. Open Electronics Workbench (EWB)/MultiSim. See your instructor for details on how to access and open the application.
2. Insert the student CD provided with the lab manual into the CD drive.
3. Select *Open* from the *File* menu.
4. Change to the CD drive and double click on CX591.ewb.
5. Complete the activity sheet for Experiment 59.

ACTIVITY SHEET EXPERIMENT 59

NAME ______________________________

DATE ______________________________

Step 5 Using the information from the schematic diagram and Ohm's Law, complete Table 59–1.

	Ebatt (V)	**R1**	**R2**	**R3**	**DS1**	**Totals**
Resistance						
Voltage						
Current						

TABLE 59–1

Using the multimeter in EWB, make circuit and component tests necessary to complete Table 59–2.

	Ebatt (V)	**R1**	**R2**	**R3**	**DS1**	**Totals**
Resistance						
Voltage						
Current						

TABLE 59–2

Compare the data from Table 59–1 to data from Table 59–2. Where does the data differ?

____

____

____

Based on the data from the two tables, determine which circuit component is bad. List the component below. Obtain your instructor's concurrence before proceeding to the next step.

____

____

Explain why the failure of this component caused the symptoms noted in Table 59–2.

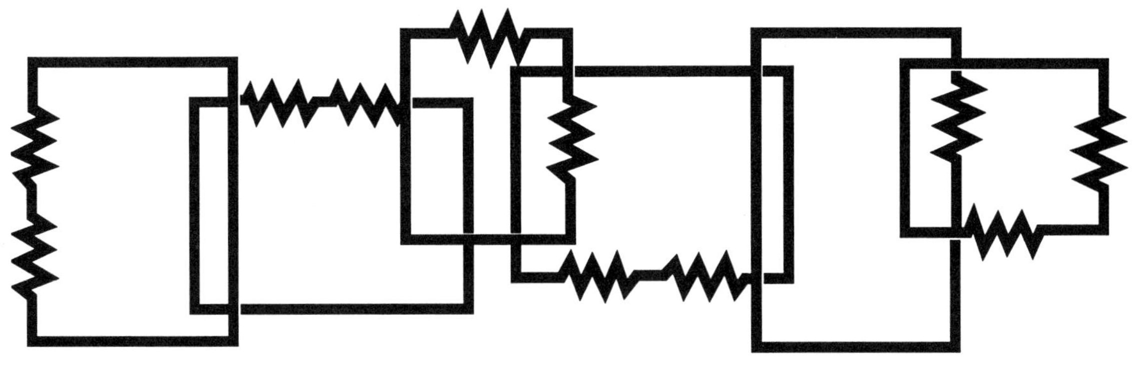

60

Parallel Circuit Measurements 1

Objectives

After completing this experiment, you will be able to:

1. Evaluate voltage distribution in a parallel circuit.
2. Determine actual resistance of individual resistors.
3. Confirm the principles of Ohm's Law.

Introduction

In Experiment 60 you will use a computer and software program to construct a Parallel resistive circuit and explore the electrical characteristics of that circuit. You will be conducting tests to measure resistance, voltage, and current. The test equipment you will use in the computer application program is very similar to what you will use on the lab workbench or out in a maintenance work shop.

Procedure

1. Open Electronics Workbench (EWB)/MultiSim. See your instructor for details on how to access and open the application.
2. Place three resistors from the parts bin onto the work surface.
3. Configure the resistors as shown in Figure 60–1.

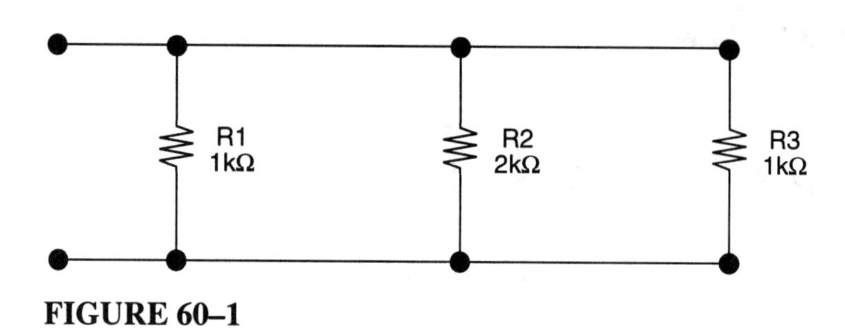

FIGURE 60–1

4. Move the multimeter from the tool bar onto the work surface and set it up to function as an ohmmeter.
5. Measure the value of each resistor by connecting the meter leads so that the meter is in parallel with the component to be tested.
6. Record the measured value of each resistor in the experiment activity sheet.
7. Disconnect the meter from the circuit.
8. Add a 10 volt battery to the circuit as shown in Figure 60–2.

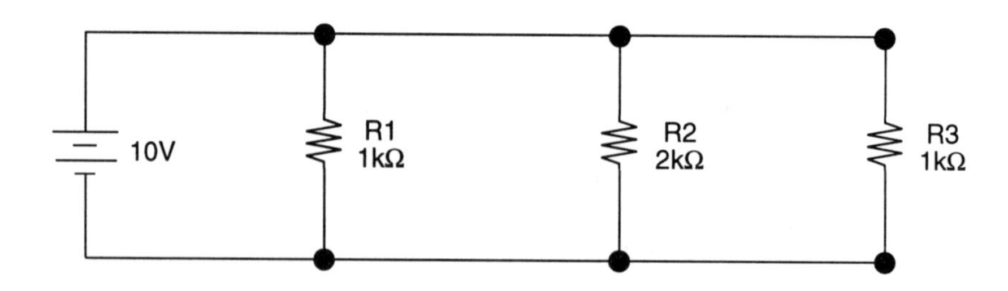

FIGURE 60–2

9. Select the multimeter and set it up to measure voltage.
10. Measure the voltage drop across each resistor by connecting the meter leads so that the meter is in parallel with the component to be tested. Record the results in the experiment activity sheet.
11. Set the multimeter to measure current.

12. Measure the current through each component by connecting the meter in series with each component. See Figure 60-3 for an illustration of how to connect the meter.

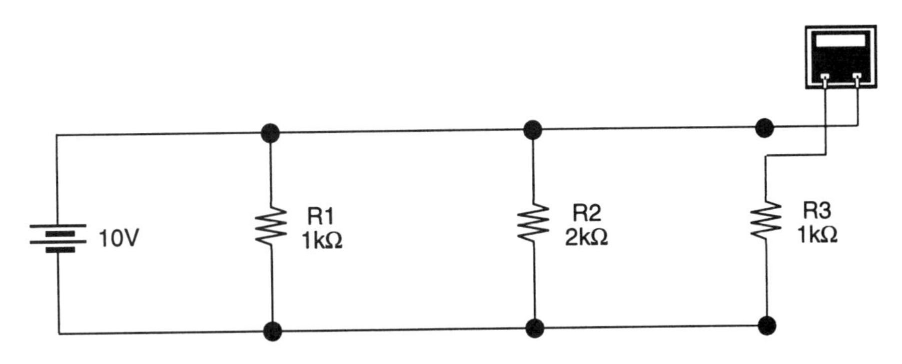

FIGURE 60–3

13. Record the results in the experiment activity sheet.
14. Verify test results by calculating the values of voltage and current for the circuit in Figure 60–2.

ACTIVITY SHEET EXPERIMENT 60

NAME ______________________________

DATE ______________________________

Step 7 R1= ____________________ R2= ____________________

R3= ____________________

Step 11 E_{batt}= ____________________ V_{R1}= ____________________

V_{R2}= ____________________ V_{R3}= ____________________

Step 14 I_{R1}= ____________________ I_{R2}= ____________________

I_T= ____________________

Step 15 Using the information from steps 6, 10, and 13, in the space below, show the Ohm's Law formulas and calculations to verify the current and voltage values measured in steps 10 and 13.

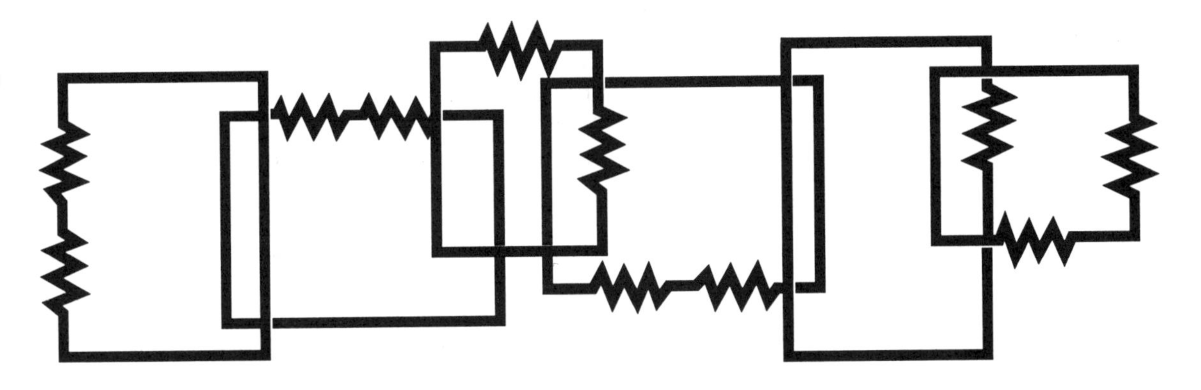

61

Troubleshooting Parallel Circuits 2

Objectives

After completing this experiment, you will be able to:

1. Recognize malfunctions in a parallel circuit.
2. Locate a failed component in a parallel circuit.

Introduction

The purpose of this experiment is to develop and practice troubleshooting skills. You will be using Electronics Workbench (EWB) and a prebuilt, prefaulted circuit. You will use the test equipment in the EWB application to make voltage, current, and resistance checks to locate and identify a failed component.

Procedure

1. Open Electronics Workbench (EWB)/MultiSim. See your instructor for details on how to access and open the application.
2. Insert the student CD provided with the lab manual into the CD drive.
3. Select *Open* from the *File* menu.
4. Change to the CD drive and double click on CX611.ewb.
5. Complete the activity sheet for Experiment 61.

ACTIVITY SHEET EXPERIMENT 61

NAME ______________________________

DATE ______________________________

Step 5 Using the information from the schematic diagram and Ohm's Law, complete Table 61–1.

	Ebatt (V)	**R1**	**R2**	**R3**	**DS1**	**Totals**
Resistance						
Voltage						
Current						

TABLE 61–1

Using the multimeter in EWB, make circuit and component tests necessary to complete Table 61–2.

	Ebatt (V)	**R1**	**R2**	**R3**	**DS1**	**Totals**
Resistance						
Voltage						
Current						

TABLE 61–2

Compare the data from Table 61–1 to data from Table 61–2. Where does the data differ?

____

____

____

Based on the data from the two tables, determine which circuit component is bad. List the component below. Obtain your instructor's concurrence before proceeding to the next step.

____

____

Explain why the failure of this component caused the symptoms noted in Table 61–2.

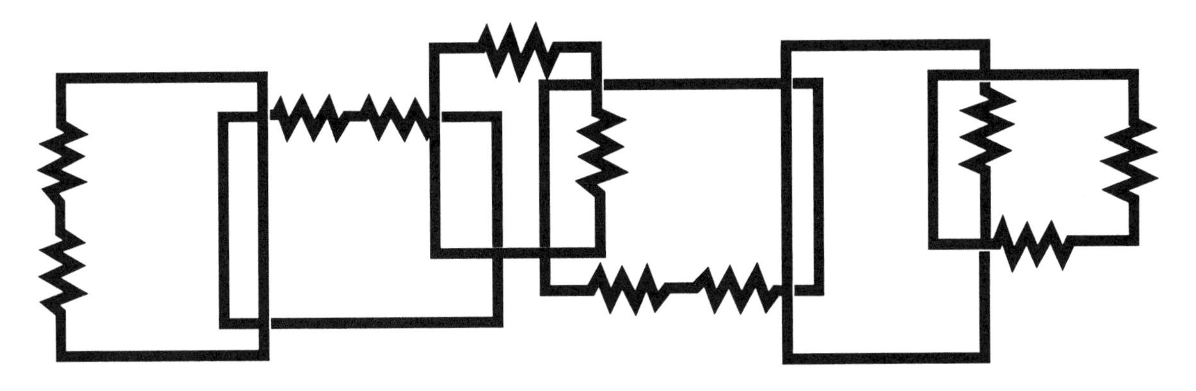

62

Troubleshooting Parallel Circuits 3

Objectives

After completing this experiment, you will be able to:

1. Recognize malfunctions in a parallel circuit.
2. Locate a failed component in a parallel circuit.

Introduction

The purpose of this experiment is to develop and practice troubleshooting skills. You will be using Electronics Workbench (EWB) and a prebuilt, prefaulted circuit. You will use the test equipment in the EWB application to make voltage, current, and resistance checks to locate and identify a failed component.

Procedure

1. Open Electronics Workbench (EWB)/MultiSim. See your instructor for details on how to access and open the application.
2. Insert the student CD provided with the lab manual into the CD drive.
3. Select *Open* from the *File* menu.
4. Change to the CD drive and double click on CX621.ewb.
5. Complete the activity sheet for Experiment 62.

ACTIVITY SHEET EXPERIMENT 62

NAME ______________________________

DATE ______________________________

Step 5 Using the information from the schematic diagram and Ohm's Law, complete Table 62–1.

	Ebatt (V)	**R1**	**R2**	**R3**	**DS1**	**Totals**
Resistance						
Voltage						
Current						

TABLE 62–1

Using the multimeter in EWB, make circuit and component tests necessary to complete Table 62–2.

	Ebatt (V)	**R1**	**R2**	**R3**	**DS1**	**Totals**
Resistance						
Voltage						
Current						

TABLE 62–2

Compare the data from Table 62–1 to data from Table 62–2. Where does the data differ?

____

____

____

Based on the data from the two tables, determine which circuit component is bad. List the component below. Obtain your instructor's concurrence before proceeding to the next step.

____

____

Explain why the failure of this component caused the symptoms noted in Table 62–2.

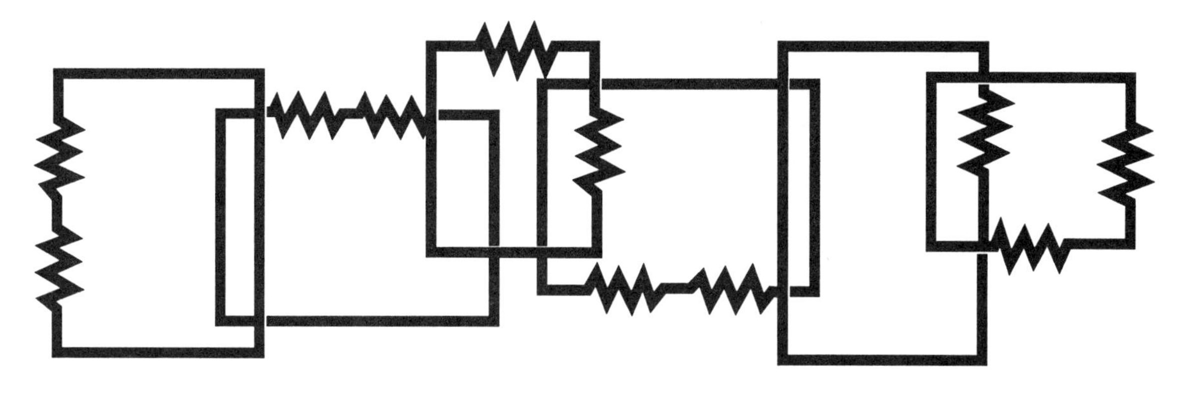

63

Series-Parallel Circuit Measurements 1

Objectives

After completing this experiment, you will be able to:

1. Evaluate voltage distribution in a series-parallel circuit.
2. Determine actual resistance of individual resistors.
3. Confirm the principles of Ohm's Law.

Introduction

In Experiment 63 you will use a computer and software program to construct a series–parallel resistive circuit and explore the electrical characteristics of that circuit. You will be conducting tests to measure resistance, voltage, and current. The test equipment you will use in the computer application program is very similar to what you will use on the lab workbench or out in a maintenance work shop.

Procedure

1. Open Electronics Workbench (EWB)/MultiSim. See your instructor for details on how to access and open the application.
2. Place three resistors from the parts bin onto the work surface.
3. Configure the resistors as shown in Figure 63–1.

FIGURE 63–1

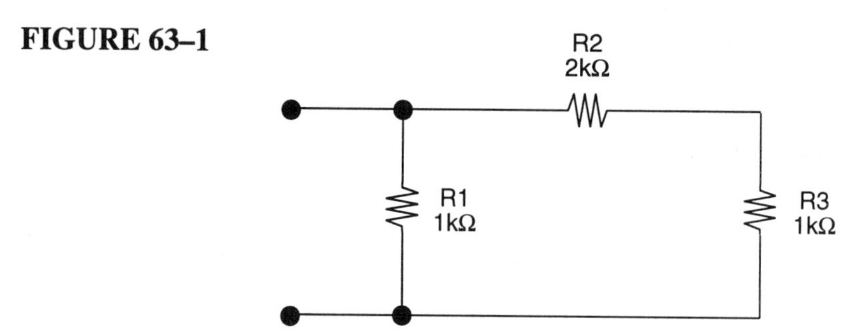

4. Move the multimeter from the tool bar onto the work surface and set it up to function as an ohmmeter.
5. Measure the value of each resistor by connecting the meter leads so that the meter is in parallel with the component to be tested.
6. Record the measured value of each resistor in the experiment activity sheet.
7. Disconnect the meter from the circuit.
8. Add a 10 volt battery to the circuit as shown in Figure 63–2.

FIGURE 63–2

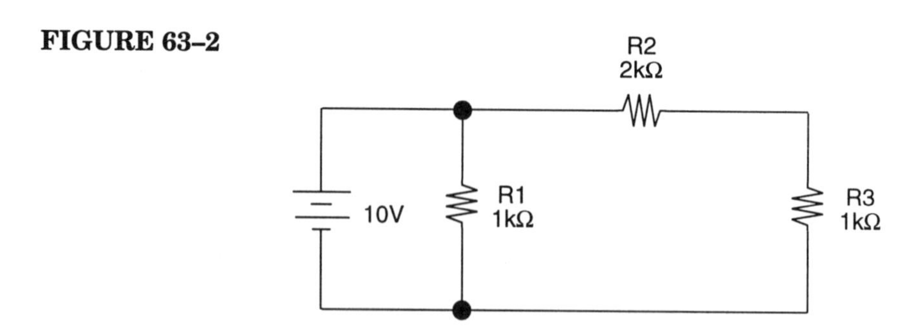

9. Select the multimeter and set it up to measure voltage.
10. Measure the voltage drop across each resistor by connecting the meter leads so that the meter is in parallel with the component to be tested. Record the results in the experiment activity sheet.
11. Set the multimeter to measure current.

12. Measure the current through each component by connecting the meter in series with each component. See Figure 63–3 for an illustration of how to connect the meter.

FIGURE 63–3

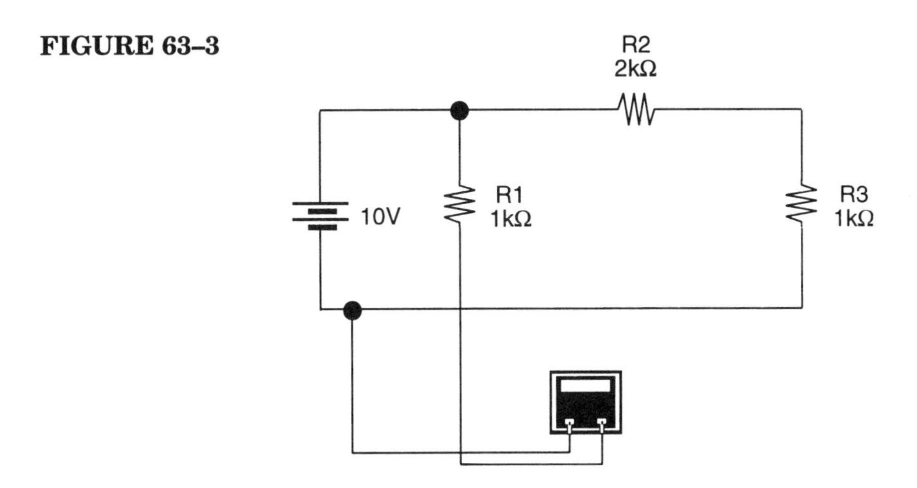

13. Record the results in the experiment activity sheet.
14. Verify test results by calculating the values of voltage and current for the circuit in Figure 63–2.

ACTIVITY SHEET EXPERIMENT 63

NAME ______________________________

DATE ______________________________

Step 6 R1= ____________________ R2= ____________________

R3= ____________________

Step 10 E_{batt}= ____________________ V_{R1}= ____________________

V_{R2}= ____________________ V_{R3}= ____________________

Step 13 I_{R1}= ____________________ I_{R2}= ____________________

I_T= ____________________

Step 14 Using the information from steps 6, 10, and 13, in the space below, show the Ohm's Law formulas and calculations to verify the current and voltage values measured in steps 10 and 13.

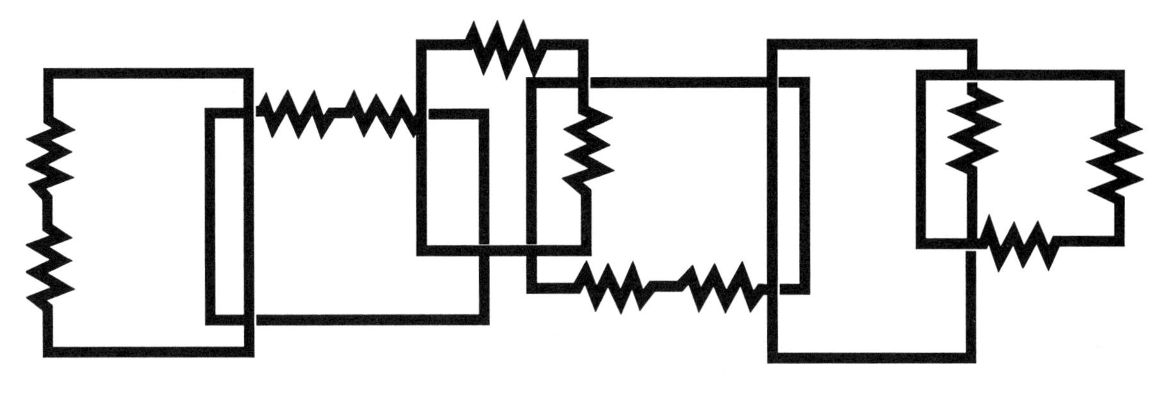

64

Troubleshooting Series-Parallel Circuits 2

Objectives

After completing this experiment, you will be able to:

1. Recognize malfunctions in a series-parallel circuit.
2. Locate a failed component in a series-parallel circuit.

Introduction

The purpose of this experiment is to develop and practice troubleshooting skills. You will be using Electronics Workbench (EWB) and a prebuilt, prefaulted circuit. You will use the test equipment in the EWB application to make voltage, current, and resistance checks to locate and identify a failed component.

Procedure

1. Open Electronics Workbench (EWB)/MultiSim. See your instructor for details on how to access and open the application.
2. Insert the student CD provided with the lab manual into the CD drive.
3. Select *Open* from the *File* menu.
4. Change to the CD drive and double click on CX641.ewb.
5. Complete the activity sheet for Experiment 64.

ACTIVITY SHEET EXPERIMENT 64

NAME ______________________________

DATE ______________________________

Step 5 Using the information from the schematic diagram and Ohm's Law, complete Table 64–1.

	Ebatt (V)	**R1**	**R2**	**R3**	**DS1**	**Totals**
Resistance						
Voltage						
Current						

TABLE 64–1

Using the multimeter in EWB, make circuit and component tests necessary to complete Table 64–2.

	Ebatt (V)	**R1**	**R2**	**R3**	**DS1**	**Totals**
Resistance						
Voltage						
Current						

TABLE 64–2

Compare the data from Table 64–1 to data from Table 64–2. Where does the data differ?

____

____

____

Based on the data from the two tables, determine which circuit component is bad. List the component below. Obtain your instructor's concurrence before proceeding to the next step.

____

____

Explain why the failure of this component caused the symptoms noted in Table 64–2.

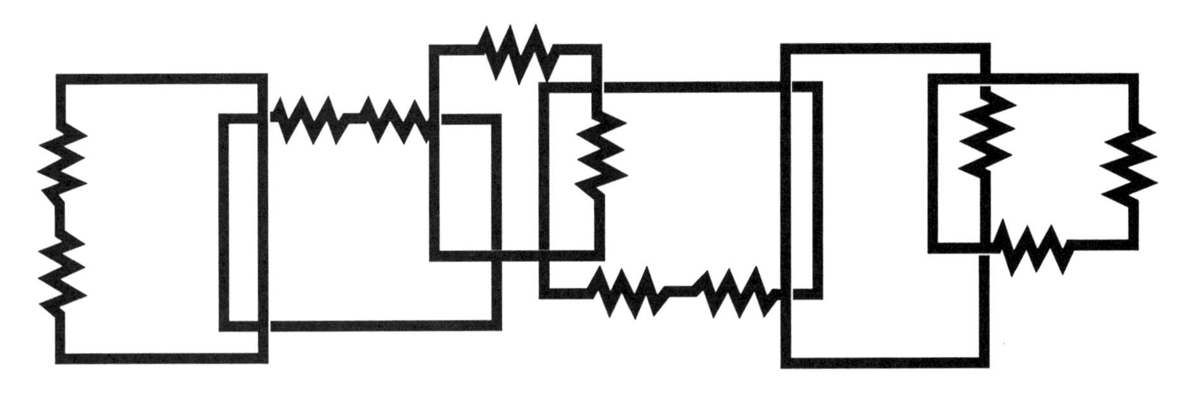

65

Troubleshooting Series-Parallel Circuits 3

Objectives

After completing this experiment, you will be able to:

1. Recognize malfunctions in a series-parallel circuit.
2. Locate a failed component in a series-parallel circuit.

Introduction

The purpose of this experiment is to develop and practice troubleshooting skills. You will be using Electronics Workbench (EWB) and a prebuilt, prefaulted circuit. You will use the test equipment in the EWB application to make voltage, current, and resistance checks to locate and identify a failed component.

Procedure

1. Open Electronics Workbench (EWB)/MultiSim. See your instructor for details on how to access and open the application.
2. Insert the student CD provided with the lab manual into the CD drive.
3. Select *Open* from the *File* menu.
4. Change to the CD drive and double click on CX651.ewb.
5. Complete the activity sheet for Experiment 65.

ACTIVITY SHEET EXPERIMENT 65

NAME ______________________________

DATE ______________________________

Step 5 Using the information from the schematic diagram and Ohm's Law, complete Table 65–1.

	Ebatt (V)	**R1**	**R2**	**R3**	**DS1**	**Totals**
Resistance						
Voltage						
Current						

TABLE 65–1

Using the multimeter in EWB, make circuit and component tests necessary to complete Table 65–2.

	Ebatt (V)	**R1**	**R2**	**R3**	**DS1**	**Totals**
Resistance						
Voltage						
Current						

TABLE 65–2

Compare the data from Table 65–1 to data from Table 65–2. Where does the data differ?

____

____

____

Based on the data from the two tables, determine which circuit component is bad. List the component below. Obtain your instructor's concurrence before proceeding to the next step.

____

____

Explain why the failure of this component caused the symptoms noted in Table 65–2.

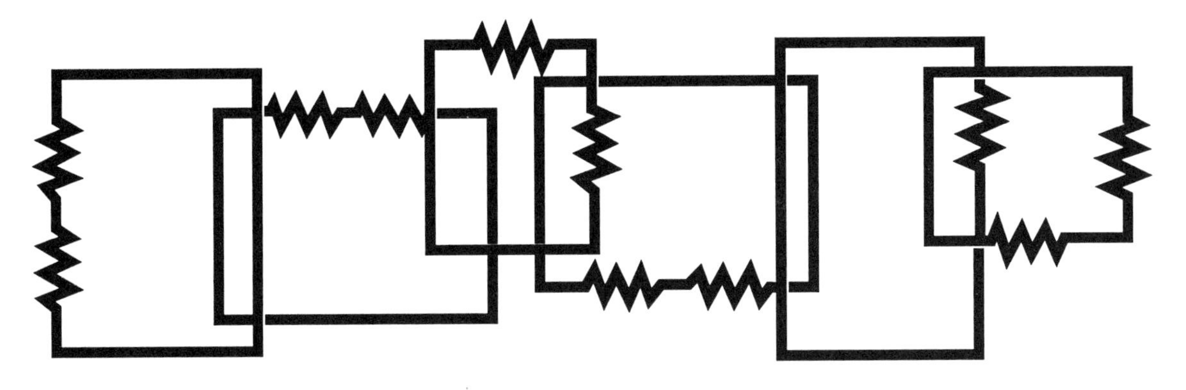

66

Loaded Voltage-Divider Circuit Measurements 1

Objectives

After completing this experiment, you will be able to:

1. Evaluate voltage distribution in a voltage-divider circuit.
2. Determine actual resistance of individual resistors.
3. Confirm the principles of Ohm's Law.

Introduction

In Experiment 66 you will use a computer and software program to construct a voltage-divider circuit and explore the electrical characteristics of that circuit. You will be conducting tests to measure resistance, voltage, and current. The test equipment you will use in the computer application program is very similar to what you will use on the lab workbench or out in a maintenance work shop.

Procedure

1. Open Electronics Workbench (EWB)/MultiSim. See your instructor for details on how to access and open the application.
2. Place five resistors and a battery from the parts bin onto the work surface.
3. Configure the resistors as shown in Figure 66–1.

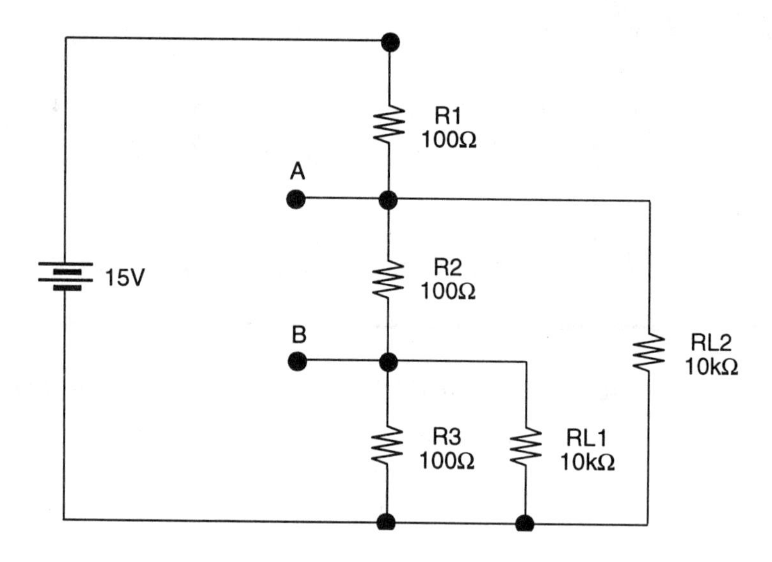

FIGURE 66–1

4. Move the multimeter from the tool bar onto the work surface and set it up to function as an ohmmeter.
5. Disconnect the battery.
6. Measure the value of each resistor by connecting the meter leads so that the meter is in parallel with the component to be tested.
7. Record the measured value of each resistor in the experiment activity sheet.
8. Disconnect the meter from the circuit.
9. Reconnect the battery.
10. Select the multimeter and set it up to measure voltage.
11. Measure the voltage drop across each resistor by connecting the meter leads so that the meter is in parallel with the component to be tested. Record the results in the experiment activity sheet.
12. Set the multimeter to measure current.

13. Measure the current through each component by connecting the meter in series with each component. See Figure 66–2 for an illustration of how to connect the meter.

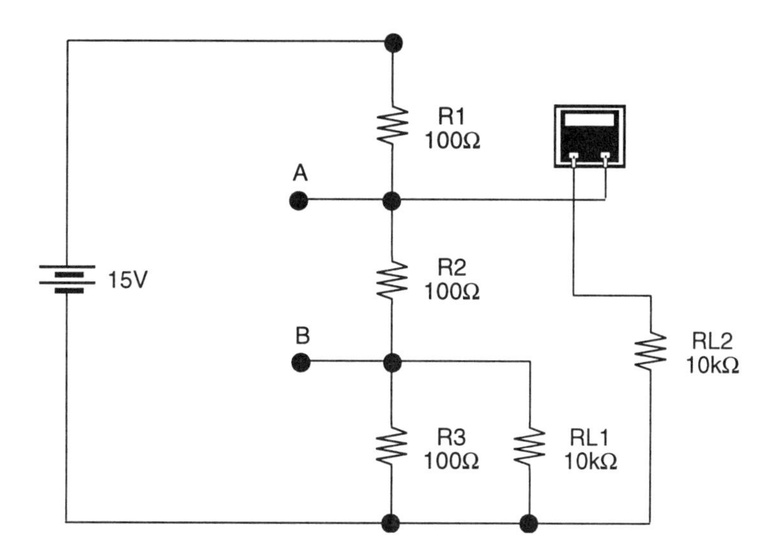

FIGURE 66–2

14. Record the results in the experiment activity sheet.
15. Verify test results by calculating the values of voltage and current for the circuit in Figure 66–2.

ACTIVITY SHEET EXPERIMENT 66

NAME ______________________________

DATE ______________________________

Step 6

R1= ____________________ R2= ____________________

R3= ____________________ RL1= ____________________

RL2= ____________________

Step 10

E_{batt}= ____________________ V_{R1}= ____________________

V_{R2}= ____________________ V_{R3}= ____________________

V_{RL1}= ____________________ V_{RL2}= ____________________

Step 13

I_{R1}= ____________________ I_{R2}= ____________________

I_{R3}= ____________________ I_{RL1}= ____________________

I_{RL2}= ____________________ I_T= ____________________

Step 14

Using the information from steps 6, 10, and 13, in the space below, show the Ohm's Law formulas and calculations to verify the current and voltage values measured in steps 10 and 13.

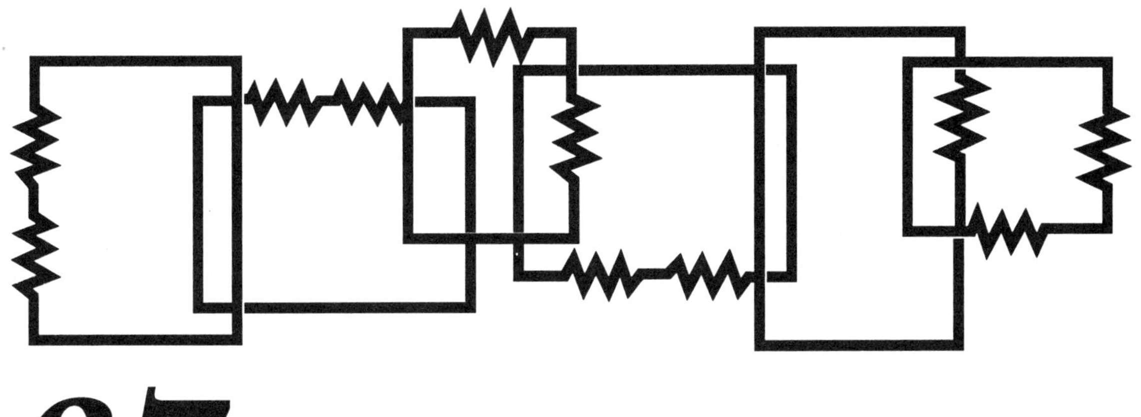

67

Troubleshooting Loaded Voltage-Divider Circuits 1

Objectives

After completing this experiment, you will be able to:

1. Recognize malfunctions in a loaded voltage-divider circuit.
2. Locate a failed component in a loaded voltage-divider circuit.

Introduction

The purpose of this experiment is to develop and practice troubleshooting skills. You will be using Electronics Workbench (EWB) and a prebuilt, prefaulted circuit. You will use the test equipment in the EWB application to make voltage, current, and resistance checks to locate and identify a failed component.

Procedure

1. Open Electronics Workbench (EWB)/MultiSim. See your instructor for details on how to access and open the application.
2. Insert the student CD provided with the lab manual into the CD drive.
3. Select *Open* from the *File* menu.
4. Change to the CD drive and double click on CX671.ewb.
5. Complete the activity sheet for Experiment 67.

ACTIVITY SHEET EXPERIMENT 67

NAME ______________________________

DATE ______________________________

Step 5 Using the information from the schematic diagram and Ohm's Law, complete Table 67–1.

	Ebatt	**R1**	**R2**	**R3**	**RL1**	**RL2**	**DS1**	**DS2**	**Totals**
Resistance									
Voltage									
Current									

TABLE 67–1

Using the multimeter in EWB, make circuit and component tests necessary to complete Table 67–2.

	Ebatt	**R1**	**R2**	**R3**	**RL1**	**RL2**	**DS1**	**DS2**	**Totals**
Resistance									
Voltage									
Current									

TABLE 67–2

Compare the data from Table 67–1 to data from Table 67–2. Where does the data differ?

____

____

____

Based on the data from the two tables, determine which circuit component is bad. List the component below. Obtain your instructor's concurrence before proceeding to the next step.

____

____

Explain why the failure of this component caused the symptoms noted in Table 67–2.

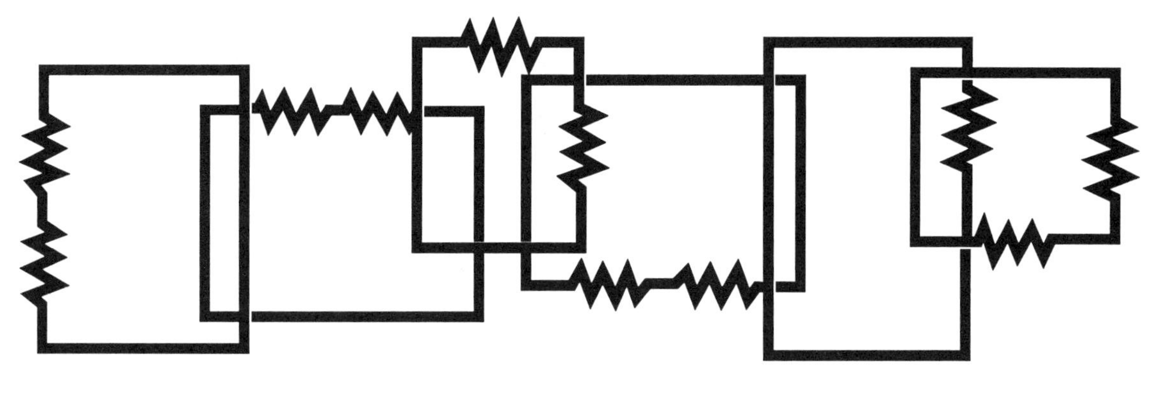

68

Troubleshooting Loaded Voltage-Divider Circuits 2

Objectives

After completing this experiment, you will be able to:

1. Recognize malfunctions in a loaded voltage-divider circuit.
2. Locate a failed component in a loaded voltage-divider circuit.

Introduction

The purpose of this experiment is to develop and practice troubleshooting skills. You will be using Electronics Workbench (EWB) and a prebuilt, prefaulted circuit. You will use the test equipment in the EWB application to make voltage, current, and resistance checks to locate and identify a failed component.

Procedure

1. Open Electronics Workbench (EWB)/MultiSim. See your instructor for details on how to access and open the application.
2. Insert the student CD provided with the lab manual into the CD drive.
3. Select *Open* from the *File* menu.
4. Change to the CD drive and double click on CX681.ewb.
5. Complete the activity sheet for Experiment 68.

ACTIVITY SHEET EXPERIMENT 68

NAME ______________________________

DATE ______________________________

(Note: Ebatt is a term often used in engineering documents to indicate "source voltage.")

Step 5 Using the information from the schematic diagram and Ohm's Law, complete Table 68–1.

	Ebatt	**R1**	**R2**	**R3**	**RL1**	**RL2**	**DS1**	**DS2**	**Totals**
Resistance									
Voltage									
Current									

TABLE 68–1

Using the multimeter in EWB, make circuit and component tests necessary to complete Table 68–2.

	Ebatt	**R1**	**R2**	**R3**	**RL1**	**RL2**	**DS1**	**DS2**	**Totals**
Resistance									
Voltage									
Current									

TABLE 68–2

Compare the data from Table 68–1 to data from Table 68–2. Where does the data differ?

____

____

____

Based on the data from the two tables, determine which circuit component is bad. List the component below. Obtain your instructor's concurrence before proceeding to the next step.

____

____

Explain why the failure of this component caused the symptoms noted in Table 68–2.

____

____

____

____

____

____

____

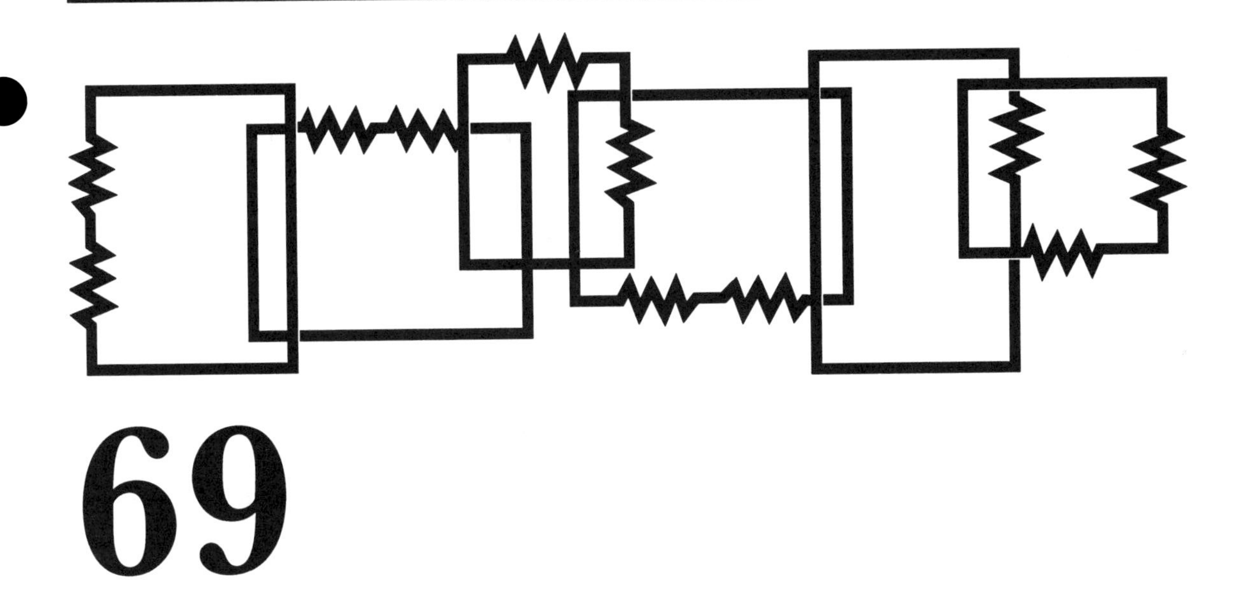

69

RC Time Constants

Objectives

After completing this experiment, you will be able to:

1. Measure charge and discharge slopes in an active RC circuit.
2. Calculate the time constant of an RC circuit.
3. Recognize failed component symptoms in an active RC circuit.

Introduction

In Experiment 69 you will use a computer and software program to construct an active RC circuit and explore the electrical characteristics of that circuit. You will be conducting tests to measure resistance, voltage, and current. You will insert failed components to observe their effect on circuit performance. The test equipment you will use in the computer application program is very similar to what you will use on the lab workbench or out in a maintenance work shop.

Procedure

1. Open Electronics Workbench (EWB)/MultiSim. See your instructor for details on how to access and open the application.
2. Insert the student CD provided with the lab manual into the CD drive.
3. Select *Open* from the *File* menu.
4. Change to the CD drive and double click on CX691.ewb.
5. Using the components on the work bench surface, assemble the circuit shown in Figure 69–1.

FIGURE 69–1

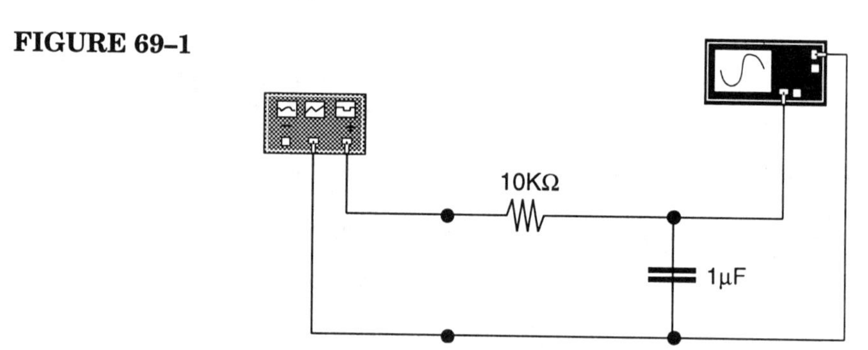

6. Remove the activity sheet for Experiment 69.
7. Set the function generator to 100 Hz, 50% Duty cycle, 10 V Amplitude, Offset 10.
8. Set the oscilloscope to Time Base 5.00ms/div, Channel 10 V/Div.
9. Measure and record minimum voltage of the waveform, maximum voltage of the waveform, time of positive slope, time of negative slope.
10. Using the values of R and C, calculate the time constant of the circuit.
11. Replace the 10kΩ resistor with the one labeled *open*.
12. Measure and record the time of the positive and negative slope.
13. Replace the open resistor with the shorted resistor.
14. Measure and record the time of the positive and negative slope.
15. Replace the shorted resistor with the 10kΩ resistor.
16. Replace the 1μF capacitor with the one labeled *open*.
17. Measure and record the time of the positive and negative slope.
18. Replace the open capacitor with the shorted capacitor.
19. Measure and record the time of the positive and negative slope.

20. Explain the results obtained with the open and the shorted resistor.

21. Explain the results obtained with the open and shorted capacitor.

ACTIVITY SHEET EXPERIMENT 69

NAME ______________________________

DATE ______________________________

Step 9 Minimum Voltage= ___________ Maximum Voltage= ___________

Positive slope time= ___________ Negative slope time= ___________

Step 10 Calculated time constant = ___________

Frequency = ___________

Step 12 Positive slope time= ___________ Negative slope time= ___________

Step 14 Positive slope time= ___________ Negative slope time= ___________

Step 16 Positive slope time= ___________ Negative slope time= ___________

Step 17 Positive slope time= ___________ Negative slope time= ___________

Step 19 Positive slope time= ___________ Negative slope time= ___________

Step 20 Explain the results obtained in steps 12 and 14.

____

____

____

____

Step 21 Explain the results obtained in steps 17 and 19.

____

____

____

____

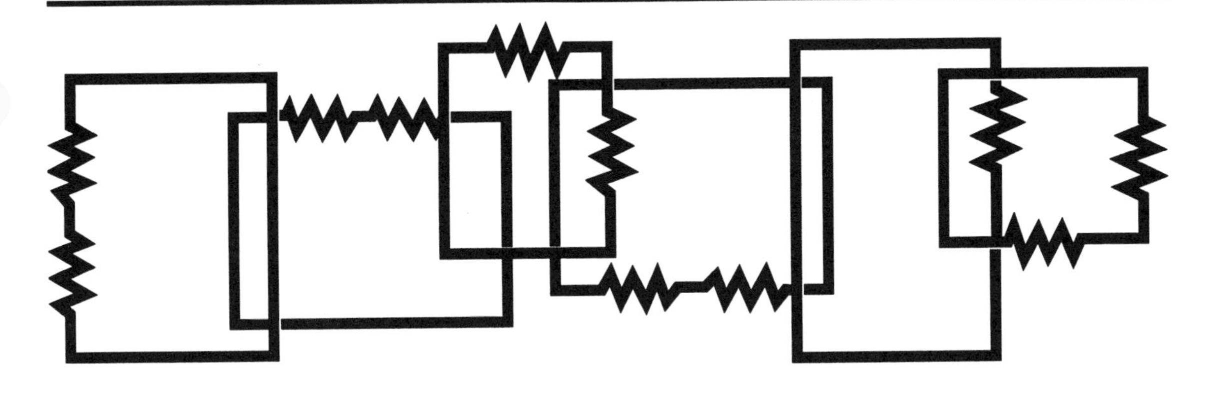

70

RL Circuits

Objectives

After completing this experiment, you will be able to:

1. Observe the signal characteristics of a RL circuit.
2. Recognize failed component symptoms in an active RL circuit.

Introduction

In Experiment 70 you will use a computer and software program to construct an active RL circuit and explore the electrical characteristics of that circuit. You will be conducting tests to measure voltage and frequency. You will insert failed components to observe their effect on circuit performance. The test equipment you will use in the computer application program is very similar to what you will use on the lab workbench or out in a maintenance work shop.

Procedure

1. Open Electronics Workbench (EWB)/MultiSim. See your instructor for details on how to access and open the application.
2. Insert the student CD provided with the lab manual into the CD drive.
3. Select *Open* from the *File* menu.
4. Change to the CD drive and double click on CX701.ewb.
5. Using the components on the work bench surface, assemble the circuit shown in Figure 70–1.

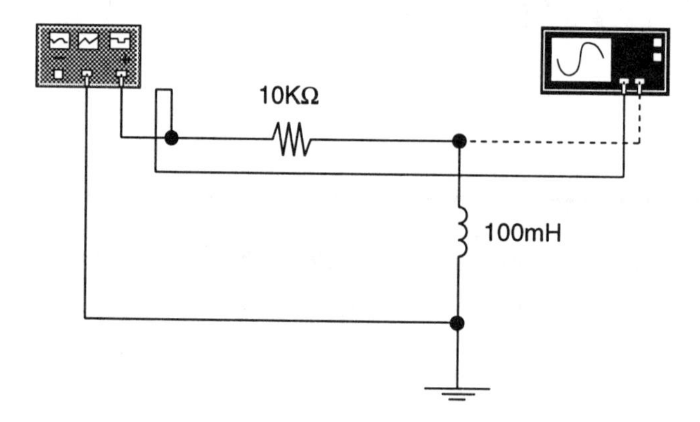

FIGURE 70–1

6. Remove the activity sheet for Experiment 70.
7. Set the function generator to 10k Hz, 50% Duty cycle, 4 V Amplitude, Offset 0.
8. Set the oscilloscope to Time Base 0.02ms/div, Channel A & B 5 V/Div.
9. Using the oscilloscope, measure and record the voltage across the 10kΩ resistor.
10. Using the resistor value and voltage drop across the resistor, determine the circuit current.
11. Determine the reactance of the inductor.
12. Replace the 10kΩ resistor with the one labeled *open*.
13. Using the oscilloscope, measure and record the voltage across the resistor and inductor.
14. Replace the open resistor with the shorted resistor.
15. Using the oscilloscope, measure and record the voltage across the resistor and inductor.
16. Replace the shorted resistor with the 10kΩ resistor.

17. Replace the 100mH inductor with the one labeled *open*.
18. Using the oscilloscope, measure and record the voltage across the resistor and inductor.
19. Replace the open capacitor with the shorted inductor.
20. Using the oscilloscope, measure and record the voltage across the resistor and inductor.
21. Explain the results obtained with the open and the shorted resistor.
22. Explain the results obtained with the open and shorted capacitor.

ACTIVITY SHEET EXPERIMENT 70

NAME ______________________________

DATE ______________________________

Step 9 V_R= ____________________

Step 10 Calculated circuit current = ____________________

Step 11 X_L = ____________________

Step 13 V_R= ____________________ V_L= ____________________

Step 15 V_R= ____________________ V_L= ____________________

Step 18 V_R= ____________________ V_L= ____________________

Step 20 V_R= ____________________ V_L= ____________________

Step 21 Explain the results obtained in steps 13 and 15.

____

____

____

____

Step 22 Explain the results obtained in steps 18 and 20.

____

____

____

____

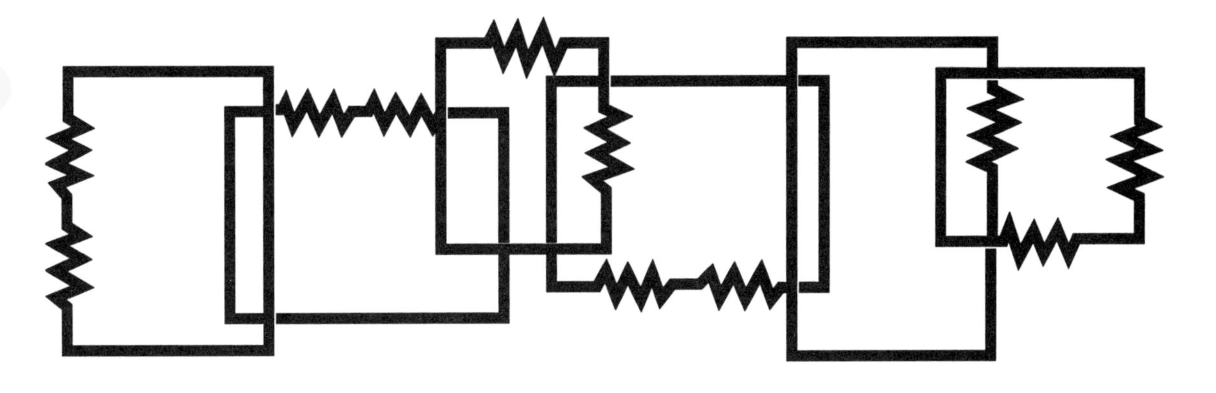

71

RLC Circuits

Objectives

After completing this experiment, you will be able to:

1. Observe the signal characteristics of series and parallel RLC circuits.
2. Recognize failed component symptoms in an active RLC circuits.

Introduction

In Experiment 71 you will use a computer and software program to construct an active RLC circuit and explore the electrical characteristics of that circuit. You will be conducting tests to measure voltage, current, and frequency. You will insert failed components to observe their effect on circuit performance. The test equipment you will use in the computer application program is very similar to what you will use on the lab workbench or out in a maintenance work shop.

Procedure

1. Open Electronics Workbench (EWB)/MultiSim. See your instructor for details on how to access and open the application.
2. Insert the student CD provided with the lab manual into the CD drive.
3. Select *Open* from the *File* menu.
4. Change to the CD drive and double click on CX711.ewb.
5. Assume R1 = 10kΩ and L1 = 1mH. Calculate the value of C1 for resonance at 1.5 kHz.
6. Using the components on the work bench surface, assemble the circuit shown in Figure 71–1 using component values from step 5.

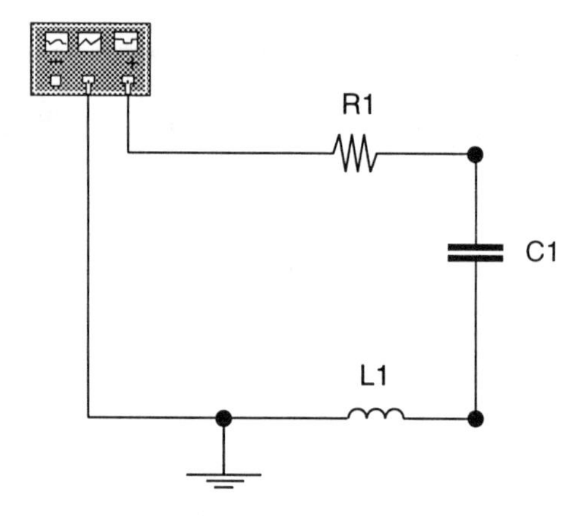

FIGURE 71–1

7. Remove the activity sheet for Experiment 71.
8. Set the function generator to 10 kHz, 50% Duty cycle, 4 V Amplitude, Offset 0.
9. Set the Bode plotter for "Log", Horizontal - F=200, I=200, Vertical - F=10.1 Mhz, I=.1 khz. Use the arrows to move the cursor to the start of the slope.
10. Measure and record the rolloff frequency.
11. Replace the 10kΩ resistor with the one labeled *open*.
12. Measure and record the rolloff frequency.
13. Measure and record the voltage across R1, C1, and L1.
14. Replace the open resistor with the shorted resistor.
15. Measure and record the rolloff frequency.

16. Measure and record the voltage across R1, C1, and L1.
17. Replace the shorted resistor with the 10kΩ resistor.
18. Replace the 1mH inductor with the one labeled *open.*
19. Measure and record the rolloff frequency.
20. Measure and record the voltage across R1, C1, and L1.
21. Replace the open inductor with the shorted inductor.
22. Measure and record the rolloff frequency.
23. Measure and record the voltage across R1, C1, and L1.
24. Replace the capacitor with the one labeled *open.*
25. Measure and record the rolloff frequency.
26. Measure and record the voltage across R1, C1, and L1.
27. Replace the open capacitor with the shorted capacitor.
28. Measure and record the rolloff frequency.
29. Measure and record the voltage across R1, C1, and L1.
30. Reconfigure the circuit to appear as shown in Figure 71–2.

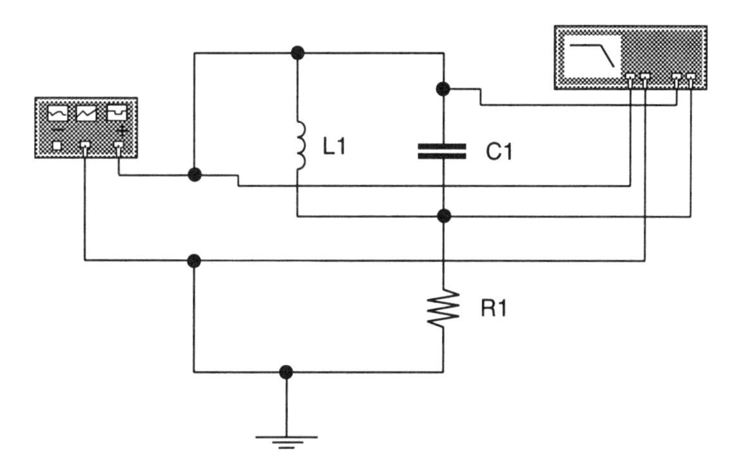

FIGURE 71–2

31. Measure and record the pass frequency.
32. Replace the 10kΩ resistor with the one labeled *open.*
33. Measure and record the pass frequency.
34. Measure and record the voltage across R1, C1, and L1.
35. Replace the open resistor with the shorted resistor.
36. Measure and record the pass frequency.
37. Measure and record the voltage across R1, C1, and L1.
38. Replace the shorted resistor with the 10kΩ resistor.
39. Replace the 1mH inductor with the one labeled *open.*
40. Measure and record the pass frequency.
41. Measure and record the voltage across R1, C1, and L1.
42. Replace the open inductor with the shorted inductor.
43. Measure and record the pass frequency.
44. Measure and record the voltage across R1, C1, and L1.
45. Replace the capacitor with the one labeled *open.*
46. Measure and record the pass frequency.
47. Measure and record the voltage across R1, C1, and L1.

48. Replace the open capacitor with the shorted capacitor.
49. Measure and record the pass frequency.
50. Measure and record the voltage across R1, C1, and L1.
51. Explain the results obtained with the open and the shorted resistors.
52. Explain the results obtained with the open and shorted capacitors.
53. Explain the results obtained with the open and shorted inductors.

ACTIVITY SHEET EXPERIMENT 71

NAME ______________________________

DATE ______________________________

Step 10 $F_{Rolloff}$= ____________________

Step 12 $F_{Rolloff}$= ____________________

Step 13 V_{R1}= ____________________ , V_{C1}= ____________________ ,

V_{L1}= ____________________

Step 15 $F_{Rolloff}$= ____________________

Step 16 V_{R1}= ____________________ , V_{C1}= ____________________ ,

V_{L1}= ____________________

Step 19 $F_{Rolloff}$= ____________________

Step 20 V_{R1}= ____________________ , V_{C1}= ____________________ ,

V_{L1}= ____________________

Step 22 $F_{Rolloff}$= ____________________

Step 23 V_{R1}= ____________________ , V_{C1}= ____________________ ,

V_{L1}= ____________________

Step 25 $F_{Rolloff}$= ____________________

Step 26 V_{R1}= ____________________ , V_{C1}= ____________________ ,

V_{L1}= ____________________

Step 28 $F_{Rolloff}$= ____________________

Step 29 V_{R1}= ____________________ , V_{C1}= ____________________ ,

V_{L1}= ____________________

Step 31 F_{PASS}= ____________________

Step 33 F_{PASS}= ____________________

Step 34 V_{R1}= ____________________ , V_{C1}= ____________________ ,

V_{L1}= ____________________

Step 36 $F_{PASS}=$ ____________________

Step 37 $V_{R1}=$ ____________________, $V_{C1}=$ ____________________,

$V_{L1}=$ ____________________

Step 40 $F_{PASS}=$ ____________________

Step 41 $V_{R1}=$ ____________________, $V_{C1}=$ ____________________,

$V_{L1}=$ ____________________

Step 43 $F_{PASS}=$ ____________________

Step 44 $V_{R1}=$ ____________________, $V_{C1}=$ ____________________,

$V_{L1}=$ ____________________

Step 46 $F_{PASS}=$ ____________________

Step 47 $V_{R1}=$ ____________________, $V_{C1}=$ ____________________,

$V_{L1}=$ ____________________

Step 49 $F_{PASS}=$ ____________________

Step 50 $V_{R1}=$ ____________________, $V_{C1}=$ ____________________,

$V_{L1}=$ ____________________

Step 51 Explain what happens when open components exist in the circuit.

Step 52 Explain what happens when shorted components exist in the circuit.

____

____

____

____

____

____

____

____

Step 53 For the circuit in Figure 71–2, what would you do to determine if the inductor or capacitor were shorted?

____

____

____

____

____

____

____

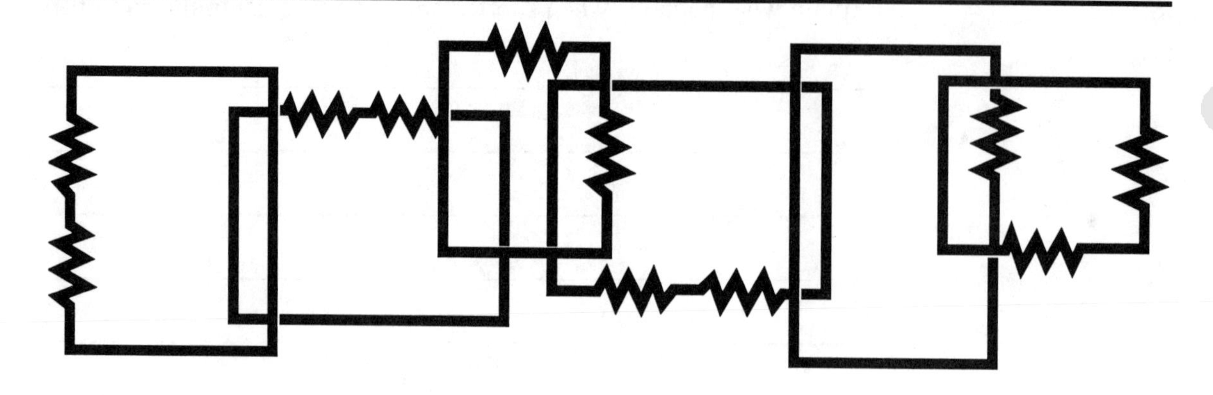

72

Troubleshooting RLC Circuits 1

Objectives

After completing this experiment, you will be able to:

1. Recognize malfunctions in a parallel RLC circuit.
2. Locate a failed component in a parallel RLC circuit.

Introduction

The purpose of this experiment is to develop and practice troubleshooting skills. You will be using Electronics Workbench (EWB)/MultiSim and a prebuilt, prefaulted circuit. You will use the test equipment in the simulation application to make resonant frequency, voltage, current, and resistance checks to locate and identify a failed component.

Procedure

1. Open Electronics Workbench (EWB)/MultiSim. See your instructor for details on how to access and open the application.
2. Insert the student CD provided with the lab manual into the CD drive.
3. Select *Open* from the *File* menu.
4. Change to the CD drive and double click on CX721.ewb.
5. Using the component values shown in CX721.ewb, calculate the resonant frequency of the circuit.
6. Remove the activity sheet for Experiment 72.
7. Set the Bode plotter for "Log", Horizontal - F=200, I=200, Vertical - F=10.1 Mhz, I=.1 khz. Use the arrows to move the cursor to the start of the slope.
8. Set the function generator to output 6 kHz at 5 V.
9. Complete the activity sheet for Experiment 72.

ACTIVITY SHEET EXPERIMENT 72

NAME ______________________________

DATE ______________________________

(Note: In this lab activity you will be responsible for determining and selecting the appropriate test equipment for obtaining accurate measurements.)

Step 9 Calculated resonant frequency = ____________________

Input frequency = ____________________

Measured resonant frequency = ____________________

V_{C1}= ____________________ , V_{L1}= ____________________ ,

V_{R1} = ____________________

R_1= ____________________ , I_{C1}= ____________________ ,

I_{L1}= ____________________ , I_{R1}= ____________________

List the failed component. State the failure (open or short).

____

____

____

____

Explain why the failed component caused the noted symptoms.

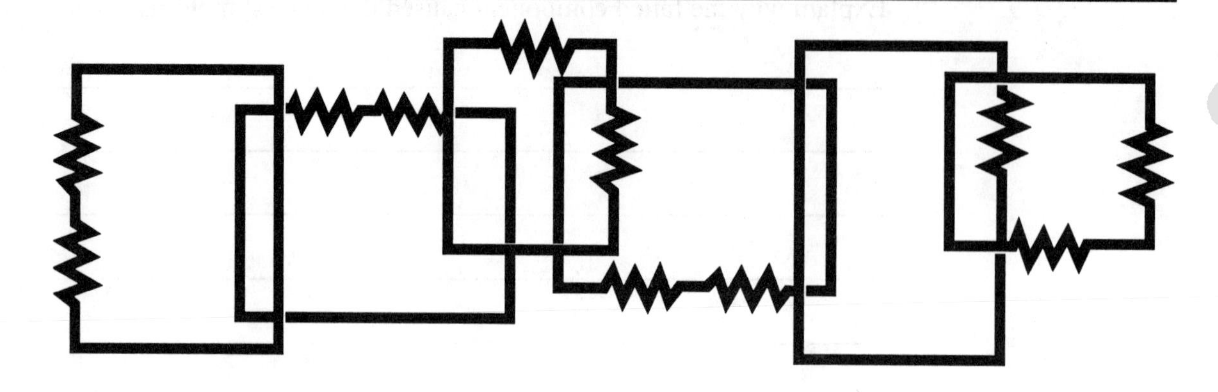

73

Troubleshooting RLC Circuits 2

Objectives

After completing this experiment, you will be able to:

1. Recognize malfunctions in a series RLC circuit.
2. Locate a failed component in a series RLC circuit.

Introduction

The purpose of this experiment is to develop and practice troubleshooting skills. You will be using Electronics Workbench (EWB)/MultiSim and a prebuilt, prefaulted circuit. You will use the test equipment in the simulation application to make resonant frequency, voltage, current, and resistance checks to locate and identify a failed component.

Procedure

1. Open Electronics Workbench (EWB)/MultiSim. See your instructor for details on how to access and open the application.
2. Insert the student CD provided with the lab manual into the CD drive.
3. Select *Open* from the *File* menu.
4. Change to the CD drive and double click on CX731.ewb.
5. Using the component values shown in CX731.ewb, calculate the resonant frequency of the circuit.
6. Remove the activity sheet for Experiment 73.
7. Set the Bode plotter for "Log", Horizontal - F=200, I=200, Vertical - F=10.1 Mhz, I=.1 khz. Use the arrows to move the cursor to the start of the slope.
8. Set the function generator to output 6 kHz at 5 V.
9. Complete the activity sheet for Experiment 73.

ACTIVITY SHEET EXPERIMENT 73

NAME ______________________________

DATE ______________________________

(Note: In this lab activity you will be responsible for determining and selecting the appropriate test equipment for obtaining accurate measurements.)

Step 9

Calculated resonant frequency = ____________________

Input frequency = ____________________

Measured resonant frequency = ____________________

V_{C1}= ____________________ , V_{L1}= ____________________ ,

V_{R1} = ____________________

R_1= ____________________ , I_{C1}= ____________________ ,

I_{L1}= ____________________ , I_{R1}= ____________________

List the failed component. State the failure (open or short).

____

____

____

____

Explain why the failed component caused the noted symptoms.

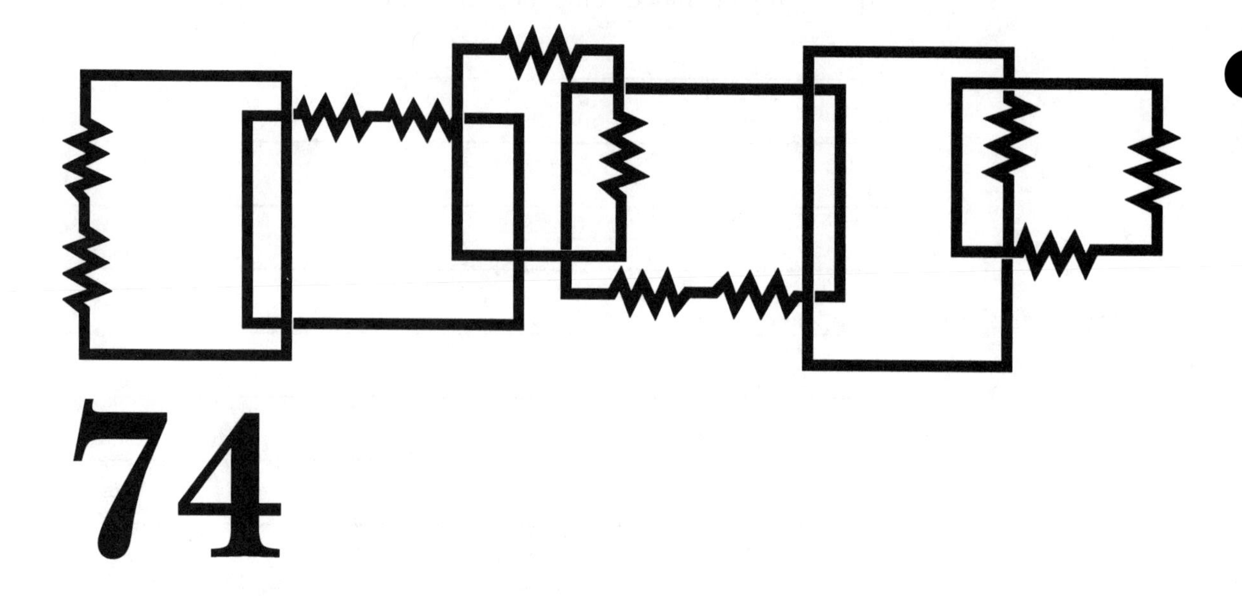

74

The Bridge Rectifier

Objectives

After completing this experiment, you will be able to:

1. Observe the signal characteristics of a bridge rectifier circuit.
2. Recognize failed component symptoms in a bridge rectifier circuit.

Introduction

In Experiment 74 you will use a computer and software program to explore the electrical characteristics of a bridge rectifier circuit. You will be conducting tests to determine the effects of component failure. You will insert failed components to observe their effect on circuit performance. The test equipment you will use in the computer application program is very similar to what you will use on the lab workbench or out in a maintenance work shop.

Procedure

1. Open Electronics Workbench (EWB)/MultiSim. See your instructor for details on how to access and open the application.
2. Insert the student CD provided with the lab manual into the CD drive.
3. Select *Open* from the *File* menu. Change to the CD drive and double click on CX741.ewb.
4. Remove the activity sheet for Experiment 74.
5. Activate the circuit and record the output wave form.
6. Select D1.
7. Click on *Circuit.*
8. Click on *Fault.*
9. Select *Open.*
10. Select the box labeled "1."
11. Click *Accept.*
12. Activate the circuit.
13. Draw the signal shown on the oscilloscope.
14. Restore the diode.
15. Select D2.
16. Click on *Circuit.*
17. Click on *Fault.*
18. Select *Open.*
19. Select the box labeled "1."
20. Click *Accept.*
21. Activate the circuit.
22. Draw the signal shown on the oscilloscope.
23. Restore the diode.
24. Select D3.
25. Click on *Circuit.*
26. Click on *Fault.*
27. Select *Open.*
28. Select the box labeled "1."
29. Click *Accept.*
30. Activate the circuit.

31. Draw the signal shown on the oscilloscope.
32. Restore the diode.
33. Select D4.
34. Click on *Circuit.*
35. Click on *Fault.*
36. Select *Open.*
37. Select the box labeled “1.”
38. Click *Accept.*
39. Activate the circuit.
40. Draw the signal shown on the oscilloscope.
41. Restore the diode.

ACTIVITY SHEET EXPERIMENT 74

NAME ______________________________

DATE ______________________________

Step 5

Step 13

Step 22

Step 31

Step 40

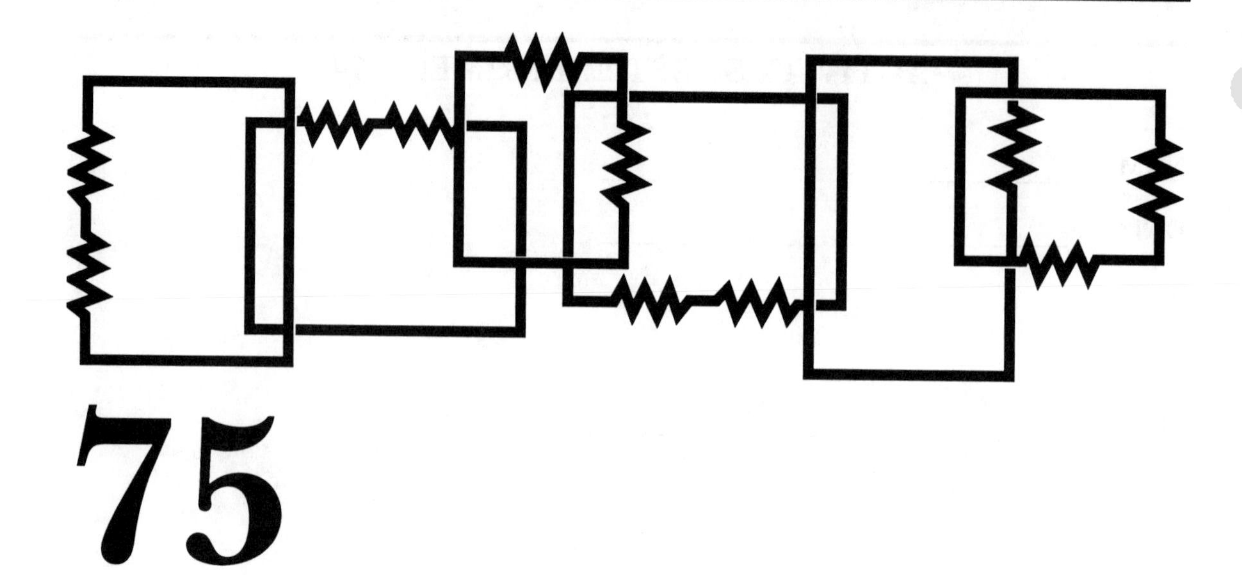

75

Troubleshooting Power Supplies 1

Objectives

After completing this experiment, you will be able to:

1. Recognize malfunctions in a power supply circuit.
2. Locate a failed component in a power supply circuit.

Introduction

The purpose of this experiment is to develop and practice troubleshooting skills. You will be using Electronics Workbench (EWB)/MultiSim and a prebuilt, prefaulted circuit. You will use the test equipment in the simulation application to make voltage, current, and resistance checks to locate and identify a failed component.

Procedure

1. Open Electronics Workbench (EWB)/MultiSim. See your instructor for details on how to access and open the application.
2. Insert the student CD provided with the lab manual into the CD drive.
3. Select *Open* from the *File* menu.
4. Change to the CD drive and double click on CX751.ewb.
5. Remove the activity sheet for Experiment 75.
6. Complete the activity sheet for Experiment 75.

ACTIVITY SHEET EXPERIMENT 75

NAME ______________________________

DATE ______________________________

(Note: In this lab activity you will be responsible for determining and selecting the appropriate test equipment for obtaining accurate measurements.)

Step 6 Use the test equipment to perform electrical tests of the circuit. List each test you make in the space provided below.

____

____

____

____

____

____

____

List the failed component. State the failure (open or short).

____

____

____

____

Explain why the failed component caused the noted symptoms.

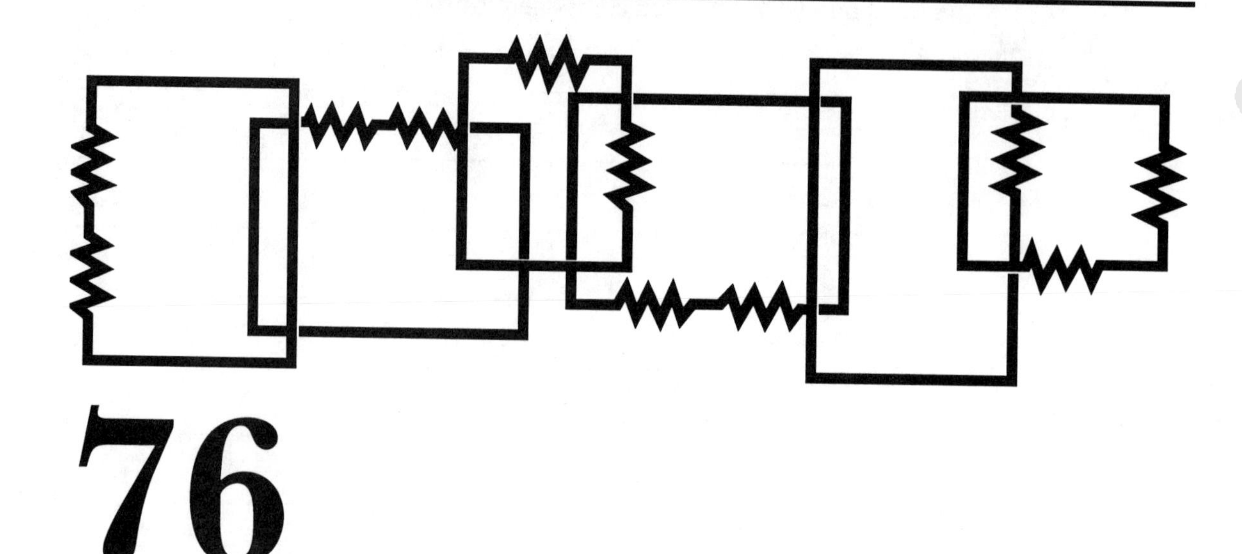

76

Troubleshooting Power Supplies 2

Objectives

After completing this experiment, you will be able to:

1. Recognize malfunctions in a power supply circuit.
2. Locate a failed component in a power supply circuit.

Introduction

The purpose of this experiment is to develop and practice troubleshooting skills. You will be using Electronics Workbench (EWB)/MultiSim and a prebuilt, prefaulted circuit. You will use the test equipment in the simulation application to make voltage, current, and resistance checks to locate and identify a failed component.

Procedure

1. Open Electronics Workbench (EWB)/MultiSim. See your instructor for details on how to access and open the application.
2. Insert the student CD provided with the lab manual into the CD drive.
3. Select *Open* from the *File* menu.
4. Change to the CD drive and double click on CX761.ewb.
5. Remove the activity sheet for Experiment 76.
6. Complete the activity sheet for Experiment 76.

ACTIVITY SHEET EXPERIMENT 76

NAME ______________________________

DATE ______________________________

(Note: In this lab activity you will be responsible for determining and selecting the appropriate test equipment for obtaining accurate measurements.)

Step 6 Use the test equipment to perform electrical tests of the circuit. List each test you make in the space provided below.

____

____

____

____

____

____

____

List the failed component. State the failure (open or short).

____

____

____

____

Explain why the failed component caused the noted symptoms.

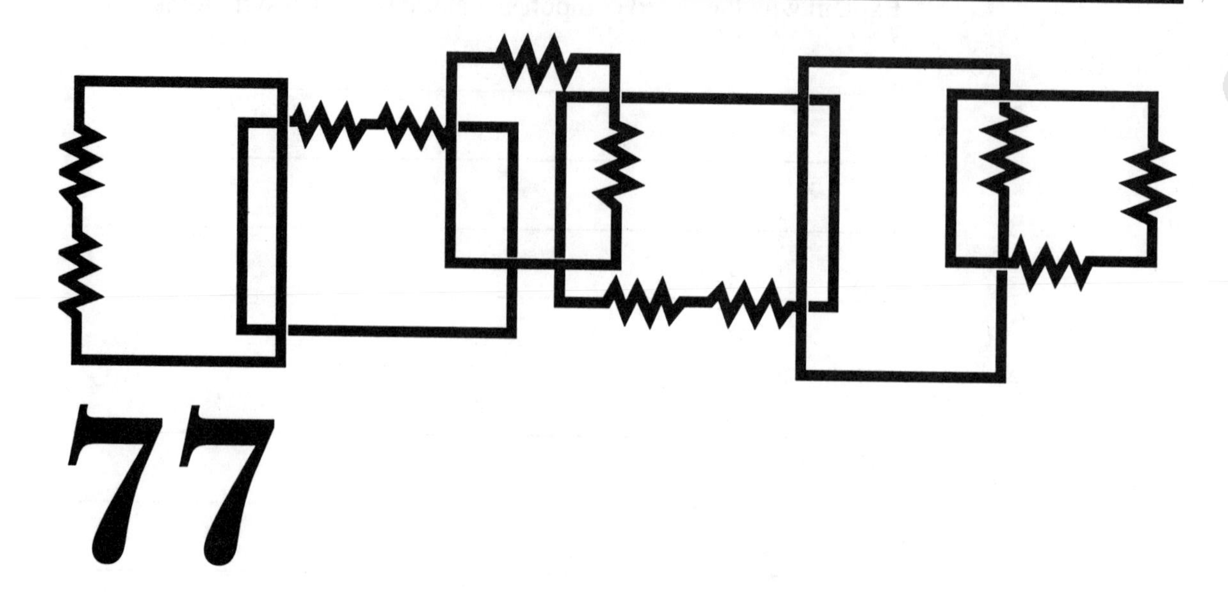

77

Troubleshooting Power Supplies 3

Objectives

After completing this experiment, you will be able to:

1. Recognize malfunctions in a power supply circuit.
2. Locate a failed component in a power supply circuit.

Introduction

The purpose of this experiment is to develop and practice troubleshooting skills. You will be using Electronics Workbench (EWB)/MultiSim and a prebuilt, prefaulted circuit. You will use the test equipment in the simulation application to make voltage, current, and resistance checks to locate and identify a failed component.

Procedure

1. Open Electronics Workbench (EWB)/MultiSim. See your instructor for details on how to access and open the application.
2. Insert the student CD provided with the lab manual into the CD drive.
3. Select *Open* from the *File* menu.
4. Change to the CD drive and double click on CX771.ewb.
5. Remove the activity sheet for Experiment 77.
6. Complete the activity sheet for Experiment 77.

ACTIVITY SHEET EXPERIMENT 77

NAME ________________________________

DATE ________________________________

(Note: In this lab activity you will be responsible for determining and selecting the appropriate test equipment for obtaining accurate measurements.)

Step 6 Use the test equipment to perform electrical tests of the circuit. List each test you make in the space provided below.

____

____

____

____

____

____

____

List the failed component. State the failure (open or short).

____

____

____

____

Explain why the failed component caused the noted symptoms.

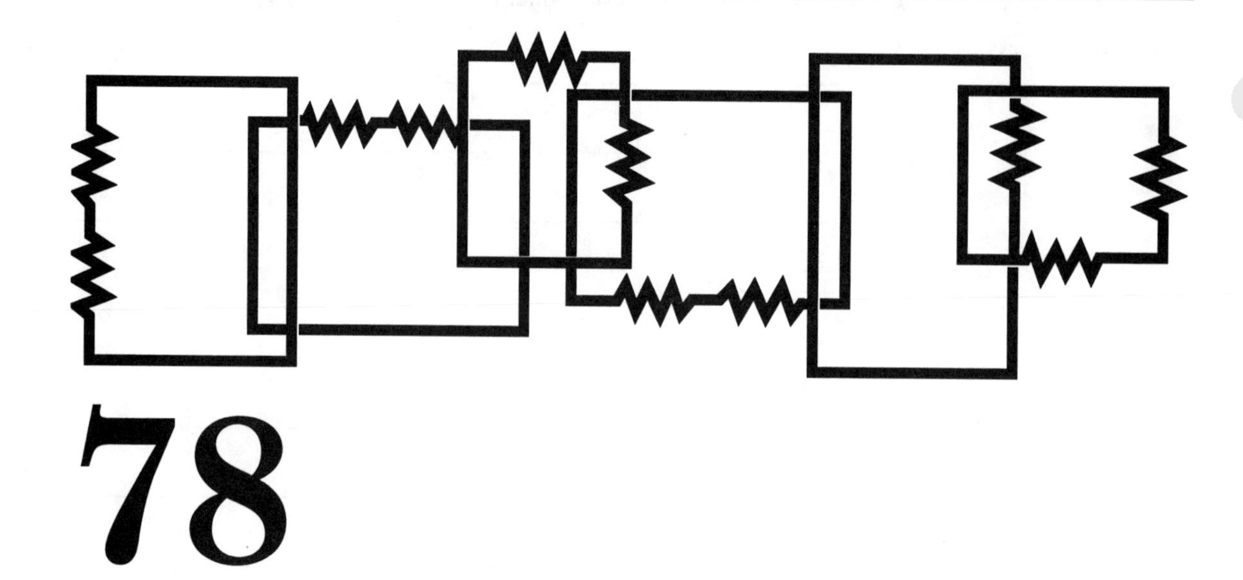

78

Troubleshooting a Common-Emitter Amplifier

Objectives

After completing this experiment, you will be able to:

1. Recognize malfunctions in a C-E amplifier.
2. Locate a failed component in a C-E amplifier circuit.

Introduction

The purpose of this experiment is to develop and practice troubleshooting skills. You will be using Electronics Workbench (EWB)/MultiSim and a prebuilt, prefaulted circuit. You will use the test equipment in the simulation application to make voltage, current, and resistance checks to locate and identify a failed component.

Procedure

1. Open Electronics Workbench (EWB)/MultiSim. See your instructor for details on how to access and open the application.
2. Insert the student CD provided with the lab manual into the CD drive.
3. Select *Open* from the *File* menu.
4. Change to the CD drive and double click on CX781.ewb.
5. Remove the activity sheet for Experiment 78.
6. Complete the activity sheet for Experiment 78.

ACTIVITY SHEET EXPERIMENT 78

NAME ______________________________

DATE ______________________________

(Note: In this lab activity you will be responsible for determining and selecting the appropriate test equipment for obtaining accurate measurements.)

Step 6 Use the test equipment to perform electrical tests of the circuit. List each test you make in the space provided below.

____

____

____

____

____

____

____

List the failed component. State the failure (open or short).

____

____

____

____

Explain why the failed component caused the noted symptoms.

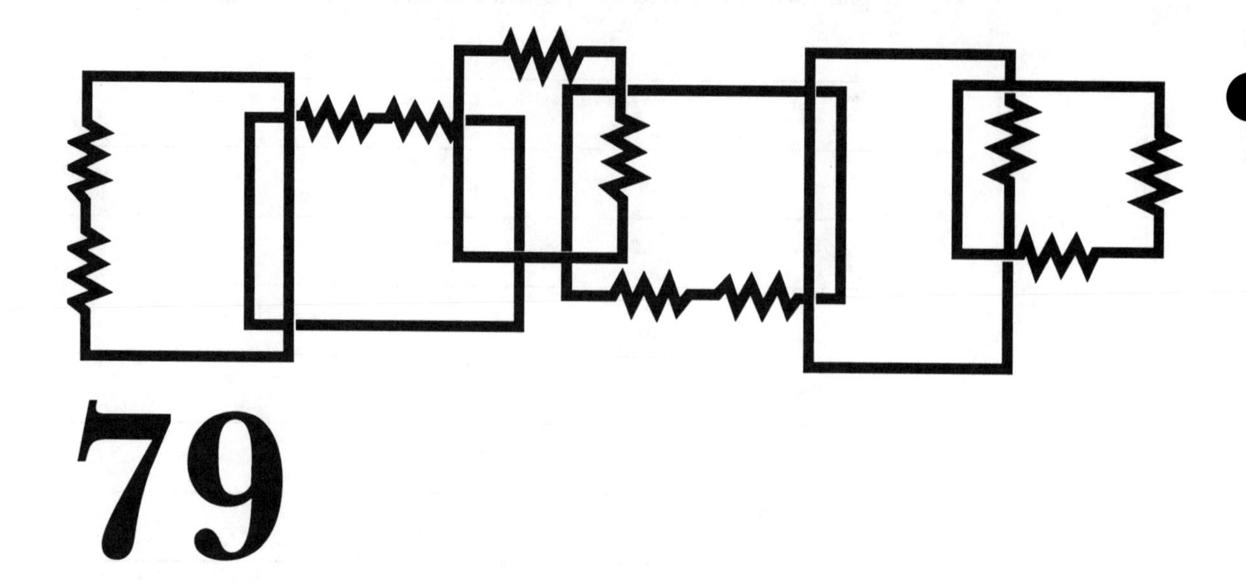

79

Troubleshooting a Common-Collector Amplifier

Objectives

After completing this experiment, you will be able to:

1. Recognize malfunctions in a C-C amplifier.
2. Locate a failed component in a C-C amplifier circuit.

Introduction

The purpose of this experiment is to develop and practice troubleshooting skills. You will be using Electronics Workbench (EWB)/MultiSim and a prebuilt, prefaulted circuit. You will use the test equipment in the simulation application to make voltage, current, and resistance checks to locate and identify a failed component.

Procedure

1. Open Electronics Workbench (EWB)/MultiSim. See your instructor for details on how to access and open the application.
2. Insert the student CD provided with the lab manual into the CD drive.
3. Select *Open* from the *File* menu.
4. Change to the CD drive and double click on CX791.ewb.
5. Remove the activity sheet for Experiment 79.
6. Complete the activity sheet for Experiment 79.

ACTIVITY SHEET EXPERIMENT 79

NAME ______________________________

DATE ______________________________

(Note: In this lab activity you will be responsible for determining and selecting the appropriate test equipment for obtaining accurate measurements.)

Step 6 Use the test equipment to perform electrical tests of the circuit. List each test you make in the space provided below.

____

____

____

____

____

____

____

List the failed component. State the failure (open or short).

____

____

____

____

Explain why the failed component caused the noted symptoms.

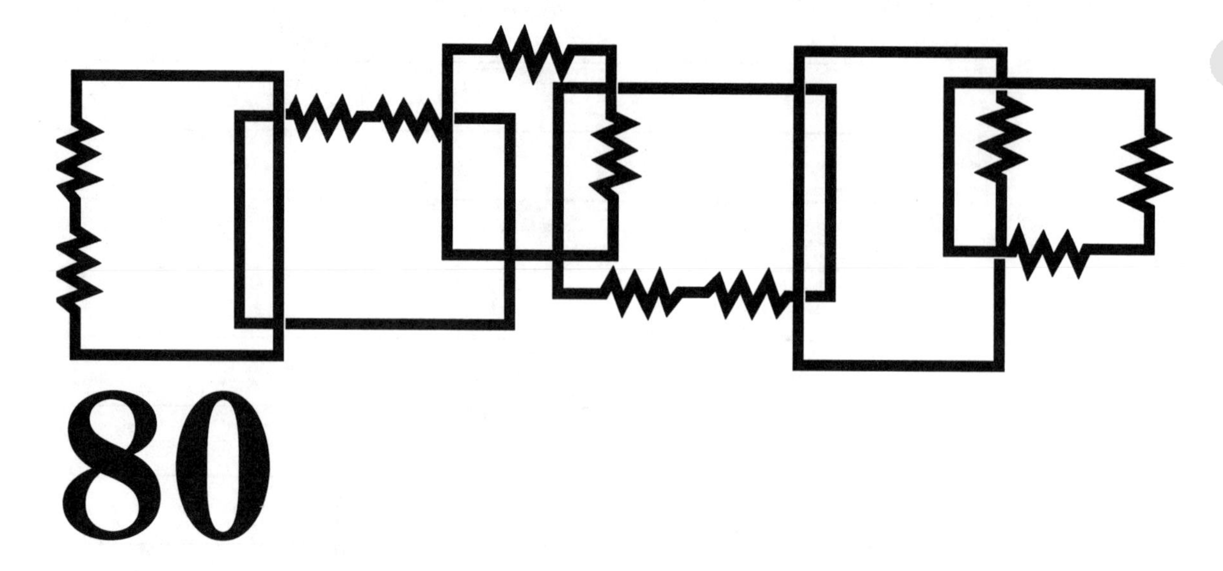

80

Troubleshooting a Common-Base Amplifier

Objectives

After completing this experiment, you will be able to:

1. Recognize malfunctions in a C-B amplifier.
2. Locate a failed component in a C-B amplifier circuit.

Introduction

The purpose of this experiment is to develop and practice troubleshooting skills. You will be using Electronics Workbench (EWB)/MultiSim and a prebuilt, prefaulted circuit. You will use the test equipment in the simulation application to make voltage, current, and resistance checks to locate and identify a failed component.

Procedure

1. Open Electronics Workbench (EWB)/MultiSim. See your instructor for details on how to access and open the application.
2. Insert the student CD provided with the lab manual into the CD drive.
3. Select *Open* from the *File* menu.
4. Change to the CD drive and double click on CX801.ewb.
5. Remove the activity sheet for Experiment 80.
6. Complete the activity sheet for Experiment 80.

ACTIVITY SHEET EXPERIMENT 80

NAME ______________________________

DATE ______________________________

(Note: In this lab activity you will be responsible for determining and selecting the appropriate test equipment for obtaining accurate measurements.)

Step 6 Use the test equipment to perform electrical tests of the circuit. List each test you make in the space provided below.

List the failed component. State the failure (open or short).

Explain why the failed component caused the noted symptoms.

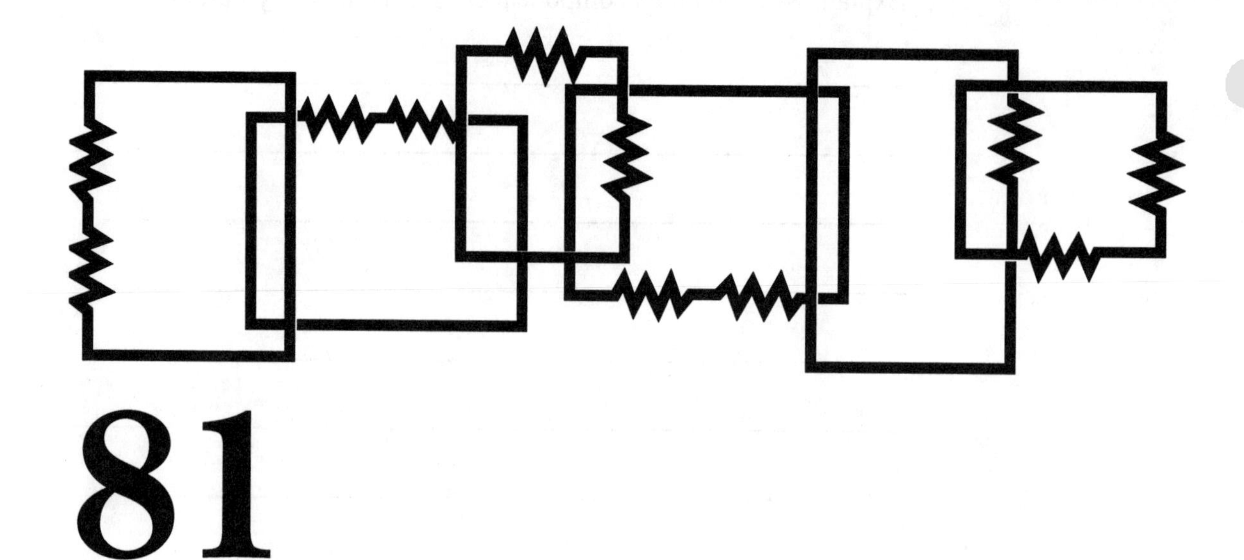

81

Vocabulary Exercise Chapter 1

Objectives

After completing this experiment, you will be able to:

1. Recognize terms relating to the electronics industry.
2. Discuss different roles of electronics technicians.

Introduction

Vocabulary is an important part of the learning process. It is essential for you to understand and speak the language of electronics. This vocabulary exercise is designed to help you become familiar with some of the terms used in the electronics industry.

Procedure

1. Remove the blank crossword puzzle.
2. Read each clue.
3. Complete the crossword puzzle by filling in the appropriate boxes.
4. Refer to your textbook for answers, if necessary.

CLUES

Across

2. Devices for entertainment, information, and safety. (2 words)
4. Typically power, motor, and process control equipment. (2 words)
5. Assembles and tests prototype devices and equipment. (2 words)
7. Demonstrated the relationship between electricity and magnetism.
8. Sends and receives information. (2 words)
12. Use of components to control the flow of electricity.
13. Repairs failed equipment.
14. Tests pre-production and production items for quality compliance.
15. Manipulates large groups of information. (2 words)

Down

1. Discovered the relationship between electricity and magnetism.
3. Health care and life-support equipment. (2 words)
6. First computer game company.
9. Unit of magnetic field strength.
10. First television picture tube.
11. First large-scale electronic digital computer.

ACTIVITY SHEET EXPERIMENT 81

NAME ______________________________

DATE ______________________________

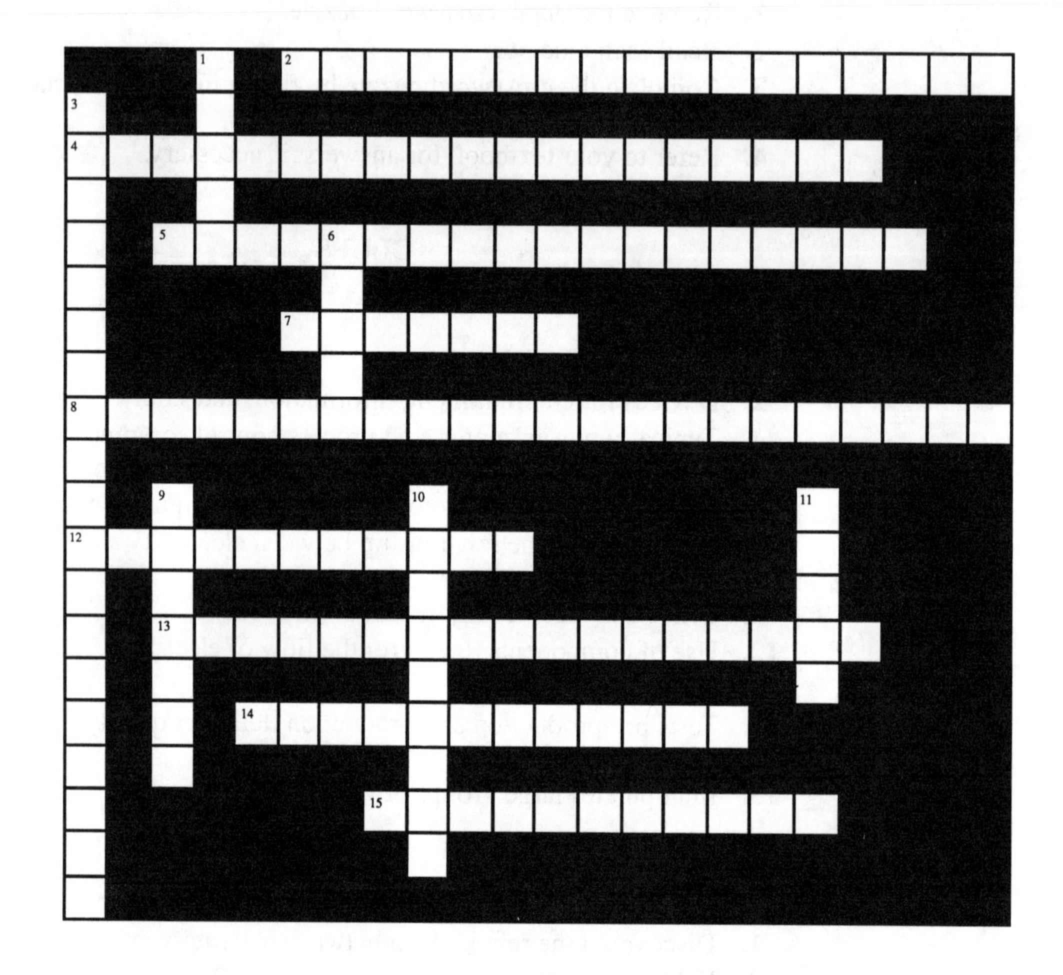

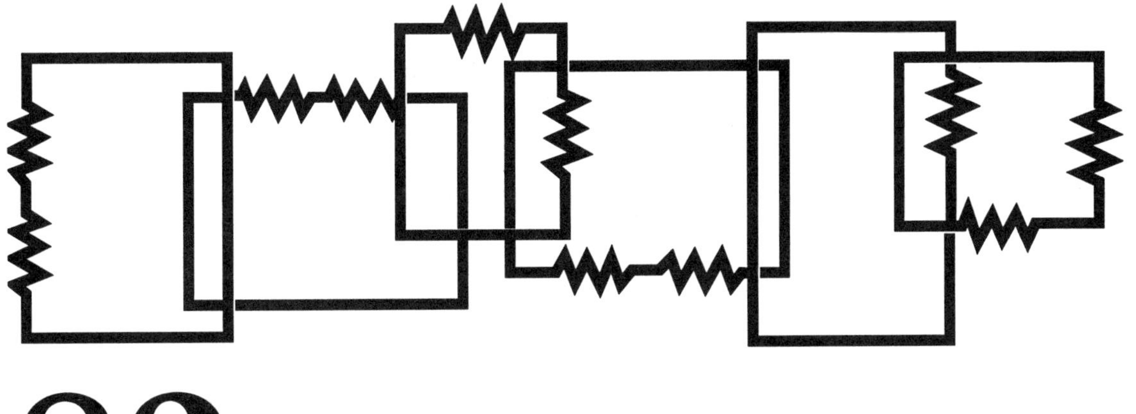

82

Vocabulary Exercise Chapter 2

Objectives

After completing this experiment, you will be able to:

1. Recognize terms relating to voltage and current.
2. Define terms relating to voltage and current.

Introduction

Vocabulary is an important part of the learning process. It is essential for you to understand and speak the language of electronics. This vocabulary exercise is designed to help you become familiar with some of the terms used in the electronics industry.

Procedure

1. Remove the blank crossword puzzle.
2. Read each clue.
3. Complete the crossword puzzle by filling in the appropriate boxes.
4. Refer to your textbook for answers, if necessary.

CLUES

Across

2. A break in the circuit path.
5. Materials that block current flow.
7. The force or pressure exerted on electrons to cause current flow.
8. Smallest particle of a compound still retaining its chemical characteristics.
10. The outermost electron-occupied shell. (2 words)
12. Unit of measurement for current.
13. Atom which has lost one or more of its electrons. (2 words)
14. Relationship between voltage and current. (2 words)
15. The flow of electrons through a conductor. (2 words)
17. An electron that is not in any orbit around a nucleus. (2 words)
18. Device used to measure current flow.

Down

1. A substance consisting of only one type of atom.
3. Atom with an excess of electrons.
4. A no-resistance path for current flow.
6. Electron orbital paths.
9. Balanced atom. (2 words)
10. Device used to measure voltage.
11. Materials that pass current easily.
16. The smallest particle into which an element can be divided without losing its identity.

ACTIVITY SHEET EXPERIMENT 82

NAME ______________________________

DATE ______________________________

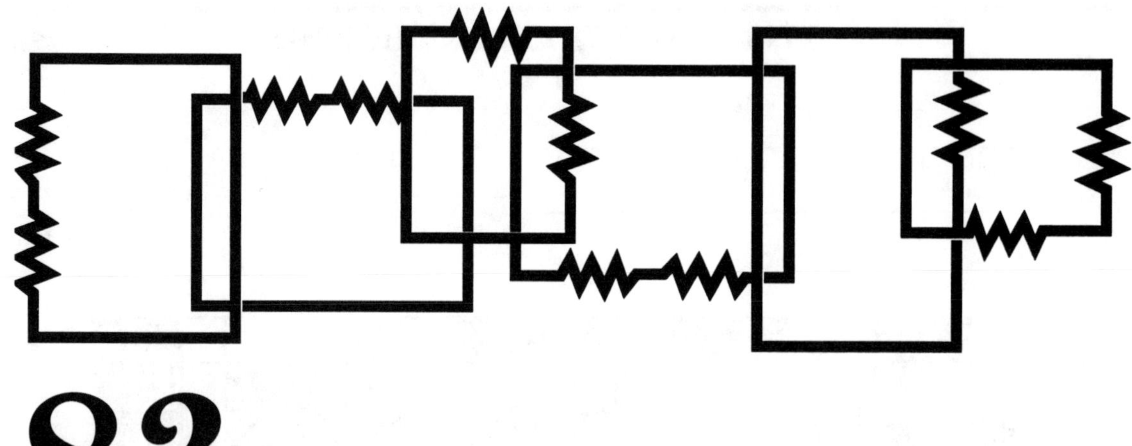

83

Vocabulary Exercise Chapters 3 and 4

Objectives

After completing this experiment, you will be able to:

1. Recognize terms relating to resistance, resistors, and power.
2. Define terms relating to resistance, resistors, and power.

Introduction

Vocabulary is an important part of the learning process. It is essential for you to understand and speak the language of electronics. This vocabulary exercise is designed to help you become familiar with some of the terms used in the electronics industry.

Procedure

1. Remove the blank crossword puzzle.
2. Read each clue.
3. Complete the crossword puzzle by filling in the appropriate boxes.
4. Refer to your textbook for answers, if necessary.

CLUES

Across

1. Measure of a material's resistance to current flow.
7. Unit of measurement for resistance.
8. Release of electrical energy in the form of heat.
10. Printed circuit board. (abbr.)
12. Resistive temperature detector (abbr.)
13. Device that converts energy from one form to another.
15. Light-dependent resistor.
16. A three-terminal variable resistor.
18. Permissible deviation.
21. Temperature-sensitive semiconductor.
22. Proportional variance.
23. Opposes the flow of current.

Down

2. Nonuniform variance.
3. Method of indicating value and tolerance of resistors.
4. A method of mounting components on a PCB.
5. Single in-line package. (abbr.)
6. The rate at which work is performed.
9. Thin-film detector. (abbr.)
11. A device whose resistance changes when heated.
14. A two-terminal variable resistor.
15. Relationship between voltage and current.
17. Device for measuring power.
19. Device for measuring resistance.
20. The unit of energy.

ACTIVITY SHEET EXPERIMENT 83

NAME ______________________________

DATE ______________________________

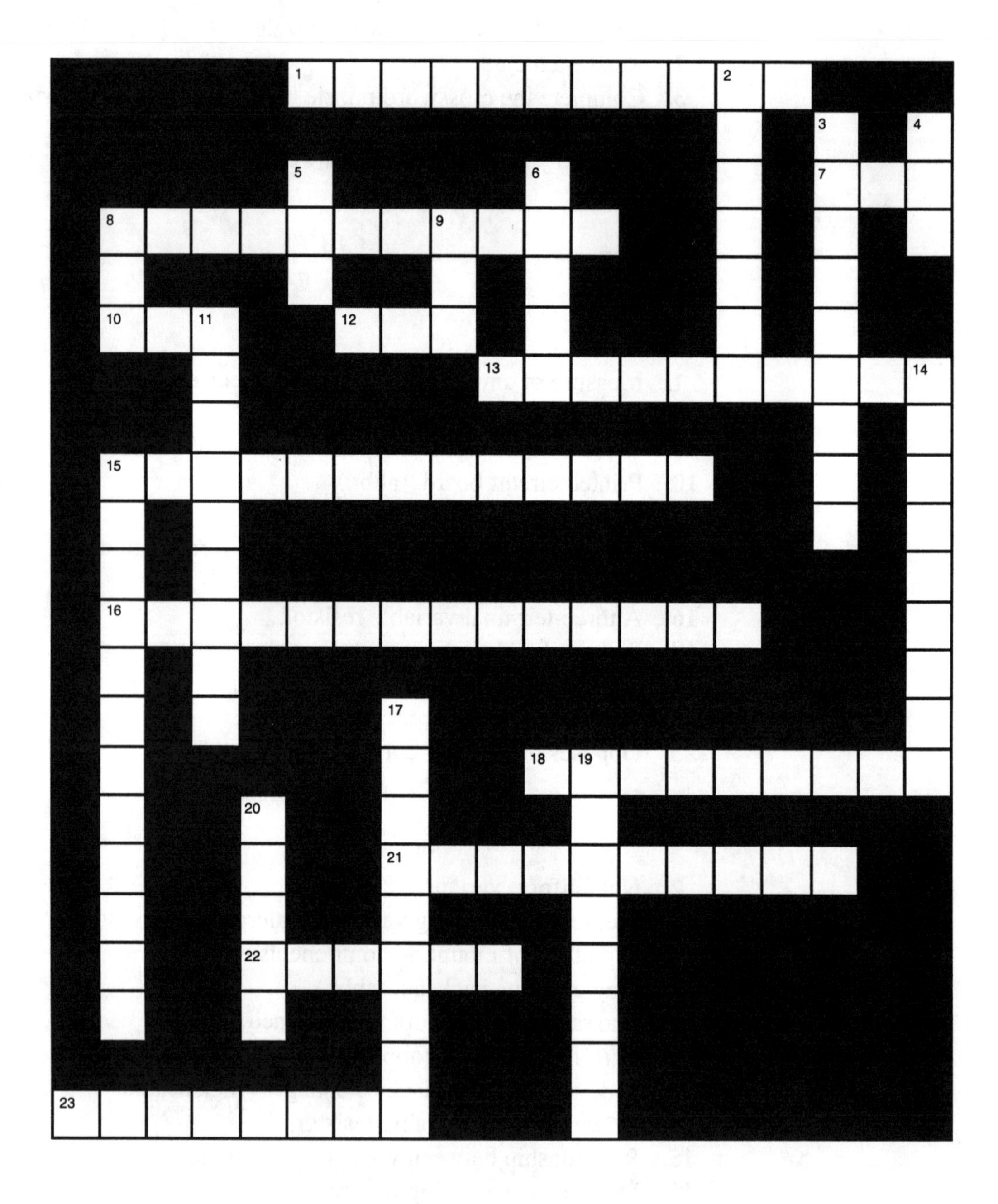

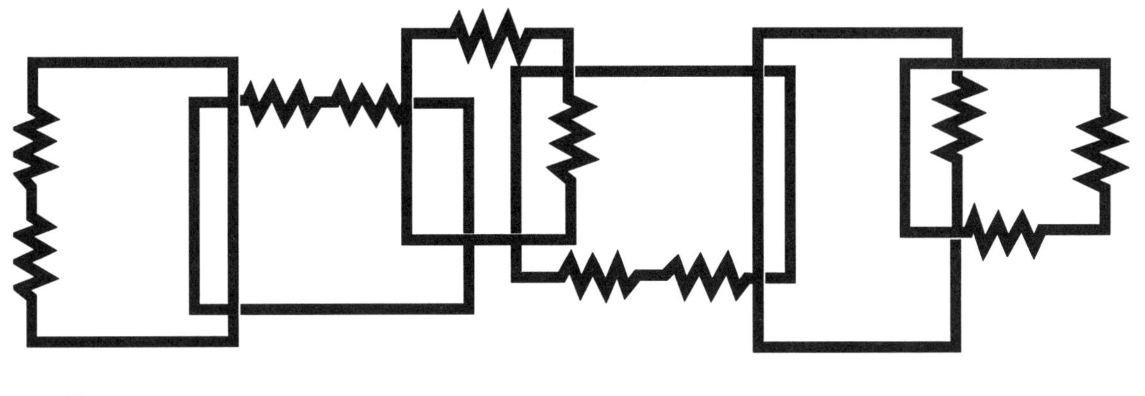

84

Vocabulary Exercise Chapters 5 through 10

Objectives

After completing this experiment, you will be able to:

1. Recognize terms relating to direct current (dc).
2. Define terms relating to direct current (dc).

Introduction

Vocabulary is an important part of the learning process. It is essential for you to understand and speak the language of electronics. This vocabulary exercise is designed to help you become familiar with some of the terms used in the electronics industry.

Procedure

1. Remove the blank crossword puzzle.
2. Read each clue.
3. Complete the crossword puzzle by filling in the appropriate boxes.
4. Refer to your textbook for answers, if necessary.

CLUES

Across

1. Electrically conducting liquid or paste.
4. Cell using a liquid electrolyte. (2 words)
6. Photovoltaic cell. (2 words)
9. Cell using a paste electrolyte. (2 words)
11. Reuseable fuse. (2 words)
12. Field produced by current flow.

Down

2. Opposition to output current. (2 words)
3. Temperature transducer.
5. Output current. (2 words)
6. A device that supplies electric energy to a load.
7. A stone possesing magnetic properties.
8. The application of force.
10. Resistance to motion.

ACTIVITY SHEET EXPERIMENT 84

NAME ______________________________

DATE ______________________________

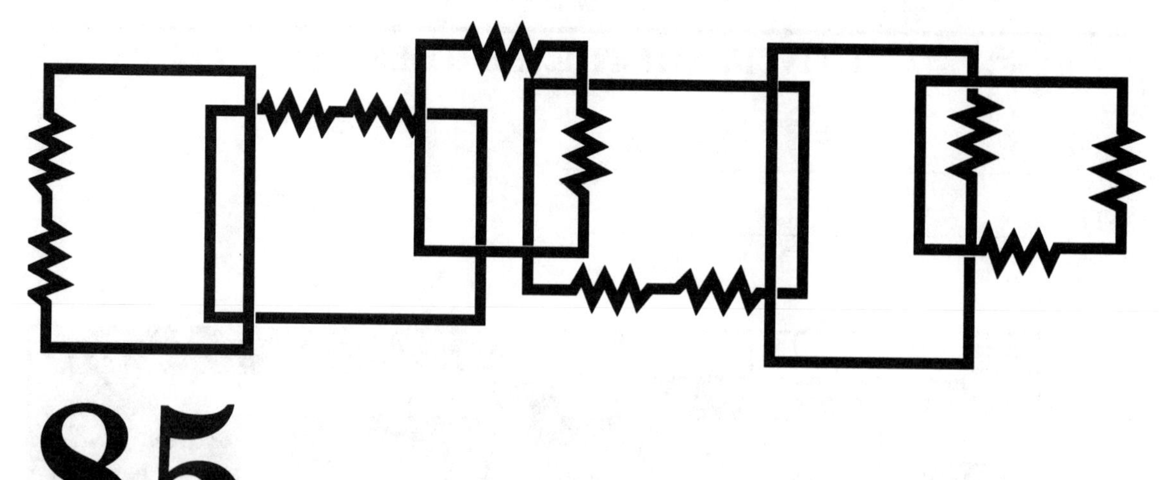

85

Vocabulary Exercise Chapters 11 and 12

Objectives

After completing this experiment, you will be able to:

1. Recognize terms relating to alternating current (ac).
2. Define terms relating to alternating current (ac).

Introduction

Vocabulary is an important part of the learning process. It is essential for you to understand and speak the language of electronics. This vocabulary exercise is designed to help you become familiar with some of the terms used in the electronics industry.

Procedure

1. Remove the blank crossword puzzle.
2. Read each clue.
3. Complete the crossword puzzle by filling in the appropriate boxes.
4. Refer to your textbook for answers, if necessary.

CLUES

Across

4. The study of triangles.
6. Physical length of one complete cycle.
7. Travels at 3×10^8 meters per second. (2 words)
10. Transmission of information between two points.
11. An inductive device used to couple electric energy from one circuit to another.
13. Highest amplitude. (2 words)
15. The time interval between two pulses. (abrr.)
17. A device that converts ac to dc.
18. Negative going transition of a pulse. (2 words)
20. Travels at 1133 feet per second.

Down

1. When the output varies in direct proportion to the input.
2. The number of times per second that a pulse is transmitted. (abbr.)
3. Positive going transition of a pulse. (2 words)
4. Any device that converts energy from one form to another.
5. Cyclic rate.
8. Electric current that has both positive and negative alternations.
9. A quantity that has both magnitude and direction.
12. Size of a signal when measured from zero.
14. Mean value.
15. Angular relationship between two waves.
16. Time required to complete one cycle.
19. Effective value.

ACTIVITY SHEET EXPERIMENT 85

NAME ______________________

DATE ______________________

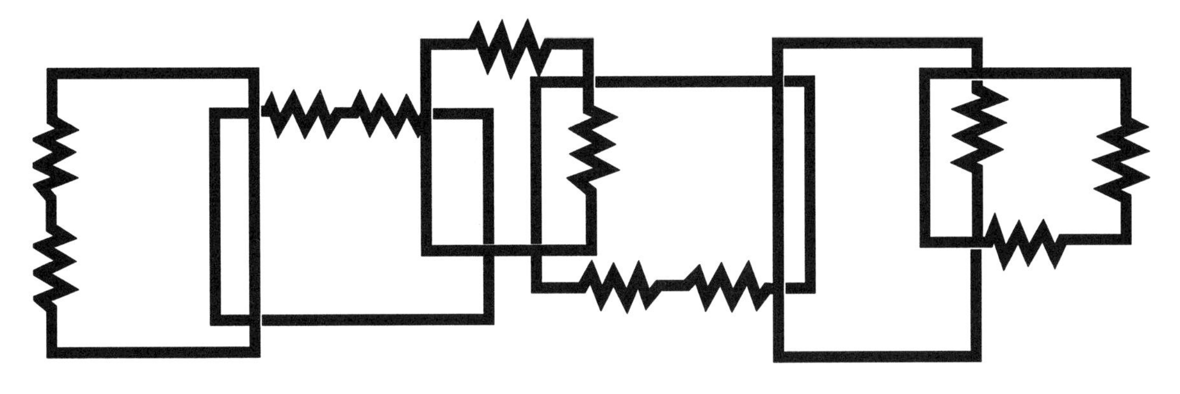

86

Vocabulary Exercise Chapters 13 and 14

Objectives

After completing this experiment, you will be able to:

1. Recognize terms relating to capacitors and capacitive circuits.
2. Define terms relating to capacitors and capacitive circuits.

Introduction

Vocabulary is an important part of the learning process. It is essential for you to understand and speak the language of electronics. This vocabulary exercise is designed to help you become familiar with some of the terms used in the electronics industry.

Procedure

1. Remove the blank crossword puzzle.
2. Read each clue.
3. Complete the crossword puzzle by filling in the appropriate boxes.
4. Refer to your textbook for answers, if necessary.

CLUES

Across

3. Blocks high frequencies.
4. Stores electrical energy in the form of an electric field.
6. Total opposition to current flow in an ac circuit.
8. Resistance presented by a capacitor.
9. Circuit whose output is proportional to the integral of its inputs.
10. Force field produced by static electrical charges.

Down

1. Unit of capacitance.
2. Circuit whose output is proportional to the rate of change of the input voltage.
3. Small undesirable flow of current through an insulator or dielectric.
4. Any device that converts energy from one form to another.
5. Reduce in amplitude.
7. Frequency selective circuit.

ACTIVITY SHEET EXPERIMENT 86

NAME ______________________________

DATE ______________________________

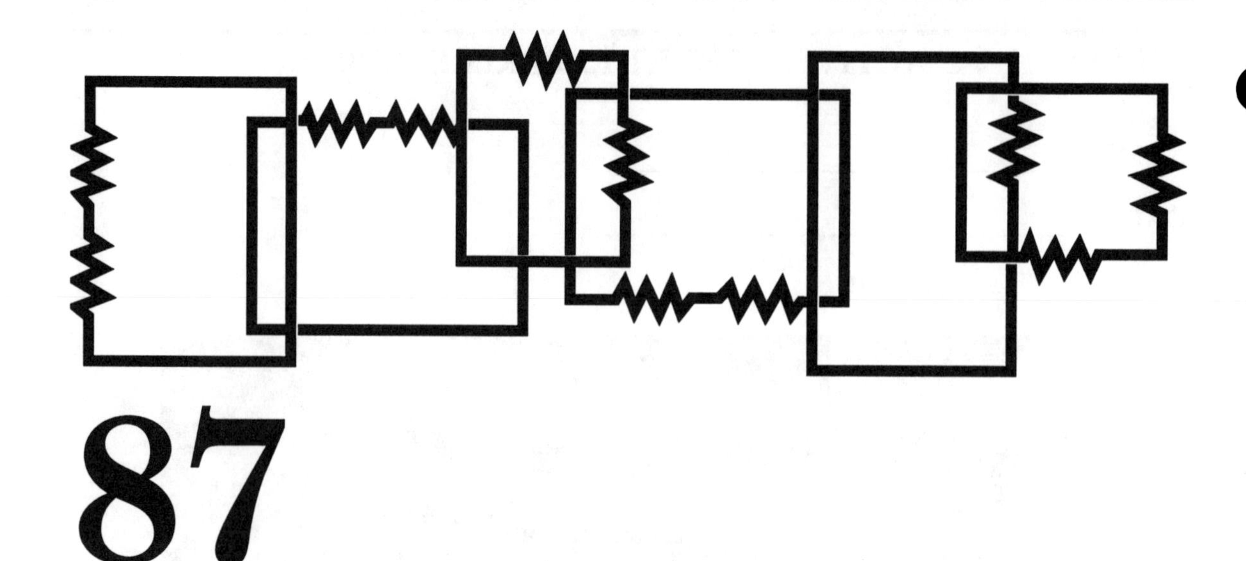

87

Vocabulary Exercise Chapters 16 and 17

Objectives

After completing this experiment, you will be able to:

1. Recognize terms relating to inductors and transformers.
2. Define terms relating to inductors and transformers.

Introduction

Vocabulary is an important part of the learning process. It is essential for you to understand and speak the language of electronics. This vocabulary exercise is designed to help you become familiar with some of the terms used in the electronics industry.

Procedure

1. Remove the blank crossword puzzle.
2. Read each clue.
3. Complete the crossword puzzle by filling in the appropriate boxes.
4. Refer to your textbook for answers, if necessary.

CLUES

Across

2. The magnetic lines of force produced by a magnet. (2 words)
4. Resistance to the flow of magnetic lines of force.
6. Unit of magnetic flux.
7. A measure of the strength of a magnetic wave.
9. A lag between cause and effect.

Down

1. Force that produces a magnetic field. (abbr.)
2. Electrostatic transducer.
3. One magnetic line of force.
4. Ability of a material to remain magnetized.
5. Property which opposes changes in current.
7. Composed of or containing iron.
8. Unit of inductance.

ACTIVITY SHEET EXPERIMENT 87

NAME ______________________________

DATE ______________________________

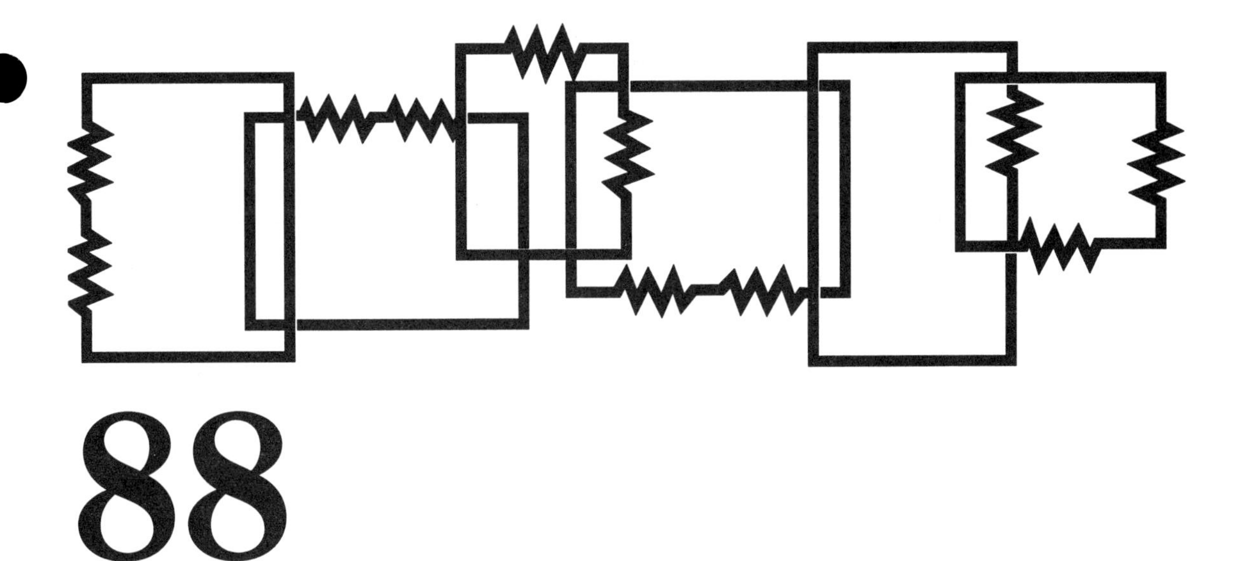

88

Vocabulary Exercise Chapters 19 through 22

Objectives

After completing this experiment, you will be able to:

1. Recognize terms relating to semiconductors and junction diodes.
2. Define terms relating to semiconductors and junction diodes.

Introduction

Vocabulary is an important part of the learning process. It is essential for you to understand and speak the language of electronics. This vocabulary exercise is designed to help you become familiar with some of the terms used in the electronics industry.

Procedure

1. Remove the blank crossword puzzle.
2. Read each clue.
3. Complete the crossword puzzle by filling in the appropriate boxes.
4. Refer to your textbook for answers, if necessary.

CLUES

Across

2. Boosting in strength.
5. Vacuum tube. (2 words)
6. The outermost electron path. (2 words)
9. Coupling from emitter to base. (2 words)
11. Cathode ray tube. (abbr.)
12. Voltage-controlled semiconductor device.
14. Three-electrode vacuum tube.
15. A positive electrode or terminal.
17. I_C
18. Bipolar junction transistor. (abbr.)
19. Base-emitter junction forward biased, base-collector junction reverse biased. (2 words)
22. At rest.
23. A small layer on either side of a PN junction. (2 words)
24. The space between two orbital shells. (2 words)
25. Gate, source, and drain.
26. The effect of transferring resistance.

Down

1. A diode's internal barrier voltage.
3. Breakdown voltage.
4. The smallest particle of a substance.
5. Mutual conductance.
7. A negative electrode or terminal.
8. An electron orbital path.
10. Field effect transistor. (abbr.)
13. I_E
16. Two-state circuits. (2 words)
18. I_B
20. Majority carrier in a N-type material.
21. Force that causes a diode to operate in a certain manner. (2 words)

ACTIVITY SHEET EXPERIMENT 88

NAME ____________________

DATE ____________________

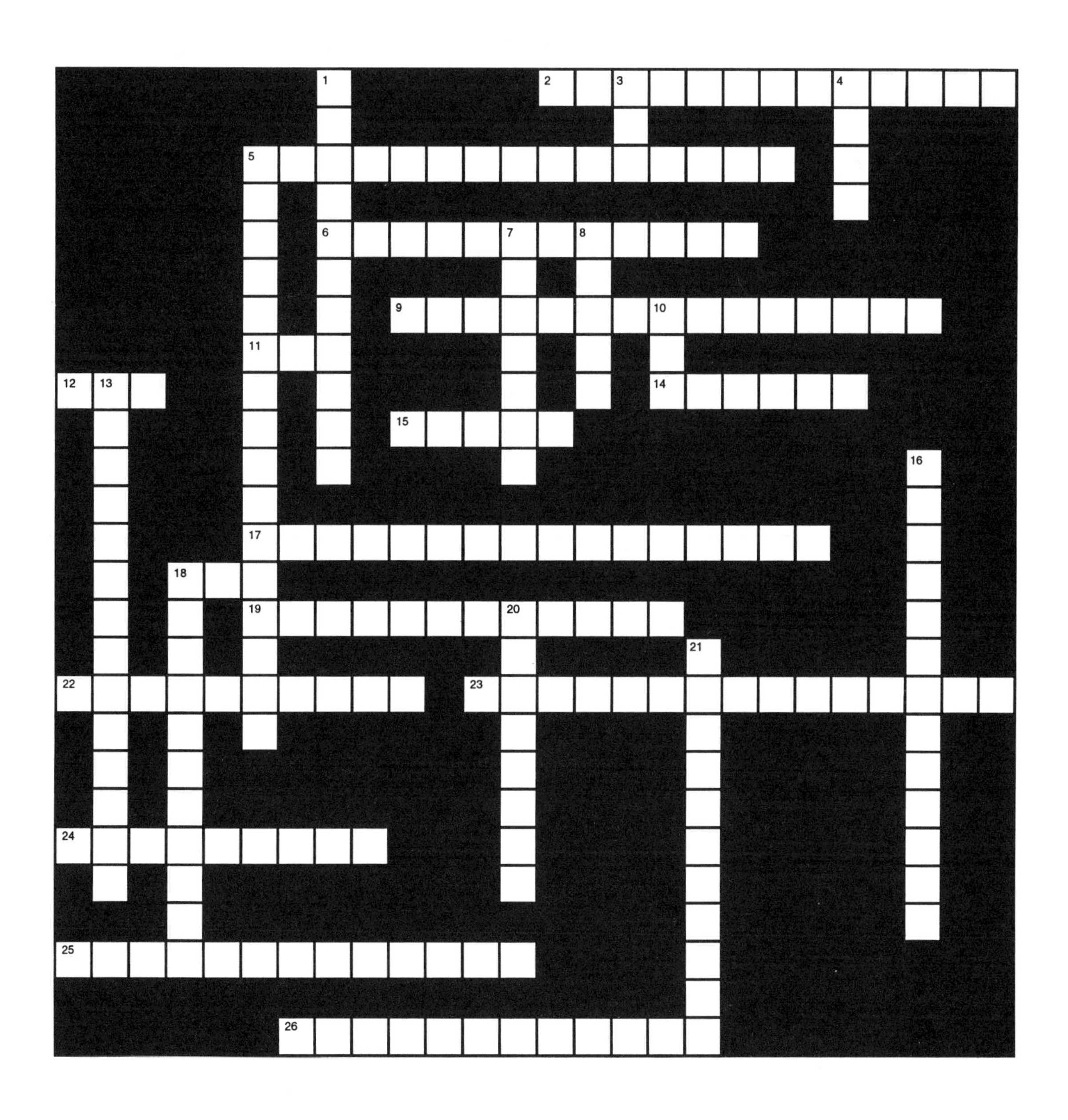

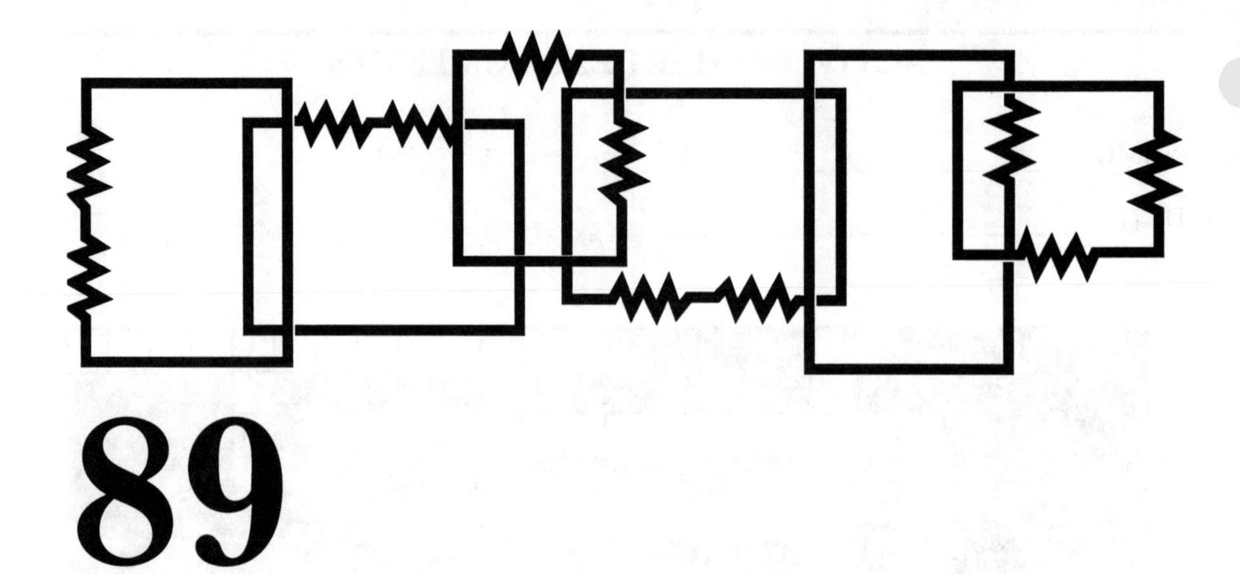

89

Vocabulary Exercise Chapter 23

Objectives

After completing this experiment, you will be able to:

1. Recognize terms relating to operational amplifiers.
2. Define terms relating to operational amplifiers.

Introduction

Vocabulary is an important part of the learning process. It is essential for you to understand and speak the language of electronics. This vocabulary exercise is designed to help you become familiar with some of the terms used in the electronics industry.

Procedure

1. Remove the blank crossword puzzle.
2. Read each clue.
3. Complete the crossword puzzle by filling in the appropriate boxes.
4. Refer to your textbook for answers, if necessary.

CLUES

Across

1. Second step in op-amp circuit troubleshooting.
5. First step in op-amp circuit troubleshooting.
6. Amplification of differential to common mode gain.
7. Ability to attenuate out of band signals.
10. Ground for voltage but not for current.
11. Third step in op-amp troubleshooting.
12. Converts dc into repeating output signals.

Down

2. Maximum rate of change.
3. Output follows input.
4. Summing amplifier. (2 words)
8. Ratio of differential to common-mode gain. (abbr.)
9. Special type of high-gain amplifier.

ACTIVITY SHEET EXPERIMENT 89

NAME ______________________

DATE